Numerical Methods

for Science and Engineering

Numerical Methods
for Science and Engineering

B. Ravindra Reddy
Professor of Mathematics & Additional Controller of Examinations,
JNTU College of Engineering, Hyderabad.

G. Shanker Rao
Former HOD of Mathematics, Govt. Girraj P.G College, and
Former Faculty of Mathematics College of Engineering,
Osmania University, Hydearabad.

V. Bhikshma
Professor Department of Civil Engineering,
College of Engineering, Osmania University, Hyderabad.

BSP BS Publications

An Imprint of **BSP Books Pvt. Ltd.**

4-4-309/316, Giriraj Lane, Sultan Bazar,
Hyderabad - 500 095.

Published by:

BSP **BS Publications**
An Imprint of **BSP Books Pvt., Ltd.**
4-4-309/316, Giriraj Lane, Sultan Bazar,
Hyderabad - 500 095
Phone : 040 - 23445600
e-mail: info@bspbooks.net
Website: www.bspbooks.net

ISBN: 978-93-88305-89-1 (Hardback)

Dedication

We wholeheartedly dedicate this book to Prof. B. Sathyanarayana

Who is popularly known as Battu

Preface

The book is designed as a self-contained comprehensive class room text for the B.sc and B.Tech students of most of the Indian universities.

The topics covered are Errors, difference operators, methods of solving linear algebraic and transcendental equations, interpolation with equal intervals, Interpolation with unequal intervals inverse interpolation, numerical Integration, numerical differentiation, central difference interpolation formulas and methods of solving ordinary differential equations. The book also introduces the methods of solving linear algebraic equations. Guass –Jacobi and Gauss Seidel methods.

Professor Battu Sathyanarayana has more than 30 years of experience of teaching. His areas of interest include catalysis, bio-inorganic chemistry, Nano materials chemistry. He has published many research papers in National and International magazines and guided many research scholars in obtaining Ph.D degree. He is a life member of Indian science congress association. He is associated with several UGC committees. He is the nodal officer wi-Fi enabled network project of Osmania University sponsored by Ministry of Electronics and Information Technology. He is president of Osmania University Teachers Association and Chairman of Telangana State Federation of University Teachers Associations. This book is dedicated to him.

We wish to thank Mr. Vasudeva Rao, Sales manager for his efforts in promoting the book.

The authors wish to express thanks to Sri. Naresh Davergave, Production manager BSP books for his interest and support in bringing out the book without errors.

We also thank sincerely our publisher Sri Anil Shah of BSP books for publishing the book.

-Authors

Contents

CHAPTER 8

CHAPTER 9

CHAPTER 10

CHAPTER 11

CHAPTER 12

CHAPTER 1

Errors in Numerical Calculations

1.1 INTRODUCTION

Numerical methods are very powerful and popular tools for solving a variety of engineering, mathematical and scientific problems using the four basic arithmetical operations. In this chapter we introduce Numerical techniques are used to solve problems involving higher order polynomials. They are used in solving transcendental equations. The numerical methods are also used in solving equations involving several variables. The techniques employed in numerical analysis are times approximate. Therefore the results (i.e., outcomes) obtained by numerical methods have some errors.

Let X_E denote an exact number and α be a number that differs slightly from X and is used in place of X in calculations, then α is called **an approximate number**.

If α is less than X then it is called a minor approximation of X , and if α is greater than X, then it is called a major approximation of X

Definition: Let x be an exact number and α be the approximate number of x, then the difference between x and α is called the **error of α**.

It is denoted by E and is given by

$$E = X - \alpha \qquad \qquad .....(1.1)$$

If $X > \alpha$, then the error is positive, and if $X < \alpha$, then the error is negative.

1.2 ABSOLUTE ERROR

The absolute error E_A of an approximate number α is the absolute value of the difference between the corresponding exact number x and the number α.

$$\therefore E_A = |X_E - X_A|$$

It is also denoted by ΔX

1.3   LIMITING ABSOLUTE ERROR

Definition: The limiting absolute error of an approximate number is any number that is not less than the absolute error of that number.

Thus if Δ_α is the limiting absolute error of an approximate number α which takes the place of the exact number X_E, then

$$|X_E - \alpha| \leq \Delta_\alpha$$

The exact number X lies within the range

$$\alpha - \Delta_\alpha \leq X_E \leq \alpha + \Delta_\alpha$$

We can write $X = \alpha \pm \Delta_\alpha$

Note: The absolute error does not suffice to describe the accuracy of a measurement or a computation. An essential point in the accuracy of the measurements is the absolute error related to unit length. It is called the relative error.

1.4   RELATIVE ERROR

The relative error E_R of an approximate number α is the ratio of the absolute error E_A of the number to the modulus of the corresponding exact number x.

From the definition we have

Relative error $E_R = \dfrac{E_A}{X}$

We can also write $E_R = \dfrac{E_A}{X} = \delta X$

1.5   THE LIMITING RELATIVE ERROR

The limiting relative error of a given approximate number α, is any number not less than the relative error of that number. It is denoted by δ_α.

By definition we have $\delta \leq \delta_\alpha$

That is $\dfrac{\Delta}{|X|} \leq \delta_\alpha$, when $\Delta \leq |x| \, \delta_\alpha$

Thus for limiting relative absolute error of a number α we can take $\Delta_\alpha = x|\, \delta_\alpha|$, from which, knowing the relative error δ_α we can obtain for the exact number. Since the exact number lies between $\alpha \, (1-\delta_\alpha)$ and $\alpha \, (1+\delta_\alpha)$

We can write $X = (1 \pm \delta_\alpha)$

If α is an approximate number taking the place of an exact number X, and Δ_α is the limiting absolute error of α taking

$$X > 0,\ \alpha > 0 \text{ and } \Delta_\alpha < \alpha, \text{ we get}$$

$$\delta = \frac{\Delta}{X} \leq \frac{\Delta_\alpha}{u - \Delta_\alpha} \Rightarrow \delta_\alpha = \frac{\Delta_\alpha}{u - \Delta_\alpha}$$

For the limiting relative error of the number α.

Similarly we can show that $\Delta_\alpha = \dfrac{\delta_\alpha}{\alpha - \delta_\alpha}$

Note: If Δ_α is very much less than α, and δ_α is very much less than 1, we can take

$$\delta_\alpha \approx \frac{\Delta_\alpha}{\alpha} \text{ and } \Delta_\alpha \approx \alpha \delta_\alpha$$

1.6 PERCENTAGE ERROR

The percentage error E_P is defined by

$$E_P = E_R \times 100$$

1.7 SOURCES OF ERRORS

The errors in mathematical solution of problems are of five types.

1. Errors involved in the statement of the problems
2. Errors stemming from the presence of infinite processes in mathematical analysis
3. Errors due to numerical parameters whose value can only be determined approximately
4. Errors associated with the system of numeration
5. Errors due to operations involving approximate numbers

In this section we discuss two types of errors namely truncation error and computational errors.

The errors which are inherent in the numerical methods employed for finding numerical solutions are known as **truncation errors**

The truncation error arises due to the replacement of an infinite process such as summation or integration by a finite one. These errors are caused by using approximate formulae in computation

Trigonometric functions are computed of by summing series.

$S = \sum_{r=0}^{\infty} a_r x^r$ is replaced by the finite sum $\sum_{r=0}^{n} a_r x^r$

For example consider $e^x = 1 + \dfrac{x}{1!} + \dfrac{x^2}{2!} + \dfrac{x^3}{3!} + \ldots$

Is summed t n terms. Suppose we wish to calculate $e^{\frac{1}{3}}$. We might begin by creating an error by specifying $e^{0.3333}$

So that the propagated error $= e^{0.3333} - e^{\frac{1}{3}} =- 0.0000465196$

Then we might truncate the series after 5 term, leading to the truncation error

$$= - (\frac{(0.3333)^5}{5!}+\frac{(0.3333)^6}{6!} + ...) =-0.0000362750$$

Finally we might sum with the rounded values:

$$1 + 0.3333 + 0.0555 + 0.0062 + 0.00005 = 1.3955$$

Where the propagated error from the rounding's is -0.00002963

The error is called **inherent error**

When performing computations with approximate numbers, we naturally   carry the errors of original detain to the final result. In this   respect errors of operation are inherent

1.8   SIGNIFICANT DIGITS, THE NUMBER OF CORRECT DIGITS

If α is a positive number it can be represented as a terminating   or non-terminating decimal as follows.

$$\alpha =\alpha_m 10^m + \alpha_{m-1}10^{m-1} + \alpha_{m-2}10^{m-2} +...+ \alpha_{m-n+1}10^{m-n+1}+ ... \qquad .....(1.2)$$

where α_i are digits of the number α [i.e $\alpha_i = 0,1,2,3,..,9$]

$\alpha_m \neq 0$ is called the leading digit m is the highest power of ten in the expansion. It is an integer

For example consider the number 5214. 73. It can be written as follows:

$$5214. 73 = 5 \cdot 10^3 +2 \cdot 10^2 +1 \cdot 10^2 +4\cdot 10 + 7\cdot10^{-1}+ 3\cdot10^{-2} +...$$

1.9   SIGNIFICANT DIGITS

A **significant digit** of an approximate number is any non-zero digit, in its decimal representation or any zero lying between significant digits or   used as placeholder, to indicate a retained place. All the other zeros of the approximate number that serve

Only to fix the position of the decimal point are not be considered as significant digits.

For example consider the number 0.007040. The first three zeros are not significant digits, since they serve only to fix the position of decimal point and indicate the place values of the other digits. The other two zeros are significant digits since the first lies between the digits 7 and 4 and the second shows that we retain the decimal place 10^{-6} in the approximate number. If the last digits of 0.007040, then the number must be written as 0.00704. From this point of view the numbers 0.007040 and 0.00704 are not the same because

The former has four significant digits and the latter has only three. When writing large numbers, the zeros on the right can serve both to indicate the significant digits and to fix the place values of other digits. This can lead to misunderstanding when the numbers are written in the ordinary way.

1.10 CORRECT DIGITS

In this section we now introduce the notion of correct digits of an approximate numbers.

Definition: If the first n significant digits of an approximate number are correct if the absolute error of the number does not one half unit in the nth place counting from left to right

If α is an approximate number given by (1.2) which takes the place of an exact number X we know that

$$| X - \alpha | \le \frac{1}{2}10^{m-n+1}$$

Then by definition the first n digits

$\alpha_m,\ \alpha_{m-1}, \alpha_{m-2}, \cdots , \alpha_{m-n+1}$ of this number are correct for example consider the exact number X = 25.97. Then with respect to X, the number α = 26 .00 is an approximation correct to three digits, since

$$| X - \alpha | = |25.97 - 26.00| = 0.03 \le \frac{1}{2}10^{m-n+1}$$

$$\Rightarrow |25.97 - 26.00| = 0.03 \le \frac{1}{2}\times (0.1)$$

1.11 GENERAL ERROR FORMULA

In this section we derive a general formula for determining error committed in using certain functions.

Let $\qquad u = f(x_1 , x_2 , \ldots , x_n)$

Be a differentiable function in the variables $x_1 , x_2 , \ldots ,$ and x_n

Then we get $u + \Delta u = f(x_1 + \Delta x_1 , x_2 + \Delta x_2 , \ldots , x_n + \Delta x_n)$

$$\Rightarrow \Delta u = u + \Delta u - u$$

$$= f(x_1 + \Delta x_1 , x_2 + \Delta x_2 , \ldots , x_n + \Delta x_n) - f(x_1 , x_2 , \ldots , x_n)$$

Expanding the right handed side by Taylor's series we get

$$\Delta u = f(x_1, x_2, \ldots, x_n) + \sum_{i=1}^{n} \frac{\partial f}{\partial x_i} \Delta x_i + \text{terms involving } (\Delta x_i)^2 \text{ and other higher}$$

order terms which are negligible

$$- f(x_1, x_2, \ldots, x_n)$$

Neglecting squares and higher powers Δx_i we have

$$\Delta u \approx \sum_{i=1}^{n} \frac{\partial f}{\partial x_i} \Delta x_i = \frac{\partial f}{\partial x_1} \Delta x_1 + \frac{\partial f}{\partial x_2} \Delta x_2 + \ldots + \frac{\partial f}{\partial x_n} \Delta x_n.. \qquad \ldots..(1.3)$$

The above equation is the equation for the absolute error of u.

Dividing (1.2) by u, we get

$$\frac{\Delta u}{u} = \sum_{i=1}^{n} \frac{\partial f}{\partial x_i} \frac{\Delta x_i}{u} = \frac{\partial f}{\partial x_1} \frac{\Delta x_1}{u} + \frac{\partial f}{\partial x_2} \frac{\Delta x_2}{u} + \ldots + \frac{\partial f}{\partial x_n} \frac{\Delta x_n}{u} \qquad \ldots..(1.4)$$

which is the formula for finding the Relative Error hence we have

$$E_R = \frac{\Delta u}{u} = \sum_{i=1}^{n} \frac{\partial f}{\partial x_i} \frac{\Delta x_i}{u} = \frac{\partial f}{\partial x_1} \frac{\Delta x_1}{u} + \frac{\partial f}{\partial x_2} \frac{\Delta x_2}{u} + \ldots + \frac{\partial f}{\partial x_n} \frac{\Delta x_n}{u} \qquad \ldots..(1.5)$$

Remarks: Let $|\Delta x_i|$, $(I = 1, 2, \ldots, n)$ be absolute errors of the arguments of the function. Then the absolute error of the function is

$$|\Delta u| = |f(x_1 + \Delta x_1, x_2 + \Delta x_2, \ldots, x_n + \Delta x_n) - f(x_1, x_2, \ldots, x_n)|$$

Expanding by Taylor's theorem and proceeding as mentioned above we get

$$|\Delta u| = |df(x_1, x_2, \ldots, x_n)| = \left|\sum_{i=1}^{n} \frac{\partial f}{\partial x_i} \Delta x_i\right|$$

$$\leq \sum_{i=1}^{n} \left|\frac{\partial f}{\partial x_i}\right| |\Delta x_i|$$

Thus
$$|\Delta u| \leq \sum_{i=1}^{n} \left|\frac{\partial f}{\partial x_i}\right| |\Delta x_i|$$

Theorem 1: If a positive approximate number a has n correct digits in the narrow sense, the relative error δ of this number does not exceed $\left(\frac{1}{10}\right)^{n-1}$ divided by the first significant digit of the given number, or $\delta \leq \frac{1}{a_m}\left(\frac{1}{10}\right)^{n-1}$

Cor 1: If for the limiting relative error of the number a_m we can take

$$\delta_a = \frac{1}{a_m}\left(\frac{1}{10}\right)^{n-1}$$

where a_m is the first significant digit of the number a_m.

Cor 2: α has more than two correct i.e. $n \geq 2$, then for all practical purposes the following formula If the number holds

$$\delta_a = \frac{1}{2a_m}\left(\frac{1}{10}\right)^{n-1}$$

1.12 ERROR OF A SUM

Theorem: The absolute error of an algebraic sum of several approximate numbers does not exceed the sum of the absolute errors of the numbers

Proof: Let $u_1, u_2, \ldots, u_n$ denote the n numbers. Let u denote the algebraic sum of these numbers.

We have $\qquad u = \pm u_1 \pm u_2 \pm \ldots \pm u_n$

$\Rightarrow \qquad |u| = |u_1| + |u_2| + \ldots + |u_n|$

Hence, we have

$\Rightarrow \qquad |\Delta u| \leq |\Delta u_1| + |\Delta u_2| + \ldots + |\Delta u_n|$

Cor 3: For the limiting point absolute error of an algebraic we can take the sum of the limiting absolute errors of terms

$$\Delta_u = \Delta_{u_1} + \Delta_{u_2} + \ldots + \Delta_{u_n}$$

From the above Inequality is follows that the limiting absolute error of the sum cannot be less that the least accurate term, which is to say the term having the maximum absolute error.

1.13 RULES FOR THE ADDITION OF APPROXIMATE NUMBERS

(i) Find the numbers with the least member number of decimal   places and leave them unchanged

(ii) Round off the remaining numbers, retaining one or two more decimal places than those with the smallest number of decimals

(iii) Add the numbers, taking into account, taking into account all retail decimals

(iv) Round off the result, reducing it by one decimal

The rounding error of the sum does not exceed

$$\Delta_{max} \leq n \cdot \tfrac{1}{2} \cdot 10^m$$

Theorem 2: If the terms one and the same sign , the same sign ,the relative error of their sum does not exist the maximum limiting error of any of the terms

i.e. $\qquad \delta_u \leq \max(\delta_{u_1}, \delta_{u_2}, \ldots, \delta_{u_n})$

1.14   ERROR OF DIFFERENCE

Let u denote the   difference between approximate numbers u_1 and u_2

Then we have $u = u_1 - u_2$

The limiting absolute error of the difference is

$$\Delta_u = \Delta_{u_1} + \Delta_{u_2}$$

Hence the limiting absolute error of difference is equal to the sum of the limiting absolute error of the difference is diminuend

$$\delta_u = \frac{\Delta_{u_1} + \Delta_{u_2}}{E}$$

Where E is the exact value of the absolute magnitude of the difference between the numbers u_1 and u_2

Sol and the numbers with examples

Example 1: Find the percentage error in computing

$$y = 3x^2 - 6x \text{ at } x = 1, \text{ if the error in x is } 0.05$$

Solution: We have $y = 3x^2 - 6x$, $\Delta x = 0.05$

Differentiating *we* get

$$\frac{dy}{dx} = 6x - 6 \Rightarrow dy = (6x - 6) \quad dx \Rightarrow$$

$$E_A = (6x - 6) \quad \Delta x = 0, \text{ at } x = 1$$

$$E_R = \frac{E_A}{X} = 0 \qquad \text{at } X = 1$$

Example 2: The height of a tower was estimated to be 50 m

Using Theodolite But the height was 45. Calculate the absolute error, relative error and percentage error involved in the measurement

Solution: We have Actual height $= X_E = 45$ m

Estimated height $= X_A = 50$ m

Absolute error $= E_A = |X_E - X_A| = |50 - 45|$

$$= 5 \text{ m}$$

Relative Error $= E_R = \frac{E_A}{X} = \frac{5}{45} = 0.111$

Percentage error $= E_R \times 100 = 11.1$ %

Example 3: If $U = 10\ x^2y^2z^3$ and errors involved in x, y, z are 0.01, 0.02, 0.03 respectively are x = 1, y = 2, z = 3. Calculate the absolute error, relative error, and percentage relative error involved in evaluating u.

Solution: It is given that $u = 10\ x^2y^2z^3$

$$X = 1, y = 2, z = 3 \text{ and } \Delta X = 0.0\ 1, \Delta y = 0.0\ 2, \Delta z = 0.0\ 3$$

$$\text{Exact value} = u = 10\ 1^2 2^2 3^3 = 1080$$

We have absolute error

$$\Delta u = \frac{\partial u}{\partial x}\Delta x + \frac{\partial u}{\partial y}\Delta y + \frac{\partial u}{\partial z}\Delta z$$

$$= 20\ x\ y^2z^3\Delta x + 20\ x^2\ yz^3\Delta y + 30x^2y^2z^2\Delta z$$

$$= 20\ (1)\ (2^2\ )\ (3^3)\ (0.0\ 1\ ) + 20\ (1^2)\ (2\ )\ (3^3)\ (\ 0.0\ 2\ )$$

$$+ 30\ (\ 1^2)\ (2^2\ )\ (3^2\ )\ (0.03) = 140.4 = 75.6$$

The Relative Error $= E_R = \dfrac{E_A}{X} = \dfrac{75.6}{1080} = 0.07$

Percentage Error $= 100\ E_R = 100 \times 0.07 = 7.13 = 7\%$

Example 4: What is the limiting relative error if n = 3 & a_m = 3.

Solution: From the given data we have $n = 3, \alpha_m = 3$

Therefore we get

Using Cor 2

$$\delta_\alpha = \frac{1}{2\alpha_m}\left(\frac{1}{10}\right)^{n-1} = \frac{1}{2.3}\left(\frac{1}{10}\right)^{3-1} = \frac{1}{6}\left(\frac{1}{10}\right)^2 = \frac{1}{6}\%$$

Example 5: Young's modulus is determined from the deflection of a rod, a and b are the dimensions of the cross section

$$E = \frac{1}{4} \cdot \frac{l^3\ p}{a^3 bs}$$

where l = length of the rod, a and b are the dimensions of the cross section, s is the bending deflection, and p is the   load. Compute the limiting relative error in a determination of young's modulus E if p = 20 kg, δ_p = 0.1 %, a = 3mm, b = 44mm

$$\delta_b = 1\ \%, l = 50 \text{ cm}, \delta_l = 1\%, s = 2.5 \text{ cm}, \delta_s = 1\%,$$

S is the bending deflection and p is the load. Compute the limiting relative error in a determination of Young's modulus E. If p = 20 kgs,

Solution: It is given that

$$E = \frac{1}{4} \cdot \frac{l^3\ p}{a^3 bs}$$

Taking logarithms on both sides, we get

$$\text{Ln } E = 3 \ln + \ln p - 3 \ln a - \ln b - \ln s - \ln 4$$

Replacing increments by differentials, we get Relative error

$$= E_R = \frac{\Delta E}{E} = 3\,\frac{\Delta l}{l} + \frac{\Delta p}{p} - 3\,\frac{\Delta a}{a} - \frac{\Delta b}{b} - \frac{\Delta s}{s}$$

The relative error $= 3\,.0 \times 0.01 + 0.00\ 1 + 3\,.0 \times 0.01 + 0.01 + 0.01 = 0.081$

Therefore the 8 % Error

EXERCISE

1. Define the terms Absolute error and Relative Error
2. Briefly explain "Round off rule "
3. Define percentage error
4. Round off the following to three decimals
   - (i) 2.3645   [***Ans:*** 2.364]
   - (ii) 4.3455   [***Ans:*** 4.346]
5. Round off the following numbers to 4 significant digits
   - (i) 63.38257   [***Ans:*** 63.38]
   - (ii) 0.009231542   [***Ans:*** 0.009232]
   - (iii) 0.2537514 0   [***Ans:*** 0.2538]
6. If the number N is correct upto 3 significant digits, then what will be the maximum relative error
7. Round off each of the following numbers to three significant figures
   - (i) 58.56258   [Ans: 58.6]
   - (ii) 0.0039417   [Ans: 0.00394]
8. Find the percentage error in approximating in computation of $x - y$ for $x = 12.05$ and $y = 8.02$ having absolute errors

   $$\Delta x = 0.0005, \Delta y = 0.001$$

   [***Hint:*** Apply, Relative Error $= \frac{\Delta x - \Delta y}{x - y} = 0.001$]

9. If $\pi = 3.14$ instead of $\frac{22}{7}$, find the relative error and percentage error.

   [***Ans:*** 0.00093, 0.093]

10. Round-off the number 4.5126 to four significant figures and find the percentage error. [: $- 0.0088$]

11. Calculate the value of e^x at 0.75   [***Ans:*** 2.12]

12. Find the limiting absolute and relative errors of the volume of a sphere $V = \frac{1}{6}\pi d^3$ if the diameter d = 3.7 $\pm$ 0.05 cm and π = 3.14   [**Ans:** 4%]

13. Given u = xy +yz + zx, find the estimate of relative percentage error in the evaluation of u for x = 2.104

    Y= 1.935 and z = 0.845, which are the Approximate values correct to the last digit
    [**Ans:** 0.062]

14. Find the sum of following approximate numbers, correct
    To the last digits o.348, 0.1834, 345.4, 235.2 11.75, 0.0849, 0.0002435 and 0.0214
    [**Ans:** 602.2]

15. The length x and the width and y of a plate is measured accurate up to 1 cm as x = 5.43 m and y 3.82 m. Area Find the area of the plate and indicate its error
    [**Ans:** $0.925\ m^2$]

16. *Given* $f(x\ y,z) = \frac{5x\ y^2}{z^2}$: Find the relative maximum error in the absolute in the evaluation of f (x, y, z) at x = y = z = 1 is 5 and if Δ x = 0.1, Δy = 0.1, Δz = 0.1 are the of x, y, z absolute errors   [**Ans:** Hint: we have]

$$\Delta f = \frac{\partial f}{\partial x}\Delta x + \frac{\partial f}{\partial y}\Delta y + \frac{\partial f}{\partial z}\Delta z \Rightarrow$$

$$|\Delta f_{max}| = |\frac{\partial f}{\partial x}\Delta x| + |\frac{\partial f}{\partial y}\Delta y| + |\frac{\partial f}{\partial z}\Delta z|$$

$$= |\frac{5\ y^2}{z^2}\Delta x| + |\frac{10x\ y}{z^2}\Delta y| + |-\frac{5x\ y^2}{z^3}\Delta z|$$

Substituting the given values we get $|\Delta f_{max}| = 2.5$

Hence $[E_R]_{max} = 0.5]$

CHAPTER 2

Solving Algebraic and Transcendental Equations

2.1 INTRODUCTION

In this chapter we introduce the methods of solving algebraic and transcendental equations. Many scientific and engineering problems require solving the algebraic and transcendental equations. The main tool used in numerical methods is to take an approximation to an expected value and to then apply an algorithm which improves the approximation. This process is then repeated until the approximation is sufficiently close.

If f(x) is purely a polynomial in x and is of the form

$f(x) = a_0x^n + a_1x^{n-1} + a_2x^{n-2} + \ldots + a_n$, where a_0 , a_1, a_2 , $\ldots$, a_n are constants real or complex and $a_0 \neq 0$.

Then f(x) = 0 is said to be an algebraic equation and f(x) is called an algebraic function. It is a function which is a sum, difference, or product of two polynomials.

Example 1: $x^2 - 7x + 12 = 0$ is an algebraic equation.

If f(x) = 0 contains some other functions, namely trigonometrically, logarithmic, exponential etc, then f(x) = 0 is called a transcendental equation.

Example 2: cot x is a transcendental function.

Example 3: Sin x – sin 2x = 0 is a transcendental equation.

Definition (Root of an equation): A number α is called a root of an equation f(x) = 0, If f(α) = 0.

If α is a root of f(x) = 0 then it is called a zero of the function f (x)

If α is a root of f(x) = 0, then we say that α satisfies the equation f(x) = 0

If α is a root of f(x) = 0, then α may be real or complex

If α is root of an equation then, $(x-\alpha)$ is a factor of f(x) so that f(x) can be expressed as

$$f(x) = (x - \alpha)^m g(x) \qquad\qquad .....(2.1)$$

Where m is a positive integer and g(x) is a function such that $g(\alpha) \neq 0$

If $m = 2$ then α is called a double root.

If $f(x) = (x - \alpha)^m g(x)$ holds for all for m > 1, then we say that α is multiple root.

The process of finding the roots of a given equation is known as   solving the equation.

In some cases it is not possible to find the exact roots of given equation .In such a situation we try to find the approximate value of the root. The methods which we use to find the approximate roots require iteration or repetition of certain computational steps. In iteration methods we start with an approximate value say x_o such that $f(x_0)$ is near to zero and construct a sequence of numbers $x_1, x_2, \ldots, x_n$ such that f(x) tends to zero as x_n approaches a certain number α.

The   number α,   which   is   the   limit   of   the   sequence $\{x_n\}$ will be the root of the equation f(x) $=$ 0. The number x_0 is called the initial approximation of the root. $x_1, x_2, \ldots, x_n$ are called successive approximations to the root α. The process of constructing successive approximations to the number α is called an iteration process. If a sequence $\{x_n\}$ constructed by an iteration actually tends to α then we say that the iteration is convergent.

we define the root of (2.1) as follows:

Algebraically a real number α is said to be a root of the equation f(x) = 0, if $f(\alpha) = 0$

And geometrically the value of x is said to be a root of the equation f (x) = 0, where the graph of f(x) meets the x-axis.

Remarks: The choice of the initial approximation depends upon the desired degree of accuracy of the root being found and the method employed for iteration.

Suppose we have an equation   f(x) = 0 $\qquad\qquad$ .....(2.2)

where the function f(x) is defined and continuous on some finite or infinite interval $a < x < b$

we will assume that (2.1) has only isolated roots i.e. for each root of (1) there is a neighborhood which does not contain   any other roots of the equation.

Approximating the isolated real roots of the equation (2.1) has two stages.

(i) Isolating the roots and (ii) Improving the values of the approximate roots In isolating the roots we establish the smallest possible intervals [a, b] containing one and only one root of the equation (2.1). In the second stage we refine the roots obtained to specified degree of accuracy.

Intermediate value theorem: Let f(x) be a real valued continuous function of a real variable x. If f(x) has opposite signs for two values x = a and x = b .i.e. f(a) f(b) < 0 .then equation f(x) = 0 has at least one root between a and b .

Theorem 1: An algebraic equation of degree n has exactly n roots.

Theorem 2: If f(a) and f(b) are of same sign, then either there is no real root or there are an even number of real roots of the equation f(x) = 0 between a and b.

Theorem 3: (Descarte's rule of sign): The number of positive roots of an algebraic equation f(x) = 0 with real co-efficient cannot exceed the number of changes in sign of f(x) and the number of negative roots of f(x) = 0 cannot have more than the number of changes in sign of f(− x).

In this chapter we describe some numerical methods, for the solution of the form f(x) = 0, where f(x) is algebraic or transcendental or combination of both.

2.2   GRAPHICAL METHOD

Consider the equation   f(x) = 0 $\qquad\qquad$ .....(2.3)

The real roots of (2.3) can be determined approximately as the abscissas of the points of intersection of the graph of the function y = f(x) with the x − axis. If the equation (2.1) does not have nearly equal roots, this method can readily used to isolate them. It is sometimes advisable equation to replace equation (2.3) by an equivalent equation.

$$\phi\,(x) = \Psi\,(x) \qquad\qquad .....(2.4)$$

Where the functions $\phi\,(x) = \Psi\,(x)$ are simpler than f(x).

Then we construct graphs of the functions

$$Y= \phi\,(x)\ \ \text{and}\ y = \Psi\,(x)$$

and the   desired   roots are   obtained   as the abscissas of the points of intersection of these graphs.

The values of x for which y = 0 or where the line crosses the x - axis gives the root.

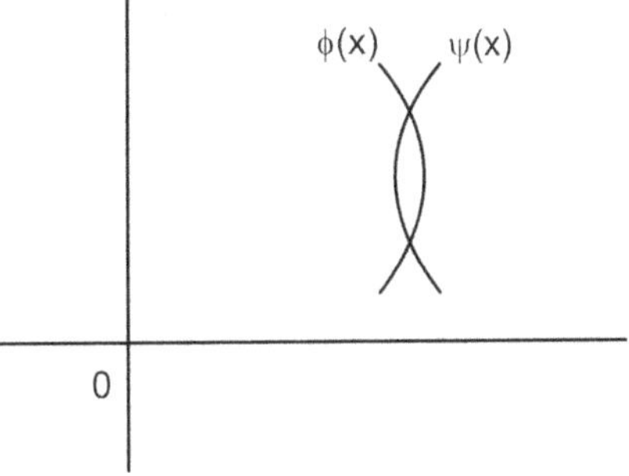

Fig. 2.1

2.3   BISECTION METHOD

Suppose we have an equation

$$f(x) = 0 \qquad\qquad .....(2.5)$$

where the function is continuous on the interval [a, b] and

f(a) f(b) < 0. In order to find a root of (lying in the interval [a, b], divide in half. If $f(\frac{a+b}{2}) = 0$. Then $\frac{a+b}{2}$ is a root of f(x) = 0.

If $f(\frac{a+b}{2}) \neq 0$, then $\frac{a+b}{2}$ is not a root of (2.5) then we choose that half $[a, \frac{a+b}{2}]$ or $[\frac{a+b}{2}, a]$, at the end points of which the function f (x) has opposite signs. The newly reduced interval be denoted by $[a_1, b_1]$ where $f(a_1) f(b_1) < 0$.

This interval is again halved and the same investigation is made.

The next approximation of the root is $\frac{a_1+b_1}{2}$

The root now lies in $[a_1 \; \frac{a_1+b_1}{2}]$ or in $[\frac{a_1+b_1}{2}, b_1]$

If $f(a_1) f(\frac{a_1+b_1}{2}) < 0$ then root lies in $[a_1, \; \frac{a_1+b_1}{2}]$

Or if $f(\frac{a_1+b_1}{2}) f(b_1) < 0$, the root lies in $[\frac{a_1+b_1}{2}, b_1]$

Denote the interval in which the root lies   by $[a_2, b_2]$

We have $f(a_2) f(b_2) < 0$.

Repeating the process of bisection , we obtain successive sub- intervals $[a_k, b_k]$ with $f(a_k) f(b_k) < 0$,and $b_k - a_k = \frac{1}{2^k}$ (b-a) so that   the root α lies between a_k and b_k. We stop the process of bisection when the desired   degree of accuracy is achieved.

Bisection method is a simple method. At any stage the approximate value of the root does not depend on its functional value but on the sign of the functional value.

The bisection method is also called Interval halving method or Bolzano's method. It known as the method of successive bisection. This method is definite convergent and the convergence is linear. It is a direct search method.

The left end points $a_1 a_2$, . . . , a_n, . . . form a   monotonic   non-decreasing bounded sequence, and the right end points $b_1, b_2, . . . , b_n, . .$ form a monotonic non-increasing   bounded sequence, therefore there exists a common limit. The method is a slowly convergent method.

$$\alpha = \lim_{n\to\infty} a_n = \lim_{n\to\infty} b_n$$

The bisection method is conveniently used to in rough approximation of the root of given equation.

Bisection method is a simple method. It is unconditionally convergent.

Advantages:

(a) The method is very simple.

(b) At any stage the approximate value of the root does not depend on its functional value but on the sign of the functional value.

(c) The method is unconditionally convergent.

(d) Disadvantages

(e) The method is very slow and laborious

(f) The method requires a large number of iterations to obtain moderately.

Example 4: Find the root of

$$x^4 + 2x^3 - x - 1 = 0$$

Lying in [0, 1] by using Bisection method

Solution: Let f (x) = Find the root of

$$x^4 + 2x^3 - x - 1 = 0$$

And $a = 0$ and $b = 1$

We   have f(a) = f(0) = $-$ 1 and f(a) = f (1) = 1

Since f(a) f(b) =   f(0) f(1) = $-$ 1   i.e.   f(a) f(b) < 0, the root of the given equation lies between 0 and 1

By bisection method

$$\therefore \quad \frac{a+b}{2} = \frac{0+1}{2} = 0.5$$

Since f(0.5) = $-$ 1.19 $\neq$ 0

0.5 is not a root of the given equation

Since f(0.5) = $-$ 1.19 and f(1) =1

The root lies   between 0.5 and 1

We have a_1 = 0.5, a_2 = 1

The next approximation of the root is $\frac{a_1+b_1}{2}$ = $\frac{0.5+1}{2}$ = 0.75

Since f(0.75) = 0.32 +0.84 $-$0.75 $-$1 = $-$ 0.59

Since f(0.75) f(1) = $-$ 0.59 (negative) < 0

Root of the equation lies between 0.75 and 1

The next approximation of the root is $\dfrac{0.75+1}{2} = 0.875$

We have f (0.875) = + 0.05 > 0

Since f(0.75) f(0.875) < 0, the root of the given equation lies in [0.75, 0.875]

The next approximation of the root is $\dfrac{0.75+0.875}{2} = 0.8125$

We have f(0.8125) = – 0.304 therefore

The next approximation of the root is $\dfrac{0.875+0.8125}{2} = 0.8438$

We have f (0.8438) = –0. 135

Since f (0.8438) f(0.875) < 0, the next approximation of the root is $\dfrac{0.8438+0.875}{2} =$ 0.8594

f(0.8594) = – 0.043 which is very near to zero

we can take the root of the given equation as $\dfrac{0.859+0.8755}{2}$

i.e. $\alpha = 0.867$

Example 5: Find the root of the equation $x^3 - 5x +1 = 0$

Using the Bisection method

Solution: Let f (x) = $x^3 - 5x +1$

We have f(0)= 1 and   f(1) = – 3

∴ f(0) f(1) = – 3 < 0 i.e. f(a)f(b) < 0 where a = 0 and b = 1

The root of the given equation lies between 0 and 1

Let α be a root of the equation.

The first approximation of the root is $\dfrac{0+1}{2} = 0.5$

Since f(0.5) = – 1.375

∴ f(0) f(0.5) = 1 (– 1.375) = – 1.375 < 0

Therefore the root of the equation lies between 0 and 0.5

The second approximation of the root is $\dfrac{0+0.5}{2} = 0.25$

f(0.25) = –0.234375 $\Rightarrow$ f(0.25).f(0) < 0

The root of the equation lies between 0 and 0.25

The third approximation of the root is $\dfrac{0+0.25}{2} = 0.125$

We have f (0. 125) = 0.37695 > 0 (positive)

Since f (0.125) f (0.25) < 0 [i.e. f (0.125) and f (0.25) are opposite in sign the root of the equation lies between 0.125 and 0.25

The fourth approximation of tho root lies between 0 125 and 0.25

The fourth approximation of the root is $\frac{0.125+0.25}{2}$ = 0.1875

∴ We get f(0.1875) = 0.06910

Since f(0.1875) and f(0.25) are opposite in sign the next approximation lies between 0.1875 and 0.25

The fifth approximation of the root = $\frac{0.1875+0.25}{2}$ = 0.21875

The value   x = 0.21875 is taken as the approximate value of the root of the given equation

Example 6: Find a real root of $x^3 - x - 1 = 0$ correct to four decimal places

Solution: Let f (x) $= x^3 - x - 1 = 0$

We have f (1.5) = 0.8750 and f (1) = −1

∴ f (1) f(1.6) < 0 i.e. f(1) and f(1.5) are opposite in sign

Therefore the root of given equation lies between 1 and 1.6

The root is $\frac{1+1.5}{2}$ = 1.25 [first approximation]

We have f(1.25) = − 0.29688 , f(1.5) = 0.8750

f(1.25) and f(1.5) are opposite in sign, therefore the root of the equation lies between 1.25 and 1. 5 the next approximation of the root is $\frac{1.25+1.5}{2}$ = 1.3750 now we have f (1.3750) = 0.22461.

Since f (1.25) and f(1.3750) are opposite in sign, the root of the equation lies between 1.25 and 1.3750. The next approximation of the root is $\frac{1..25+1.3750}{2}$ = 1.3125

We get f(1.3125) = − 0.051514

Since f(1.3125) and f(1.3750) are opposite in sign the root of the equation lies between 1.3125 and 1.3750

Therefore the next approximation of the root is $\frac{1.3125+1.3750}{2}$ = 1.3438

We have f (1.3438) = 0.082832

Since f (1.3125) and f (1.3438) are opposite in sign, the root of the equation is
$\frac{1.3125+1.34\ 38}{2}$ = 1. 1.3282

We have f (1.3282) = 0.014 898

Since f (1.3125) f(3282) < 0, the root of the equation lies between 1.3125 and 1.3282

The next approximation of the root is $\dfrac{1.3125+1.3282}{2}$ = 1.3204

We have f (1. 3204) = $-$ 0.018340

Since f (1.3204) f (1.3282) < 0 the root of the equation lies between 1.3204 and 1.3282

Hence the approximation of the root is $\dfrac{1.3204 +1.3282}{2}$ = 1.3243

Since f (1.3243) f (1.3282) < 0

The root of the equation lies between 1.3243 and 1.3282

The next approximation of the root is $\dfrac{1.3243+1.3263}{2}$ = 1.3282

$\Rightarrow$ f (1.3263) f(1.3243)< 0 $\Rightarrow$ the root of the given equation lies between. The root of the equation is

$$\dfrac{1.3243+1.3263}{2} = 1.3253$$

Since f (1.3243) f (1.3253) < 0 , the root lies between 1.3243 and 1.3253

Hence the root is $\dfrac{1.3243+3253}{2}$ = 1.3248

The root of the given equation (Correct 4 decimal places) is taken as 1.3248.

Example 7: Find the root of the equation $x - e^{\frac{1}{x}}$ = 0 by bisection method up to four approximations, given that the root lies between 1 and 2

Solution: Given that the root of the equation lies between 1 and 2

Let $\qquad$ $f(x) = x - e^{\frac{1}{x}}$,

we have a = 1, b = 2, f(a) = f(1) = 1 $- e$ = $-$ 1.718

$$f (b) = f(2) = 2 - e^{\frac{1}{2}} = 0.352$$

f(1) f(2) < 0 , therefore   the root of the equation lies between 1 and 2 the root by bisection method is $\dfrac{1+2}{2}$ = 1.5

$$f (1.5) = -0.477 \Rightarrow f (1.5) f(2) < 0$$

Therefore the root of the equation lies between 1.5 and 2 the root is $\dfrac{1.5+2}{2}$ = 1.75

now f(1.75) = 0.021 $\Rightarrow$ f(1.75) f(1.5) < 0

$\Rightarrow$ The root of the given equation lies between 1.5 and 1.75 the next approximation of the root is $\dfrac{1.5+1.75}{2} = \dfrac{1.5+1.75}{2} = 1.625$

Since f (1. 687) = $-$ 0.201, we have f (1. 687) f(1.75) < 0

Hence the root of the equation lies in [1.687, 1.75]

The next approximation of the root is $\dfrac{1.87+1.75}{2} = 1.718$

The exact root of the given equation is taken as 1.718

EXERCISE 1

1. Find the positive root of x $-$ cos x =0 , by bisection method $\qquad$ [*Ans:* 0.7388]

2. Using bisection method find the negative root of $x^3 - 4x + 9 = 0$ by bisection method $\qquad$ [*Ans:* $-$ 2.7065]

3. Find the root of $x^3 - 5x + 1 = 0$ $\qquad$ [*Ans:* 0.21875]

4. Using the bisection method find the root of $x^3 - x - 4 = 0$, in (1.75 ,2) .carry out 5 steps $\qquad$ [*Ans:* 1.793]

5. Find the approximate value of the root of   the equation $x^3 - 4x - 9 = 0$, in (2.5, 3) by using bisection method $\qquad$ [*Ans:* 2.707]

6. Find the root of $x^3 - 2x - 5 = 0$,in (2,3) by using bisection method in (2,3) carryout   five steps (2,3) $\qquad$ [*Ans:* 2.094]

7. Find the root of $x^3 - x^2$ +x $-$ 7 = 0m, by using bisection method in (2, 2.5).

$\qquad$ [*Ans:* 2.1049]

8. Find the approximate value of the root of the equation e^x= x +2 in (1, 1.4) by using bisection method $\qquad$ [*Ans:* 1.146]

9. Find a   real root of $e^x - x^3$= 0 correct upto four significant digits [*Ans:* 4,536376]

10. By using bisection method, find the approximate root of the equation sin x = $\dfrac{1}{x}$. That lies between x = 1 and x = 1.5 .Carry out   computations upto seventh stage

$\qquad$ [*Ans:* 1.11328]

2.4   ITERATION METHOD

Iteration method is one of the most important methods in numerical solution of equations. It is also known as successive approximation method

Suppose we have an equation

$$f(x) = 0 \qquad\qquad .....(2.6)$$

Where f(x) is a continuous function if it is required to find the roots of (2.4), we rewrite the equation (2.6), in the form

$$x = \phi(x) \qquad\qquad .....(2.7)$$

Let x_0 be an appropriate value of the desired root α. Substituting it for x on the right side of (2.7), we obtain a new value of x i.e. The first approximation x_1 of x is given by

$$x_1 = (x_0)$$

Now inserting x_1 in the right side member of (2.7) in place of x_0, we get a new value of x_1 i.e. x_2 and so on

Hence the successive approximations are as follows

$$x_1 = \phi(x_0)$$
$$x_2 = \phi(x_1)$$
$$x_3 = \phi(x_2)$$
$$x_{n+1} = \phi(x_n)$$

If the sequence of approximations $x_0, x_1, x_2, \ldots, x_{n+1}, \ldots$

Is convergent, then the equation (2.6) will yield the result, otherwise not.

The sequence of approximations $x_0, x_1, x_2, \ldots, x_{n+1}$, is not always convergent.

Consider $\qquad x_{n+1} = \phi(x_n)$

As n increases, the left side tends to the root α and if ϕ is continuous the right side tends to (α). Hence in the limit, we have $\alpha = \phi(\alpha)$.

Theorem 4: Let α be a root of f(x) = 0 and let I be an interval containing the point x = α. Let $\phi(x)$ and $\phi'(x)$ be continuous on I where $\phi(x)$ is defined by the equation x = $\phi(x)$ which is equivalent to f(x) = 0. Then $|\phi'(x)| < 1$ for all x ϵ I. The sequence of approximations $x_0, x_1, x_2, \ldots, x_{n+1}, \ldots$ defined by $x_{n+1} = \phi(x_n)$.

Converges to the root α, provided that the initial approximation x_0 is chosen in I
Proof. It is given that α is a root of $\alpha = \phi(\alpha)$ $\qquad\qquad .....(2.8)$

The first approximation of the root is $x_1 = (x_0)$ $\qquad\qquad .....(2.9)$

From (2.8) and (2.9) we get

$$\alpha - x_1 = \phi(\alpha) - \phi(x_0)$$

By using mean value theorem the right hand side can be written as $(\alpha - x_0)\,\phi'(x_0)$ therefore   we obtain

$$\alpha - x_1 = (\alpha - x_0)\,\phi'(\alpha_0)\ ,\ x_0 < \alpha_0 < \alpha \qquad\qquad .....(2.10)$$

$$\alpha - x_2 = (u - x_1)\,\phi'(\alpha_1)\ ,\ x_1 < \alpha_1 < \alpha \qquad\qquad .....(2.11)$$

$$\alpha - x_3 = (\alpha - x_2)\,\phi'(\alpha_2),\ x_2 < \alpha_2 < \alpha \qquad\qquad .....(2.12)$$

$$\alpha - x_{n+1} = (\alpha - x_n)\,\phi'(\alpha_n)\ ,\ x_n < \alpha_n < \alpha \qquad\qquad .....(2.13)$$

If we let $|\phi'(\alpha_i)| \le k < 1$, for all I then we have $|\alpha - x_1| \le |\alpha - x_0|$

$$|\alpha - x_2| \le |\alpha - x_1| \qquad\qquad .....(2.14)$$

Which shows that each successive approximation remains in *I*

Multiplying (2.10) to (2.13) and simplifying, we get

$$\alpha - x_{n+1} = (\alpha - x_0)\,\phi'(\alpha_0)\,\phi'(\alpha_1)...\,\phi'(\alpha_n)$$

Since, $|\phi'(\alpha_i)| < k$, the above equation becomes

$$|\alpha - x_{n+1}| \le k^{n+1}|\alpha - \alpha_0|$$

As $n \to \infty$, the right hand side tends to zero, and it follows that the sequence of approximations $x_0, x_1, x_2, \dots, x_{n+1}$, converges to the root α if $k < 1$.

Multiplying (2.10) to (2.13) and simplifying, we get

$$\alpha - x_{n+1} = (\alpha - x_0)\,\phi'(\alpha_0)\,\phi'(\alpha_1)...\,\phi'(\alpha_n)$$

Since, $|\phi'(\alpha_i)| < k$, the above equation becomes

$$|\alpha - x_{n+1}| \le k^{n+1}|\alpha - \alpha_0|$$

As n $\to \infty$, the right hand side tends to zero, and it follows that the sequence of approximations $x_0, x_1, x_2, \dots, x_{n+1}$, converges to the root α if $k < 1$.

Example 8: Find a real root of $e^x - x^3 = 0$, correct up to four significant digits

Solution: Let f (x) = $e^x - x^3 = 0$

It can be expressed as follows

$$X = e^{x/3} \qquad\qquad .....(2.15)$$

We have $\phi(x) = e^{x/3} \Rightarrow \phi'(x) = \frac{1}{3}e^{x/3} \Rightarrow |\phi'(x)| = |\frac{1}{3}e^{x/3}| < 1$ is true for x = 0 to x = 3

i.e. The criteria of convergence is satisfied. Therefore the iterative method is applicable.

Since the root of the equation lies between 0 and 3, the initial value of the root can be taken as $x_0 = \frac{0+3}{2} = 1.5$.

Applying iteration method we get

$$x_1 = \phi(x_0) = e^{\frac{x_0}{3}} = e^{\frac{1.5}{3}} = 1.64872$$

$$x_2 = \phi(x_1) = e^{\frac{x_1}{3}} = e^{\frac{1.64872}{3}} = 1.732514$$

$$x_3 = \phi(x_2) = e^{\frac{x_2}{3}} = e^{\frac{1.732514}{3}} = 1.7815875$$

$$x_4 = \phi(x_3) = e^{\frac{x_3}{3}} = e^{\frac{1.7815875}{3}} = 1.8109698$$

$$x_5 = \phi(x_4) = e^{\frac{x_4}{3}} = e^{\frac{1.8109698}{3}} = 1.8287937$$

$$x_6 = \phi(x_5) = e^{\frac{x_5}{3}} = e^{\frac{1.8287937}{3}} = 1.8396915$$

$$x_7 = \phi(x_1) = e^{\frac{x_1}{3}} = e^{\frac{1.8396915}{3}} = 1.8463865$$

$$x_8 = \phi(x_7) = e^{\frac{x_7}{3}} = e^{\frac{1.8463865}{3}} = 1.8505117$$

The root of the equation at this stage is taken as 1.8505

The given equation can be rewritten as x = 3 ln x

Here we have $\phi(x)$ = 3 ln x

$\Rightarrow |\phi'(x)| = |\frac{3}{x}| < 1 \Rightarrow x < -3$ is not applicable, we can take the root as a value > 3

Let the initial approximation of the root be taken as $x_0 = 4$

We have $x_1 = (x_0) = 3 \ln x_0 \Rightarrow x_1 = 3 \ln 4 = 4.1599831$

$x_2 = (x_1) = 3 \ln x_1 \Rightarrow x_2 = 3 \ln 4.1599831 = 4.2757396$

$x_3 = (x_2) = 3 \ln x_2 \Rightarrow x_3 = 3 \ln 4.2757396 = 4.3588713$

$x_4 = (x_3) = 3 \ln x_3 \Rightarrow x_4 = 3 \ln 4.3588713 = 4.4166395$

$x_5 = (x_4) = 3 \ln x_4 \Rightarrow x_5 = 3 \ln 4.4166395 = 4.4561373$

$x_6 = (x_5) = 3 \ln x_5 \Rightarrow x_6 = 3 \ln 4.4561373 = 4.4828469$

$x_7 = (x_6) = 3 \ln x_6 \Rightarrow x_7 = 3 \ln 4.4828469 = 4.500775$

The root of given equation can be taken as 4.500775

Example 9: Solve the equation $x^3 + x^2 - 1 = 0$, by iteration method

Solution: Let   f(x) = $x^3 + x^2 - 1$

We have f(0) = -1 and f(1) = 1, f(0) f(1) = $-1 < 0$

Therefore the root of the equation lies between 0 and 1

The given equation can be written as $x^3+x^2 = 1$ Or $x^2(x+1) = 1$ or $x^2=\dfrac{1}{x+1}$ or $x = \dfrac{1}{\sqrt{x+1}}$ we have $\phi(x) = \dfrac{1}{\sqrt{x+1}}$, $\phi'(x) = -\dfrac{1}{2}\cdot\dfrac{1}{(x+1)^{3/2}}$, $|\phi'(0)| = \dfrac{1}{2} < 1$ and $|\phi'(1)| = \dfrac{1}{2^{5/2}} < 1$ $|\phi'(x)| < 1 \ \forall \ x \in (0, 1)$ therefore the iteration can be applied. The root of the given equation lies between 0 and 1.

Taking the initial value of the root as 0.75 i.e. $x_0 = 0.75$, and applying iteration method we get

$$x_1 = (x_0) = \frac{1}{\sqrt{x_0+1}} = \frac{1}{\sqrt{0.75+1}} = \frac{1}{\sqrt{1.75}} = 0.75593$$

$$x_2 = (x_1) = \frac{1}{\sqrt{0.75593}} = 0.75465$$

$$x_3 = (x_2) = \frac{1}{\sqrt{0.75465}} = 0.75493$$

$$x_4 = (x_3) = \frac{1}{\sqrt{0.75493}} = 0.75487$$

$$x_5 = (x_4) = \frac{1}{\sqrt{0.75487}} = 0.75488$$

$$x_6 = (x_5) = \frac{1}{\sqrt{0.75488}} = 0.75488$$

Hence the root of the equation is 0.75488

Example 10: Find the root of the equation $x = \cos x$, near $x = \dfrac{\pi}{4}$, correct to three decimal process by fixed point iteration.

Solution: we have $\phi(x) = \cos x$, $x_0 = \dfrac{\pi}{4}$

Applying fixed point iteration we get

$$x_1 = \phi(x_0) = \cos x_0 = \cos \frac{\pi}{4} = 0.7071068$$

$$\Rightarrow x_2 = \phi(x_1) = \cos x_1 = \cos 0.7071068 = 0.7602446$$

$$x_3 = \phi(x_2) = \cos x_2 = \cos 0.7602446 = 0.7246675$$

$$x_4 = \phi(x_3) = \cos x_3 = \cos 0.7246675 = 0.7487199$$

$$x_5 = \phi(x_4) = \cos x_4 = \cos 0.7487199 = 0.7325608$$

$$x_6 = \phi(x_5) = \cos x_5 = \cos 0.7325608 = 0.7434642$$

$$x_7 = \phi(x_6) = \cos x_6 = \cos 0.7434642 = 0.7361282$$

$$x_8 = \phi(x_7) = \cos x_7 = \cos 0.7361282 = 0.7410736$$

$$x_9 = \phi(x_8) = \cos x_8 = \cos 0.7410736 = 0.7377441$$

$$x_{10} = \phi(x_9) = \cos x_9 = \cos 0.7377441 = 0.7399878$$

The required root of the given equation is 0.738 correct to three decimal places

Example 11: Evaluate the square root of 5 using the equation $x^2 - 5 = 0$ by applying the fixed point iteration algorithm.

Solution: Consider $x^2 - 5 = 0 \Rightarrow x^2 = 5$ dividing both sides by 5 we get

$$x = \frac{5}{x} \text{ or } x + x = \frac{5}{x} + x \Rightarrow 2x = \frac{5}{x} + x \Rightarrow x = \frac{1}{2}[\frac{5}{x} + x] \text{ which is of the form}$$

$$x = \phi(x) \text{ where } \phi(x) = \frac{1}{2}[\frac{5}{x} + x]$$

$$x = \phi(x) \text{ where } \phi(x) = \frac{1}{2}[\frac{5}{x} + x]$$

Differentiating with respect to x we get $\phi'(x) = \frac{1}{2}[-\frac{5}{x^2} + 1]$ clearly $|\phi'(x)|$ is less than 1 for all x near 2. Let $x_0 = 2$ be the initial approximation. Applying fixed point algorithm we get

$$x_1 = \phi(x_0) \Rightarrow x_1 = \frac{1}{2}[\frac{5}{x_0} + x_0] = \frac{1}{2}[\frac{5}{2} + 2] = 2.25$$

$$x_2 = \phi(x_1) \Rightarrow x_2 = \frac{1}{2}[\frac{5}{x_1} + x_1] = \frac{1}{2}[\frac{5}{2.25} + 2.25] = 2.236$$

Hence we get $\sqrt{5} = 2.236$

The results obtained may be written in the form of a table as shown in the example given below.

Example 12: Find by the method of fixed point iteration, find the root of $x^2 - 6x + 2 = 0$ which lies between 5 and 6 correct up to four significant figures.

Solution: Let $\quad f(x) = x^2 - 6x + 2 = 0$

We can write $x = \dfrac{6x-2}{x} = 6 - \dfrac{2}{x}$

We have $\phi(x) = 6 - \dfrac{2}{x} \Rightarrow \phi'(x) = \dfrac{2}{x^2}$

Then $|\phi'(x)| < 1$ in $[5,6]$.

Let $x_0 = 5$

Applying the method of fixed point iteration

$$x_{n+1} = \phi(x_n), \, n = 0,1,2,.....$$

The successive iterations (approximations) obtained are given below in the form of a table

n	x_n	$\phi(x_n)$
0	5	5.6
1	5.6	5.64286
2	5.64557	5.64574
3	5.64574	5.64575
4	5.64575	5.64575

Hence the root of the equation $x^2 - 6x + 2 = 0$ is 5.6458

EXERCISE 2

1. Solve the equation $3x - log_{10}x = 6$, to find the root by iteration method [***Ans:*** 2.108]

2. Evaluate $\sqrt{30}$ by iteration method   [***Ans:*** 5.477]

3. Find the root of $x^4 - x - 10 = 0$ by iteration method   [***Ans:*** 1.8556]

4. Find the root of $x^2 - 5x + 1 = 0$ by iteration method when the root lies very near to 0.   [***Ans:*** 0.2087]

5. Find by iteration method the root near 3.8, of the equation $2x - log_{10} x = 7$, correct to four decimal places.   [***Ans:*** 3.7893]

6. Evaluate by fixed point iteration method: $\sqrt{12}$ and $\frac{1}{\sqrt{12}}$ [***Ans:*** 3.464, 0.292]

7. Solve $x = 1 + tan^{-1} x$ by iteration method   [***Ans:*** 2.132]

8. Solve for x from $\cos x - x\, e^x = 0$ by iteration method   [***Ans:*** 0.5177]

9. Find the positive root of $3x - \sqrt{1 + \sin x} = 0$ by iteration method   [***Ans:*** 0.39185]

2.5 REGULA FALSI METHOD

Regula-falsi method is similar to bisection method in that we find two values x_0, x_1 (where $x_0 < x_1$) such that f(x_0) and f(x_1) are of opposite signs. Then the curve f(x) crosses the x -axis only once at the point x = α lying between the points x = x_0 and x = x_1

Consider the points A $(x_0, f(x_0))$ and B $(x_1, f(x_1))$ on the curve y = f (x). Then the equation of the chord AB is

$$\frac{y - f(x_0)}{x - x_0} = \frac{f(x_1) - f(x_1)}{x_1 - x_0} \qquad\qquad .....(2.16)$$

In the small interval $[x_0, x_1]$. The graph of the function can be considered a straight line. So that x – coordinate of the point of intersection of the chord joining A $(x_0, f(x_0))$ and B ($x_1, f(x_1)$) with the x – axis. Will give an approximate value of the root. So putting y = 0 in (2.16) we get

$$\frac{-f(x_0)}{x - x_0} = \frac{f(x_1) - f(x_1)}{x_1 - x_0} \qquad\qquad .....(2.17)$$

Or $\qquad\qquad$ x = $x_0 - \dfrac{f(x_0)}{f(x_1) - f(x_0)} (x_1 - x_0)$

Or $x = \dfrac{x_0 f(x_1) - x_1 f(x_0)}{f(x_1) - f(x_0)} = x_2$ (say)

If $f(x_0)$, $f(x_2)$ are of opposite signs, the root of the equation lies between x_0 and x_2, otherwise the root of the equation lies between x_2 and x_1. If the root of the equation les between x_0 and x_2 the next approximate value of the root is given by

$$x_3 = \dfrac{x_0 f(x_2) - x_2 f(x_0)}{f(x_2) - f(x_0)}$$

Otherwise $x_3 = \dfrac{x_2 f(x_1) - x_1 f(x_2)}{f(x_1) - f(x_2)}$

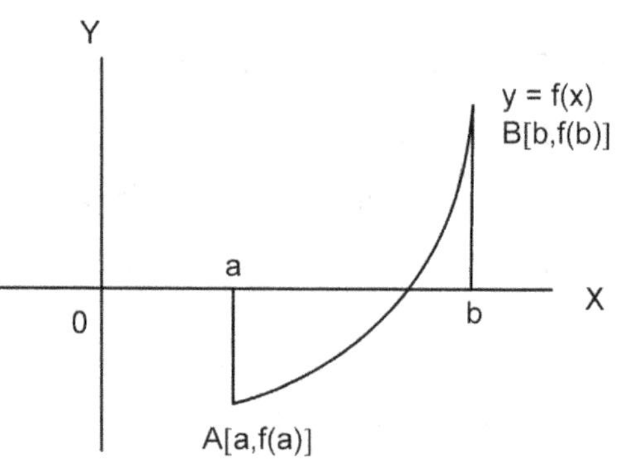

Fig. 2.2

The above process is repeatedly applied till the desired accuracy is obtained.

Geometrical interpretation: In the figure given below the curve y = f(x) between A (x = x_0) and B (x = x_1) cuts OX at Q. The chord AB cuts x-axis at P. It is clear that x = OQ is the actual value of the root. Whereas x = OP= x_2 the first approximation to the root and f (x_2) and f(x_0) are of opposite signs. So we apply the false position method in the interval (x_0, x_2) and OP_1 the next approximation to the root. The procedure is continued till the root is obtained to the desired degree of accuracy. The points of

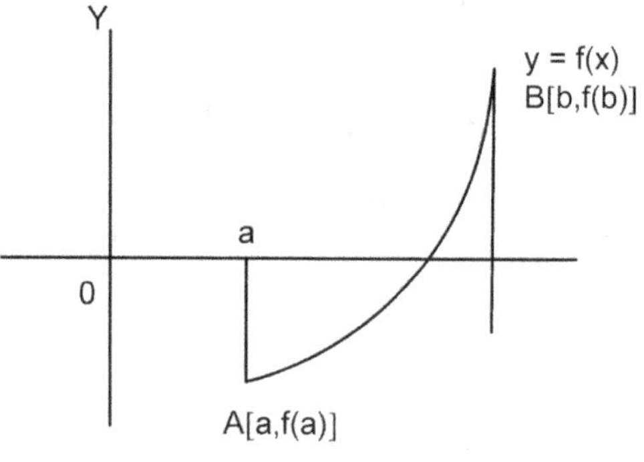

Fig. 2.3

intersection of the successive chords with x − axis , namely P_1 , P_2, P_3 , . . . tend to coincide with Q, the point where the curve y = f(x) cuts the x-axis . We get successive approximate value of the root of the equation f(x) = 0.

Example 13: Using Regula – Falsi method find the real root of the equation $e^x - 4x = 0$

Solution: Let   f(x)= $e^x - 4x$

We have f(0) = $e^0 - 4.(0) = 1$ (Positive), f (1) = $e^1 - 4 = e - 4 = -1.28$ (negative) and f(0)f(1) < 0,.

Therefore a root of the given equation lies between lies 0 and 1

Let $x_0 = 0$ and $x_1 = 1$

Let x_2 denote the next approximation of the root of the equation.

Then we have $x_2 = \dfrac{x_0 f(x_1) - x_1 f(x_0)}{f(x_1) - f(x_0)} = \dfrac{(0)(-1.128) - 1.1}{-1.128 - 1} = 0.439$

$\therefore \quad f(x_2) = f(0.439) = e^{0.439} - 4(0.439) = -0.2048$

Since $f(x_0)\, f(x_2) < 0$, the root of the equation lies between 0 and 0.439

The next approximation of the root is

$$x_3 = \dfrac{x_0 f(x_2) - x_2 f(x_0)}{f(x_2) - f(x_0)} = \dfrac{(0)(-0.2048) - 0.439 \times 1}{-0.2048 - 1} = 0.3643$$

We have $f(x_3) = f(0.3643) = e^{0.3643} - 4(0.3643) = -0.0177$

Since $f(x_0)\, f(x_3) < 0$, the root of the given equation lies between x_0 and x_3

The approximation of the root is

$$x_4 = \dfrac{x_0 f(x_3) - x_3 f(x_0)}{f(x_3) - f(x_0)} = \dfrac{0.(-0.0177) - 0.3643 \times 1}{-0.0177 - 1} = 0.35796$$

$$f(x_4) = f(0.35796) = -0.00143$$

$\therefore \quad f(x_0)\, f(x_4) < 0 \Rightarrow$ the root of the given equation lies between x_0 and x_4 .

The next approximation of the root is

$$x_5 = \dfrac{x_0 f(x_4) - x_4 f(x_0)}{f(x_4) - f(x_0)} = \dfrac{0.(-0.00143) - 0.35796 \times 1}{-0.00143 - 1} = 0.35745$$

And $f(x_5) = f(0.35745) = e^{0.35745} - 4(0.35745) = -0.000121$

$\therefore \quad f(x_0)\, f(x_4) < 0 \Rightarrow$ the root of the equation lies between x_0 and x_5

The next approximation of the root is

$$x_6 = \dfrac{x_0 f(x_5) - x_5 f(x_0)}{f(x_5) - f(x_0)} = \dfrac{0.(-0.000121) - 0.35745 \times 1}{-0.000121 - 1} = 0.35741$$

Therefore $f(x_6) = f(0.35741) = -0.0000181$

The values of x_5 and x_6 are almost equal i.e slightly differ

Hence, the root of the given equation is 0.35741.

Example 14: Find the real root of the equation $e^x \sin x = 1$, by Regula -Falsi method

Solution: Let $\quad f(x) = e^x \sin x - 1$

We have $f(2) = e^2 \sin 2 - 1 = -0.7421$, $f(3) = e^3 \sin 3 - 1 = 0.0511$ f(2) and f(3) are opposite in sign. Therefore the root of the equation lies between 2 and 3

Let $x_0 = 2$ and $x_1 = 3$ then $f(x_0) = -0.7421$, $f(x_1) = 0.0511$

The next approximate root of the equation is

$$x_2 = \dfrac{x_0 f(x_1) - x_1 f(x_0)}{f(x_1) - f(x_0)} = \dfrac{(2)(0.0511) - 3.(-0.7421)}{0..0511 + + 0.7421} = 2.93557$$

$$f(x_2) = f(2.93557) = e^{2.93557} \sin 2.93557 - 1 = -0.35538$$

Since $f(x_2)\, f(x_1) < 0$, the root lies between x_2 and x_1

The next approximation of the root is

$$x_3 = \frac{x_2 f(x_1) - x_1 f(x_2)}{f(x_1) - f(x_2)} = \frac{(2.93557)0.0511 - 3.(-0.35538)}{0..0511 + 0.35538} = 2.96199$$

$\therefore$ $f(x_3) = f(2.96199) = e^{2.96199} \sin 2.96199 - 1 = -0.000819$

Since $f(x_3)\, f(x_1) < 0$ the root of the equation lies between x_3 and x_1

The next approximation of the root is

$$x_4 = \frac{x_3 f(x_1) - x_1 f(x_3)}{f(x_1) - f(x_3)} = \frac{(2.96199)0.0511 - 3.(-0.000819)}{0..0511 + 0.000819} = 2.9625898$$

$\therefore$ $f(x_4) = e^{2.9625898} \sin 2.9625898 - 1 = -0.0001898$

Since $f(x_4)\, f(x_1) < 0$, the root of the equation lies between x_1 and x_4 the next approximation of the root is

$$x_5 = \frac{x_4 f(x_1) - x_1 f(x_4)}{f(x_1) - f(x_4)} = \frac{(2.9625898)(0.0511) - 3.(-0.00001898)}{0..0511 + 0.00001898} = 2.9626$$

Here we have $x_5 = 2.9626$ (approx)

The root of the given equation = 2.9626

Example 15: Write an algorithm for Regula-False method to evaluate root of a given equation f(x) = 0

Algorithm 1:

Step 1: Choose an interval (x_r, x_s), r < s such that $f(x_r)\, f(x_s) < 0$

Step 2: Determine the approximate root x_p in (x_r, x_s) by using the formula

$$x_p = \frac{x_r\, f(x_s) - x_s\, f(x_r)}{f(x_s) - f(x_r)}$$

Stop when the value of x_p is approximately equal in two successive stages or upto the desired degree of accuracy.

Example 16: Find the positive root of $x^2 - \log_{10}x - 12 = 0$ by Regula - Falsi method

Solution: The given equation is $x^2 - \log_{10}x - 12 = 0$

Let $\qquad f(x) = x^2 - \log_{10}x - 12$

We have $\qquad f(3) = 3^2 - \log_{10}3 - 12 = -3.4771$

$\qquad\qquad f(4) = 4^2 - \log_{10}4 - 12 = 3.3979$

$f(3)\, f(4) < 0$, therefore root of equation lies between 3 and 4

Let $\qquad x_0 = 3$ and $x_1 = 4$

The next approximation of the root is

$$x_2 = \frac{x_0 f(x_1) - x_1 f(x_0)}{f(x_1) - f(x_0)} = \frac{3(3.3979) - 4(-3.4771)}{3.3979 - (-3.4771)} = 3.5058$$

$\therefore$ f$(x_2) = (3.3979)^2 - log_{10}3.3979 - 12 = -0.2542$

Since f(x_2) and f(x_1) are opposite in sign the root of the equation lies between x_2 and x_1. The next approximation of the root is

$$x_3 = \frac{x_2 f(x_1) - x_1 f(x_2)}{f(x_1) - f(x_2)} = \frac{(3,5058(3.3979) - 4(-0.2542)}{f(x_1) - f(x_2)} = 3.5402$$

$$f(x_3) = (3.5402)^2 - log_{10}(3.5402) - 12 = -0.016$$

$$f(x_3) = (3.5402)^2 - log_{10}(3.5402) - 12 = -0.016$$

Since f(x_3) f$(x_1) < 0$, the root of the given equation lies between x_3 and x_1 the next approximation of the root is

$$x_4 = \frac{x_3 f(x_1) - x_1 f(x_3)}{f(x_1) - f(x_3)} = \frac{3.5402(3.3979) - 3(-0.016)}{3.3979 + 0.016} = 3.5424$$

We have f(x_4) = f(3.5424) = $(3.5424)^2 - log_{10}(3.5424) - 12 = -0.0007$

Since f(x_4) f$(x_1) < 0$, the root of the equation lies between x_4 and x_1

The next approximation of the root is

$$x_5 = \frac{x_4 f(x_1) - x_1 f(x_4)}{f(x_1) - f(x_4)} = \frac{3.5424(3.3979) - 4(-0.0007)}{(3.3979) - (-0.0007)} = 3.542$$

Hence the root of the given equation is 3.542.

EXERCISE 3

1. Briefly explain Regula Falsi method for determining the roots of a given equation

2. Find the real root of the equation $x^3 - 2x - 5 = 0$, using Regula -Falsi method

    [**Ans:** 2.093]

3. Evaluate the real root of the equation cos x $- xe^x = 0$, by Regula-Falsi method

    [**Ans:** 0.515201]

4. By using Regula–Falsi method, find an approximate root of the equation $x^4 - x - 10 = 0$, that lies between 1.8 and 2. Carry out three approximations

    [**Ans:** 1.8555]

5. Using Regula-Falsi method find the root of $x^3 - 4x - 9 = 0$ over (2.5, 3)

    [**Ans:** 2.706]

6. Find the root of $x^3 - 2x - 5 = 0$ by using Regula-Falsi method over (2,2.5)

[***Ans:*** 2.094]

7. Find the root of $e^x = 3x$ over (0.5, 1.0) by Regula-Falsi method   [***Ans:*** 0.619]

8. Solve by Regula-Falsi method $x^3 + x^2 + x + 7 = 0$, given that the root of the equation lies between $(-2.5, -2.0)$ [***Ans:*** -2.109]

9. Solve by Regula-Falsi method $xe^x = 2$, when the root lies between 0.5 and 1.0

[***Ans:*** 0.85]

10. Solve the equation for positive root by False position method $x\,e^x = 2$, when the root lies between 0 and 1   [***Ans:*** 0.85261]

11. Solve the equation x tan x +1 = 0 when, the root of the equation lies between 2.5 and 3   [***Ans:*** 2.798]

12. Solve the equation $x^4 - x - 1 = 0$ in (1, 1.5) by Regula Falsi method [***Ans:*** 1.22]

13. Solve the equation $x^3 - x - 1 = 0$ in (1.3, 1.43) by Regula Falsi method

[***Ans:*** 13247]

2.6   NEWTON-RAPHSON METHOD

Newton Raphson method is also an iteration method.   It is used to find the isolated roots of the equation

$$f(x) = 0 \qquad\qquad .....(2.18)$$

Let x = x_0 be an approximate value of the root of the equation (2.18) in [a, b].

If x = x_1 is the exact root of the equation (2.18), then we have

$$f(x_1) = 0 \qquad\qquad .....(2.19)$$

where x_0 and x_1 differ by a small quantity say h. Then we have

$$x_1 = x_0 + h \qquad\qquad .....(2.20)$$

Substituting in (2.19) we get f(x_0 + h) = 0

Expanding by Taylor's theorem we get $f(x_1) = 0$

$$= f(x_0) + \frac{h}{1!} f'(x_0) + \frac{h^2}{2!} f''(x_0) + \qquad\qquad .....(2.21)$$

Since h is very small, neglecting all the power's of h above the first, we get

$$0 = f(x_0) + \frac{h}{1!} f'(x_0) \Rightarrow h = -\frac{f(x_0)}{f'(x_0)}$$

Putting 'h' in 2.20 we get $x_1 = x_0 + h = x_0 - \dfrac{f(x_0)}{f'(x_0)}$

Where the above value of x_1 is a closer approximation to the root of f(x) = 0 than x_0.

Similarly if x_2 is a better approximation, starting with x_1 we get

$$x_2 = x_1 - \frac{f(x_1)}{f'(x_1)}$$

Proceeding in this way we get

$$x_{n+1} = x_n - \frac{f(x_n)}{f'(x_n)} \qquad \qquad .....(2.22)$$

The above is a general formula known as Newton-Raphson formula. Geometrically, Newton's method is equivalent replacing a small arc of the curve.

For definiteness sake let us suppose let us suppose $f''(x) > 0$ in [a, b] and f(b) > 0, where x_0 = b for which $f(x_0) f''(x) > o$

Draw a tangent line to the curve y = f(x) at the point $B_0[x_0, f(x_0)]$.

Let us take the abscissa of the point of intersection of the tangent with the x -axis as the first approximation x_1 of the root.

Again draw a tangent at $B_1[x_1, f(x_1)]$ whose abscissa of the point of intersection with the x − axis gives the second approximation x_2, of the root of f(x) = 0, and so on. The equation of the tangent at the point $B_n [x_n, f(x_n)]$, {x = 0, 1,2,...,n} is given by

$$Y - f(x_n) = f'(x_n)(x - x_n)$$

Putting y = 0, x = x_{n+1}, we get

$$x_{n+1} = x_n - \frac{f(x_n)}{f'(x_n)}, \{n = 0,1,2..\}$$

***Note:* Criterion to end the iteration :** The decision of stopping the iteration process depends on the accuracy desired by the user. If ϵ denotes the tolerable error, then the process of iteration should be terminated when'

$$|x_{n+1} - x_n| \leq \epsilon$$

Where ϵ is tolerable error.

The Newton-Raphson method uses the tangent of a curve to iteratively approximate a zero of a function f(x). It is faster than Regula–Falsi method and is sensitive to the initial approximation. The initial approximation should be nearer to the actual root of the given equation. The convergence is assured. Newton-Raphson method is second order convergent.

2.6.1 ALGORITHM FOR NEWTON-RAPHSON METHOD

Step 1 Input x_0, epsilon, maxit

[*where* x_0 is the initial guess of the root, epsilon the desired accuracy, maxit is the maximum number of iterates prescribed.

Step 2 Set I = 0

Step 3 Set $f_0 = f(x_0)$

Step 4 Compute $df_0 = f'(x_0)$

Step 5 Set $x_1 = x_0 - \dfrac{f_0}{df_0}$

Step 6 Set i = I + 1

Step 7 Check if $|(x_1 - x_0)x_1| < epsilon$, then print "root is ", x_1 and stop

Else if i < n, then set $x_0 = x_1$ and go to step 3

Step 8 Write "iterations do not converge"

Step 9 End

SOLVED   EXAMPLES

Example 17: Find the smallest positive root of the equation $x^3 - 5x + 3 = 0$ by Newton-Raphson method

Solution: Let   f (x) = $x^3 - 5x + 3$

Differentiating with respect to x we get $f'(x) = 3x^2 - 5$

We have f(−2) = $(-2)^3 - 5(-2) + 3 = 5$, $f'(-2) = 3(-2)^2 - 5 = 7$

F(−3) = $(-3)^3 - 5(-3) + 3 = -9$, f(-2) f(-3) = − 45 < 0

Therefore there is a root between −2 and −3

Also f(1) = $(1)^3 - 5(1) + 3 = -1$, f(2) = $2^3 - 5(2) + 3 = 1$

⇒ f(1) f(2) < 0 ⇒ there is a root between 1 and 2

Since f(0) = 3, we have f(0)f(1) < 0, there is a root between 0 and 1

The required smallest positive root lies between 0 and 1

Let $x_0 = 1$, be the initial approximation of the root. Then by Newton–Raphson method the next approximation is $x_1 = x_0 - \dfrac{x_0{}^3 - 5x_0 + 3}{3x_0{}^2 - 5}$ $\Rightarrow x_1 = 1 - \dfrac{(-1)}{(-2)} = 0.5$, we have $f(x_1) = (0.5)^3 - 5(0.5) + 3 = 0.625$, $f'(0.5) = -4.25$

The next approximation is

$$x_2 = x_1 - \frac{x_1{}^3 - 5x_1 + 3}{3x_1{}^2 - 5} = 0.5 - \frac{(0.5)^3 - 5(0.5) + 3}{3(0.5)^2 - 5} = 0.5 - \frac{0.625}{(-4.25)} = 0.64$$

Therefore we get $f(x_2) = (0.64)^3 - 5(0.64) + 3 = 0.062144$, $f'(x_2) = -3.7712$

The next approximation of the root is $x_3 = x_2 - \dfrac{x_2{}^3 - 5x_2 + 3}{3x_2{}^2 - 5} = 0.64 - \dfrac{0.062144}{-3.7712} = 0.6565$

$$f(x_3) = (0.6565)^3 - 5(0.6565) + 3 = 0.000446412125, \text{ and}$$

$$f'(x_3) = 3x_3^2 - 5 = 3(0.6565)^2 - 5 = -0.3.70702325$$

The next approximation of the root is $x_4 = x_3 - \dfrac{x_3^3 - 5x_3 + 3}{3r_0^2 - 5} = 0.6565 + 0.000120 = $

0.65662

Since $f(x_4) = 0.0000015976975$, $f'(x_4) = -3.70655\ 05268$

The next approximation of the root is $x_5 = x_4 - \dfrac{x_4^3 - 5x_4 + 3}{3x_4^2 - 5} = 0.65662$

$$-\frac{0.0000015976975}{-3.7065505268} = 0.656620431047$$

Therefore the root at the end of fifth iteration is 0.65662

Example 18: Find the root of 3x – cos x – 1 = 0, by Newton Raphson method, given that the root of the equation lies between 0 and 1.

Solution: We have f(x) = 3x – cos x –1, $f'(x)$ = 3 + sin x

And   f(0) = – 2 and f(1) = 1.4597

Let x_0 = 1 be the initial approximation of the root.

The next approximation of the root is

$$x_1 = x_0 - \frac{f(x_0)}{f'(x_0)} = 1 - \frac{3x_0 - \cos x_0 - 1}{3 + \sin x_0} = 1 - \frac{3.1 - \cos 1 - 1}{3 + \sin 1} = 1 - \frac{1.4597}{3.8415} = 0.62$$

Since $f(x_1) = 3x_1 - \cos x_1 - 1 = 3(0.62) - \cos (0.62) - 1 = 0.0461$

And $f'(x_1) = 3 + \sin x_1 = 3 + \sin 0.62 = 3.5810$, the next approximation of the root is

$$x_2 = x_1 - \frac{3x_0 - \cos x_0 - 1}{3 + \sin x_0} = -\frac{3.0461}{3.5810} = 0.6071$$

We have f(x_2) = $3x_2 - \cos x_2 - 1 = 3(0.6071) - \cos 0.6071 - 1 = 0$ (approx)

$$f'(x_2) = 3 + \sin 0.6071$$

The next approximation of the root is

$$x_3 = 0.62 - \frac{3x_2 - \cos x_2 - 1}{3 + \sin x_2} = 0.62 - \frac{0}{3.57057} = 0.62$$

x_2 and x_3 are equal

Hence the root of the equation is 0.62

Example 19: By applying Newton – Raphson's method twice, find the real root near 2 of the equation $x^4 - 12x + 7 = 0$

Solution: Let f (x) = $x^4 - 12x + 7$. Differentiating we get $f'(x) = 4x^3 - 12$

The initial approximation is x_0 = 2

$f(x_0) = x_0{}^4 - 12 x_0 + 7 = 2^4 - 12.(2) + 7 = -1, \quad f'(x_0) = 4x_0{}^3 - 12 = 4.2^3 - 12 = 20$

the first approximation of the root is $x_1 = x_0 - \dfrac{f(x_0)}{f'(x_0)} = 2 - \dfrac{(-1)}{20} = 2.05$

$f(x_1) = x_1{}^4 - 12x_1 + 7 = (2.05)^4 - 12(2.05) + 7 = -1, \quad f'(x_1) = 4x_1{}^3 - 12 = 4(2.05)^3 - 12$

The next approximation is

$$x_2 = x_1 - \frac{f(x_1)}{f'(x_1)} = 2.05 - \frac{x_1{}^4 - 12x_1 + 7}{4x_1{}^3 - 12} = 2.05 - \frac{(2.05)^4 - 12(2.05) + 7}{4(2.05)^3 - 12} = 2.6706$$

The root of the equation is 2.6706

Example 20: Evaluate a formula for finding the reciprocal of a number and find approximately the reciprocal of 22.

Solution: Let N the given number, then its reciprocal is $\dfrac{1}{N}$.

Let $\qquad\qquad f(x) = \dfrac{1}{x} - N$

Differentiating with respect to x we get $f'(x) = -\dfrac{1}{x^2}$

Applying Newton - Raphson's formula we get

$$x_{n+1} = x_n - \frac{f(x_n)}{f'(x_n)} \quad\Rightarrow\quad x_{n+1} = x_n - \frac{\frac{1}{x_n} - N}{-\frac{1}{x_n{}^2}} = x_n + x_n\,(-1 - Nx_n)$$

$\Rightarrow \qquad\qquad x_{n+1} = 2x_n - x_n{}^2 N = x_n\,(2 - Nx_n),\ n = 0,1,2,\dots \qquad\qquad \dots(2.23)$

Is the required formula for finding the reciprocal of N.

To find the reciprocal of 22

We have N = 22

Let $x_0 = 0.045$ be the initial approximation

Applying Newton's formula we get

$$x_1 = x_0\,(2 - N x_0) = 0.045\,(2 - 22(0.045)) = 0.0454$$

The next approximation of the root is

$$x_2 = x_1\,(2 - N x_1) = 0.0454\,(2 - 22(0.0454)) = 0.04545$$

$$x_3 = x_2\,(2 - N x_2) = 0.04545\,(2 - 22(0.04545)) = 0.0454\,5$$

Hence, the reciprocal of 22 is 0.04545

Example 21: Find the positive root of $x^2 + 2x - 2 = 0$ by Newton – Raphson Method, correct up to three significant figures.

Solution: Let $f(x) = x^2 + 2x - 2$,

Differentiating f(x) with respect to x, we get $f'(x) = 2x + 2$

We have $f(0) = -2 < 0$ and $f(1) = 1 > 0$. Hence one positive root lies between 0 and 1

Let $x_0 = 1$ be the initial approximation of the root

$$f(x_0) = x_0^2 + 2x_0 - 2 = 1^2 + 2 - 2 = \text{and } f'(x_0) = 2x_0 + 2 = 4$$

Starting with the initial value and applying Newton- Raphson method we get

$$x_1 = x_0 - \frac{x_0^2 + 2x_0 - 2}{2x_0 + 2} = 1 - \frac{1}{4} = 0.75$$

$$x_2 = x_1 - \frac{x_1^2 + 2x_1 - 2}{2x_1 + 2} = 0.75 - \frac{(0.75)^2 + 2(0.75) - 2}{2(0.75) + 2} = 0.75 - \frac{0.0625}{3.5} = 0.74$$

$$x_3 = x_2 - \frac{x_2^2 + 2x_2 - 2}{2x_2 + 2} = 0.74 - \frac{(0.74)^2 + 2(0.74) - 2}{2(0.74) + 2} = 0.74 - \frac{0.0276}{3.48} = 0.7321$$

$$x_4 = x_3 - \frac{x_3^2 + 2x_3 - 2}{2x_3 + 2} = 0.7321 - \frac{(0.7321)^2 + 2(0.7321) - 2}{2(0.74) + 2} = 0.7321 - \frac{0.00017}{2.4642} = 0.73205$$

Therefore the root of the equation is 0.732

Example 22: Find a negative root of the equation $\sin x = 1 + x^3$ by Newton's method

Solution: Let $\qquad f(x) = x^3 - \sin x + 1 \Rightarrow f'(x) = 3x^2 - \cos x$

We have $\qquad f(-1) = -1 + 0.8415 + 1 = 0.8415 > 0$

And $\qquad f(-2) = -8 + 0.9091 + 1 = -6.0907 < 0$

Therefore f(−1) f(−2)< 0

The root of the equation lies between (−1) and (−2)

Since the magnitude of f(−1) lies close to zero than f(−2) we take $x_0 = -1$

The first approximation is $x_1 = x_0 - \dfrac{f(x_0)}{f'(x_0)} = -1 - \dfrac{0.8415}{3(-1)^2 - \cos(-1)} = -1.3421$

The next approximation is given by

$$x_2 = x_1 - \frac{f(x_1)}{f'(x_1)} = -1.3421 - \frac{-2.4174 + 0.9740 + 1}{5..4037 - 0.2267} = -1.2565$$

The third approximation of the root is

$$x_3 = x_2 - \frac{f(x_2)}{f'(x_2)} = -1.2565 - \frac{-1.9838 + 1.9510}{4.7364 - 0.3091} = -1.2490$$

The fourth approximation is $x_3 = x_2 - \dfrac{f(x_2)}{f'(x_2)} = -1.2490 - \dfrac{-1.9489+1.9487}{4.6808-0.3162} = -1.2491$

Hence the required negative root is -1.249

Example 23: Using Newton's formula, establish the iterative formula

$x_{n+1} = \dfrac{1}{2}[x_n + \dfrac{N}{x_n}]$ to calculate the square root of N and compute the square root of

Solution: Let $x = \sqrt{N}$, then we have $x^2 = N$

Or $x^2 - N = 0$

Let $f(x) = x^2 - N$

Differentiating we get $f'(x) = 2x$

By Newton-Raphson's formula we have

$x_{n+1} = x_n - \dfrac{f(x_n)}{f'(x_n)} = x_n - \dfrac{x_n^2 - N}{2x_n}$ where x_n is the nth iterate, n = 0, 1, 2,...

$\therefore x_{n+1} = \dfrac{x_n^2 + N}{2x_n} \Rightarrow x_{n+1} = \dfrac{1}{2}[x_n + \dfrac{N}{x_n}]$, n = 0,1, 2,...

Hence, the iterative formula for finding the square root of the number is

$$x_{n+1} = \dfrac{1}{2}[x_n + \dfrac{N}{x_n}], \text{ n = 0,1, 2,...}$$

Now to find the square root of 8:

we have N = 8,

we know that $4 < 8 < 9$, or $\sqrt{4} < \sqrt{8} < \sqrt{9} \Rightarrow 2 < \sqrt{8} < 3$

let $x_0 = 2.5$

Substituting in Newton's iterative formula $x_{n+1} = \dfrac{1}{2}[x_n + \dfrac{N}{x_n}]$, n = 0,1 ,2 ...

we get $x_1 = \dfrac{1}{2}[x_0 + \dfrac{8}{x_0}] = \dfrac{1}{2}[2.5 + \dfrac{8}{2.5}] = 2.85$

The next approximation of the square root is

$$x_2 = \dfrac{1}{2}[x_1 + \dfrac{8}{x_1}] = \dfrac{1}{2}[2.85 + \dfrac{8}{2.85}] = 2.8285$$

The next approximation is

$$x_3 = \dfrac{1}{2}[x_2 + \dfrac{8}{x_2}] = \dfrac{1}{2}[2.8585 + \dfrac{8}{2.8585}] = 2.82\,84271$$

Hence, $\sqrt{8} = 2.82\,84271$

Example 24: Using Newton–Raphson's method establish an iterative formula for computing the cube root of N and compute the cube root of 12.

Solution: Let $\quad x = \sqrt[3]{N}$ or $x^3 = N$ or $x^3 - N = 0$

let $\quad f(x) = x^3 - N$, then $f'(x) = 3x^2$

By applying Newton's formula we have

$$x_{n+1} = x_n - \frac{f(x_n)}{f'(x_n)} = x_n - \frac{x_n^3 - N}{3x^2_n} = \frac{1}{3}\left[\frac{3x^2_n - x_n^2 + N}{x^2_n}\right] = \frac{1}{3}\left[\frac{2x^2_n + N}{x^2_n}\right]$$

$\Rightarrow \qquad x_{n+1} = \frac{1}{3}[2x_n + \frac{N}{x_n^2}]$, n = 0,1, 2...

Is the iterative formula for finding the cube root of N.

To find the cube root of 12: We have $8 < 12 < 27 \Rightarrow \sqrt[3]{8} < \sqrt[3]{12} < \sqrt[3]{27} \Rightarrow 2 < \sqrt[3]{12} < 3$

Let x_0 = 2.5 be the initial approximation of the cube root.

Applying Newton's formula for finding the cube root of a number N

i.e. $\qquad x_{n+1} = \frac{1}{3}[2x_n + \frac{N}{x_n^2}]$, n = 0, 1, ...we get

$\Rightarrow \qquad x_1 = \frac{1}{3}[2x_0 + \frac{N}{x_0^2}] = \frac{1}{3}[2(2.5) + \frac{12}{(2.5)^2}] = 2.3066$

The next approximation is

$$x_2 = \frac{1}{3}[2x_2 + \frac{N}{x_2^2}] = \frac{1}{3}[2(2.3066) + \frac{12}{(2.3066)^2}] = 2.2901$$

Hence, the cube root of 12 is 2.2901.

2.6.2 CONVERGENCE OF NEWTON-RAPHSON METHOD

$$x_{n+1} = x_n - \frac{f(x_n)}{f'(x_n)} = \phi(x_n) \text{ (say)} \qquad\qquad .....(2.24)$$

The general form of (2.26) is $x = \phi(x)$ $\qquad\qquad .....(2.25)$

We know that by the iteration method (2.25) converges if

$$|\phi'(x)| < 1$$

Here we have $\phi(x) = x - \frac{f(x)}{f'(x)}$ from (2.24)

Therefore $\phi'(x) = 1 - \frac{(f'(x)^2 - f(x)f''(x)}{[f'(x)]^2} \Rightarrow |\phi'(x)| = |\frac{f(x)f''(x)}{f'(x)]^2}|$

Hence, Newton – Raphson method converges if $|\frac{f(x)f''(x)}{f'(x)]^2}| < 1$

I.e. $\qquad$ if$| f(x) f''(x) | < f'(x)]^2$ $\qquad\qquad .....(2.26)$

If α denotes the actual root of f(x) = 0, then we can select a small interval in which f (x), $f'(x)$ and $f''(x)$ are continuous and condition (2.26) is satisfied. Hence Newton–

Raphson always converges provided the initial approximation x_0 is taken very close to the actual root α.

2.6.3 RATE OF CONVERGENCE OF NEWTON RAPHSON METHOD

Let α denote the exact value of the root of the equation f(x) = 0 and let x_n, x_{n+1} denote two successive approximations to the actual root α. If ϵ_n, ϵ_{n+1} are the corresponding errors, we have

$$x_n = \alpha + \epsilon_n \text{ and } x_{n+1} = \alpha + \epsilon_{n+1}$$

By Newton– Raphson's iterative formula

$$\alpha + \epsilon_{n+1} = \alpha + \epsilon_n - \frac{f(\alpha + \epsilon_n)}{f'(\alpha + \epsilon_n)}$$

Or
$$\epsilon_{n+1} - \epsilon_n = -\frac{f(\alpha + \epsilon_n)}{f'(\alpha + \epsilon_n)}$$

Expanding we get
$$\epsilon_{n+1} - \epsilon_n = -\frac{f(\alpha) + \epsilon_n f'(\alpha) + \frac{\epsilon_n^2}{2} f''(\alpha) + \cdots}{f'(\alpha) + \epsilon_n f''(\alpha) + \ldots}$$

Therefore we get
$$\epsilon_{n+1} = \epsilon_n - \frac{\epsilon_n f'(\alpha) + \frac{\epsilon_n^2}{2} f''(\alpha) + \cdots}{f'(\alpha) + \epsilon_n f''(\alpha) + \ldots}$$

$$= \epsilon_n - \frac{\epsilon_n [f'(\alpha) + \frac{\epsilon_n}{2} f''(\alpha) + \cdots]}{f'(\alpha) + \epsilon_n f''(\alpha) + \ldots}$$

$$= \frac{1}{2} \left[\frac{\epsilon_n^2 f''(\alpha) + \ldots}{f'(\alpha) + \frac{\epsilon_n}{f'(\alpha)} f''(\alpha) + \ldots} \right]$$

$$\Rightarrow \epsilon_{n+1} = \epsilon_n^2 \frac{f''(\alpha)}{2f'(\alpha)}$$

Hence it's clear that the error at each stage is proportional to the error in the previous stage. Therefore the Newton-Raphson method has quadratic convergence.

In this method a tangent is drawn at the point $(x_1, f(x_1))$, where $x_1 < x < x_n$. The point where the tangent cuts x – axis is the estimate of the root. This process is repeated until the root is obtained.

Advantages: This method is one of the rapid convergent methods. Near a root, the number of significant digits approximately doubles with each step. Easy to convert to multiple dimensions.

Disadvantages: Newton Raphson method requires a derivative evaluation in to a functional evaluation for each iteration. Addition to the derivative it is dependent on the initial guess. But this method is not guaranteed to converge to the root. It has poor global convergence properties Dependent on initial guess. It may encounter a zero derivatives. If the curve is very flat near the root, $f'(x)$ is nearly zero, then the iterates deviate farther from the root. Newton Raphson method is preceded by secant or bisection method to get approximate root.

2.6.4 MODIFIED NEWTON RAPHSON METHOD

There are many modified methods.

When large systems of equations are involved a modified method can be used in which the first evaluation of $x_{n+1} = x_n - \dfrac{f(x_n)}{f'(x_n)}$

$f'(x_n)$ is used for all other further extrapolations.

Consider f (x) = 0

If the derivative $f'(x)$ varies but slightly on the interval [a,b]. Then in the formula $x_{n+1} = x_n - \dfrac{f(x_n)}{f'(x_n)}$, n = 0 ,1, 2, . . . we can put $f'(x_n) \approx f'(x_0)$.

From this the root α of the equation f(x) = 0 we get the successive approximations

$$x_{n+1} = x_n - \frac{f(x_n)}{f'(x_0)},\ n = 0,1,2,\dots \qquad\qquad \dots\dots(2.27)$$

Geometrically, this method signifies that we replace the tangents at the points $B_n[x_n, f(x_n)]$ by straight lines parallel to the tangent to the curve y = f(x) at its fixed point $B_0[x_0, f(x_0)]$

The necessity to compute $f'(x_n)$ at each time is avoided if use the modified formula given by (1), when $f'(x_n)$ is complicated.

It is graphically shown in the fig. given below. The modified formula is given below.

$$x_{n+1} = x_n - \frac{f(x_n)}{f'(x_0)}\quad n = 1, 2. \dots$$

(i) The Modified Newton-Raphson method uses the fact that f(x) and u(x) := f(x)/f'(x) have the same zeros and instead approximates a zero of μ(x) That is, by defining

$$\mu(x) = \frac{f(x)}{f'(x)}$$

The Newton-Raphson update for u(x) is $x_{n+1} = x_n - \dfrac{\mu(x_i)}{\mu'(x_i)}$ it becomes

$$x_{n+1} = x_n - \frac{f(x_n)\,f'(x_n)}{(f'(x_n))^2 - f(x_n)\,f''(x_n)}\text{for}$$ effectively estimating the order of convergence, it overrelaxes or under relaxes the update, in particular converging much more quickly for roots with multiplicity greater than 1.

Due to its multiplicity, Newton – Raphson method is likely to reach the maximum number of iterations before converging. Modified Newton-Raphson computes the multiplicity and over relaxes, converging quickly using either provided derivatives or numerical differentiation.

Example 25: Applying Newton–Raphson's modified formula compute the roots of the equation

Close to 2 of the function $x^3 - x - 1 = 0$

Solution: We have $x^3 - x - 1 = 0$, $x_0 = 2$, $f'(x) = 3x^2 - 1 \Rightarrow f'(x_0) = f'(2) = 5$

The successive approximations (iterates) are given in the form of a table

x	f(x)
2	5.000
1.5455	1.1458
1.4413	0.5528
1.3910	0.3006
1.3637	0.1724
1.3480	0.1016

2.6.5   NEWTON -RAPHSON EXTENDED FORMULA (CHEBYSHEV 'S FORMULA)

Consider f (x) = 0.

Expanding f(x) by using Taylor's series and neglecting the second order terms in the neighborhood of x_0, we get $f(x) = f(x_0) + (x - x_0) f'(x_0) + .. = 0$

Neglecting the second order and the higher order terms we get

$$x = x_0 - \frac{f(x_0)}{f'(x_0)}$$

Therefore the first approximation of the root is given by

$$x_1 = x_0 - \frac{f(x_0)}{f'(x_0)} \qquad\qquad .....(2.28)$$

Again expanding f(x) by Taylor's series and neglecting the third order terms we get

$$f(x) = f(x_0) + (x - x_0)\ f'(x_0) + \frac{(x-x_0)^2}{2!} . f''(x_0) = 0 \qquad .....(2.29)$$

from (2.28) and (2.29) we get

$$f(x_0) + (x - x_0)\ f'(x_0) + \frac{1}{2}\frac{[f(x_0)]^2}{(\ f'(x_0)\ )^2}\ f''(x_0) = 0$$

Hence the Newton – Raphson extended formula is given by

$$f(x_0) + (x - x_0)\ f'(x_0) + \frac{1}{2}\frac{[f(x_0)]^2}{(\ f'(x_0)\ )^2}\ f''(x_0) = 0$$

We can also compare the methods geometrically. Instead of sliding along the tangent line, the Secant Method slides along a nearby secant line. The Secant Method has some advantages over the Newton Method. It is more stable, less subject to the wild gyrations that can afflict the Newton Method. since To use the Secant Method, we do not need the derivative,

The convergence speed of the Newton-Raphson method is very high in the most of cases. Using Taylor series, it can be shown that the error is approximately proportional to the square of the previous error. In other words the Newton-Raphson method is quadratic convergent.

The algorithm of the Newton-Raphson method is very simple. It is given below as follows:

Step 1: Choose x_0 as a starting point.

Step 2: Let $x = x_0$

$$x = x_0 - \frac{f(x_0)}{f(x_0)}$$

Step 3: if $|x\text{-}x0| < e$ then let root = x, else x_0 = x; go to step 2.

Step 4: End.

e: Acceptable approximated error

A multiple root occurs when a certain function has two or more roots which they are in the same position. For example, assume the function:

$$f(x) = (x + 2)(x - 5)(x - 5)$$

This function has a multiple root in the position of x = 5. In this situation most of the numerical formula face major difficulties for finding roots. For example when even multiple root occurs the function is tangential to the x axis in the position of the root. Therefore, many bracketing methods don't work because the sign of the function is not changed near the position of the root. The Newton –Raphson and secant method face difficulties, too. In these methods, the convergence speed decreases near a multiple root. To overcome these defect, Ralston and Rabinowitz (1978) suggested a modification for the Newton-Raphson method. They defined a new function as follows:

$$\eta = \frac{f(x)}{f'(x)}$$

It can be shown that the roots of this new function are as same as the old function. The only difference is that the multiple roots in old function have been changed to single roots in the new function. Using this function, the multiple roots can be found by employing conventional numerical method without any major difficulties. For example the Newton-Raphson formula is changed to a new formula as follows:

$$x_{i+1} = x_i - \frac{\eta(x_i)}{\eta'(x_i)}$$

Or
$$X_{i+1} = X_i - \frac{f(x_i)f'(x_i)}{\left[f'(x_i)\right] - f(x_i)f''(x_i)}$$

Step 1: Choose x0 as a starting point. N-r

Step 2: Let

$$x = x_0 - \frac{f(x_0)}{f'(x_0)}$$

Step 3: if |x-x0|< e then let root = x, else x0 = x; go to step 2.

Step 4: End.

e: Acceptable approximated error.

Step 1: Choose x0 as a starting point.

Step 2: Let x = g(x0).

Step 3: if |x-x0|< e then let root = x, else x0 = x; go to step 2.

Step 4: End.

e: Acceptable approximated error.

Step 1: Choose two starting points x0 and x1.

Step 2: Let

$$X_2 = X_1 - \frac{(x_0 - x_1)f(x_1)}{f(x_0) - f(x_1)}$$

Step 3: if |x2-x1|<e then let root = x2, else x0 = x1; x1 = x2; go to step 2.

Step 4: End.

e: Acceptable approximated error

This leads to the following algorithm. Algorithm (Secant Method) Let: $\mathbb{R} \to \mathbb{R}$ be a continuous function. The following algorithm computes an approximate solution $x*$ to the equation $(x) = 0$. The Secant Method converges at least linearly, with asymptotic rate constant 2/3.

2.7 SECANT METHOD

The Secant Method is one of the most popular methods of the many variants of the Newton Method. We start with two estimates of the root x_0 and x_1. The iterative formula, for n ≥ 1, is. The Secant Method has some advantages over the Newton Method. It is more stable, less subject to the wild gyrations that can afflict the Newton

Method. (The differences are not great, since the geometry is nearly the same.) To use the Secant Method, we do not need the derivative.

The convergence speed of the Newton-Raphson method is very high in the most of cases. Using Taylor series, it can be shown that the error is approximately proportional to the square of the previous error. In other words the Newton-Raphson method is quadratic ally convergent.

The algorithm of the Newton-Raphson method is very simple as follows:

Algorithm 2: Computing root of the Newton-Raphson method :

Step 1: Choose x0 as a starting point.

Step 2: Let $x = x_0 - \dfrac{f(x_0)}{(f'(x_0)}$

Step 3: if |x-x0|< e then let root = x, else x0 = x ; go to step 2.

Step 4: End.

 e: Acceptable approximated error

A multiple root occurs when a certain function has two or more roots which they are in the same position.

For example, consider the function $f(x) = (x+1)(x-5)^2$ which has a multiple root in the position of x = 5. In this situation most of the numerical formula face major difficulties for finding roots. For example when even multiple root occurs the function is tangential to the x axis in the position of the root. Therefore, many bracketing methods don't work because the sign of the function is not changed near the position of the root. The Newton –Raphson and secant methods face difficulties, too. In these methods, the convergence speed decreases near a multiple root. To overcome these defect, Ralston and Rabinowitz (1978) suggested a modification for the Newton-Raphson method. They defined a new function as follows:

$$\eta(x) = \frac{f(x_1)}{f'(x)}$$

It can be shown that the roots of this new function are as same as the old function. The only difference is that the multiple roots in old function have been changed to single roots in the new function. Using this function, the multiple roots can be found by employing conventional numerical method without any major difficulties. For example the Newton-Raphson formula is changed to a new formula as follows:

$$X_{i+1} = X_i - \frac{\eta(x_i)}{\eta'(x_i)}$$

Or
$$X_{i+1} = X_i - \frac{f(x_1)f'(x_1)}{[f'(x_i)]^2 - f(x_i)f''(x_i)}$$

Step 1: Choose x0 as a starting point. N-r

Step 2: Let

$$x = x_0 - \frac{f(x_0)}{f'(x_0)}$$

Step 3: if |x-x0| < e then let root = x, else x0 = x ; go to step 2.

Step 4: End.

e: Acceptable approximated error.

It can be shown that if in the area of search, this method is convergent. The algorithm of simple one point iteration method is very easy as follows:

Step 1: Choose x0 as a starting point.

Step 2: Let x = g(x0).

Step 3: if |x-x0| < e then let root = x, else x0 = x ; go to step 2.

Step 4: End.

e: Acceptable approximated error.

Although the Newton-Raphson method is very powerful to solve non-linear equations, evaluating of the function derivative is the major difficulty of this method. To overcome this deficiency, the secant method starts the iteration by employing two starting points and approximates the function derivative by evaluating of the slope of the line passing through these points. The secant method has been shown in Fig. 1. As it is illustrated in Fig. 1, the new guess of the root of the function f(x) can be found as follows:

Step 1: Choose two starting points x0 and x1.

Step 2: Let

$$X_2 = X_1 - \frac{(x_0 - x_1)f(x_1)}{f(x_0) - f(x1)}$$

Step 3: if |x2-x1| < e then let root = x2, else x0 = x1; x1 = x2 ; go to step 2.

Step 4: End.

e: Acceptable approximated error.

In the secant method, it is not necessary that two starting points to be in opposite sign. Although the Newton-Raphson method is very powerful to solve non-linear equations, evaluating of the function derivative is the major difficulty of this method. To overcome this deficiency, the secant method starts the iteration by employing two starting points and approximates the function derivative by evaluating ot the slope of the line passing through these point the algorithm of the secant method can be written as follows:

Algorithm 3: Computing a root of an equation f(x) = 0 by Secant method

Step 1: Choose two starting points x0 and x1

Step 2: Let $\qquad x_2 = x_1 - \dfrac{(x_0-x_1)\, f(x_1)}{f(x_0)-f(x_1)}$

Step 3: if $|x_2 - x_1| < e$ then let root = x_2, else $x_0 = x_1$; $x_1 = x_2$; go to step 2.

Step 4: End

e: Acceptable approximated error.

SECANT METHOD

The Secant Method is one of the most popular methods of the many variants of the Newton Method. We start with two estimates of the root $x_0 ,and\ x_1$. The iterative formula, for n ≥ 1, is. The Secant Method has some advantages over the Newton Method. It is more stable, less subject to the wild gyrations that can afflict the Newton Method. (The differences are not great, since the geometry is nearly the same.) To use the Secant Method, we do not need the derivative, which can be expensive to calculate. The Secant Method, when it is working well, which is most of the time, is fast.

MULLER'S METHOD

Muller's method is a generalization of the secant method. Instead of starting with two initial values and then joining them with a straight line in secant method, Mullers method starts with three initial approximations to the root and then join them with a second degree polynomial (a parabola), then the quadratic formula is used to find a root of the quadratic for the next approximation. That is if x_0, x_1 and x_2 are the initial approximations then x_3 is obtained by solving the quadratic which is obtained by means of x_0, x_1 and x_2. Then two values among x_0, x_1 and x_2 which are close to x_3 are chosen for the next iteration Muller's method is a root finding algorithm. The   method is based on Secant method and is used for solving equations of the form f(x) = 0.

Muller's method uses three points, constructs the parabola through these three points, and takes the intersection of the x -axis with the parabola to be the next approximation.

To find the root we Start with three initial guesses x_0, x_1 and x_2 then get the values $f(x_0)$, $f(x_1)$ and $f(x_2)$. We compute $h_0 = x_1 - x_0$, $h_1 = x_2 - x_1$

$$\delta_0 = \frac{f(x_1)-f(x_0)}{x_1-x_0} = \frac{f(x_1)-f(x_0)}{h_0} \text{ and } \delta_1 = \frac{f(x_2)-f(x_1)}{x_2-x_1} = \frac{f(x_2)-f(x_1)}{h_1}$$

$$a = \frac{\delta_1-\delta_0}{h_1+h_0}, \quad b=ah_1 + \delta_1 \text{ and } c = f(x_2)$$

The next approximation of the root is

$$x_3 = x_2 + \frac{-2c}{b \pm \sqrt{b^2 - 4ac}}$$

EXERCISE 4

1. Find by Newton -Raphson method the real root of $3x - \cos x - 1 = 0$ correct to three significant figures   [*Ans:* 0.0607]

2. Find all the roots of $\cos x - x^2 - x = 0$ to five decimal places by applying Newton – Raphson method   [*Ans :* 0.55001 and – 1.25115]

3. Compute $\sqrt{27}$ and $\sqrt{12}$ [*Ans:* 5.1961, 3.46412]

4. Find the roots of the equation $x - e^{-x} = 0$ by Newton – Raphson method
                                                                                  [*Ans:* 0.5671]

5. Find the real root of $x + \log x - 2 = 0$ by Newton –Raphson method [*Ans:* 1.557]

6. Find a real root of $x \tan x + 1 = 0$ using Newton – Raphson's method [*Ans:* 2.798]

7. Find real root of $y = x + log_{10}x - 2 = 0$ [*Ans:* 1.75557]

8. Using Newton's method compute $\sqrt{41}$ correct to four decimal places,
                                                                                  [*Ans:* 6.4032]

9. Find a real root of the equation $e^x \sin x - 1 = 0$ by Newton- Raphson method
                                                                                  [*Ans:* 0.58853]

10. Find by Newton's method the real root of $xe^x - 1 = 0$ [*Ans:* 0.853]

11. Find the square root of 10   [*Ans:* 3.162]

12. Find the reciprocal of 22 using Newton-Raphson method   [*Ans:* 0.0454]

13. Find the reciprocal of 18   [*Ans:* .0.0555]

14. Using Newton – Raphson find the root of the equation $x + log_{10}x = 3.375$

                                                             [*Ans:* 2.911]

15. Equation $x - \dfrac{x^3}{3} + \dfrac{x^5}{10} - \dfrac{x^7}{42} + .. = 0.4431135$ [*Ans:* 0.47683]

CHAPTER 3

Differential Operators

3.1 INTRODUCTION

In this chapter we introduce differential operators and their application in numerical analysis.

3.2 FORWARD DIFFERENCE OPERATOR

Suppose a function f(x) is defined for the equally spaced values   a, a+ h, a+ 2h, ..., a+ (n-1)h,... of the argument x,. The values f(a), f(a+ h) , f(a+2h), . . . i.e. are called entries Then the forward difference operator Δ is defined by the equation

$$\Delta f(x) = f(x + h) - f(x)$$

Where $\Delta f(x)$ is called the first difference of f(x). It is itself a function of x and h is called the step length or interval of differencing. The second difference of f(x) denoted by $\Delta^2 f(x)$ and is defined as follows.

$$\Delta^2 f(x) = \Delta[\Delta f(x)] = \Delta [f(x + h) - f(x)] = \Delta f(x + h) - \Delta f(x)$$

In general $\Delta^n f(x) = \Delta^{n-1} f(x + h) - \Delta^{n-1} f(X)$

The symbol Δ is called the forward difference operator

From the above definition we have

$$\Delta^2 f(x) = \Delta[\Delta f(x)] = \Delta f(x + h) - \Delta f(x) = f(x + h + h) - f(x + h) - [f(x + h) - f(x)]$$

$$= f(x + 2h) - 2 f(x + h) + f(x)$$

$$\Delta^3 f(x) = \Delta[\Delta^2 f(x)] = \Delta f(x + 2h) - 2\Delta f(x + 2h) + \Delta f(x)$$

$$= f(x + 3h) - f(x + 2h) - 2f(x + 2h) + 2 f(x + h) + f(x + h) - f(x)$$

$$= f(x+3h) - f(x + 2h) - 2f(x + 2h) + 2 f(x + h) + f(x + h) - f(x)$$

$$= f(x + 3h) - 3f(x + 2h) + 3 f(x + h) - f(x)$$

Using the definition we can write

$$\Delta f(a) = f(a + h) - f(a)$$

$$\Delta f(a + h) = f(a + 2h) - f(a + h)$$

$$\Delta f(a + 2h) = f(a + 3h) - f(a + 2h)$$

$$\cdots \qquad \cdots \qquad \cdots$$

3.2.1   ALTERNATIVE DEFINITION

Let the function $y = f(x)$ be a given function of x taking the values $y_0, y_1, y_2, \ldots, y_n$ corresponding to the equally spaced values (i.e. arguments) $x_0, x_1, x_2, \ldots, x_n$ of the independent variable x.

Then we define the first differences of y as follows

$$\Delta y_0 = y_1 - y_0$$

$$\Delta y_1 = y_2 - y_1$$

$$\Delta y_2 = y_3 - y_2$$

$$\cdots \quad \cdots \quad \cdots$$

$$\Delta y_{n-1} = y_n - y_{n-1}$$

The higher differences of y

The second differences of y are defined as follows

$$\Delta^2 y_0 = \Delta[\Delta y_0] = \Delta[y_1 - y_0] = \Delta y_1 - \Delta y_0 = y_2 - y_1 - [y_1 - y_0] = y_2 - 2y_1 + y_0$$

$$\Delta^2 y_1 = \Delta[\Delta y_1] = \Delta[y_2 - y_1] = \Delta y_2 - \Delta y_1 = y_3 - y_2 - [y_2 - y_1] = y_3 - 2y_2 + y_1$$

$$\cdots \qquad \cdots \qquad \cdots \qquad \cdots$$

$$\Delta^2 y_{n-1} = \Delta[\Delta y_{n-1}] = \Delta[y_n - y_{n-1}] = \Delta y_n - \Delta y_{n-1} = y_{n+1} - y_n - [y_n - y_{n-1}] = y_{n+1} - 2y_n + y_{n-1}$$

Where Δ^2 is called the second order forward difference operator

In general we define

$$\Delta^n y_i = \Delta^{n-1} y_{i+1} - \Delta^{n-1} y_i$$

Note : In general the arguments $x_0, x_1, x_2, \ldots, x_n$ need not be equally spaced

In this chapter we take them to be equally spaced

If $x_0, x_1, x_2, \ldots, x_n$ are equally spaced with h as the interval of differencing we have

$$x_1 - x_0 = x_2 - x_1 = x_3 - x_2 = \ldots, x_n - x_{n-1} = h$$

$$x_1 = x_0 + h, x_2 = x_1 + h = x_0 + 2h, x_3 = x_2 + h = x_0 + 3h, \ldots, x_n = x_0 + nh$$

The operatorΔ obeys the following laws:

1. $\Delta\ [f(x) + g(x)\} = \Delta f(x) + \Delta g(x)$

   Proof: $\Delta\ [f(x) + g(x)\} = f(x + h) + g(x + h) - [f(x) + g(x)\} = [f(x + h) - f(x)\ ] + [\ g(x + h) - g(x)] = \Delta f(x) + \Delta g(x)$

2. $\Delta\ [c\ f(x)] = c\ \Delta\ f(x)$ where c is a constant

   Proof: $\Delta\ [c\ f(x)] = c\ f(x + h) - c\ f(x) = c\ [f(x + h\ ) - f(x)] = c\ \Delta\ f(x)$

Theorem 1: The nth differences of a polynomial of degree n are constant and the $(n + 1)^{th}$ differences are zero

Proof: Let $f(x) = \sum_{i=0}^{n} a_i x^i$ be a polynomial of degree n

   For n = 1 we have $f(x) = a_1 x + a_0$ and $\Delta f(x) = a_1 (x + h) + a_0 - [a_1 x + a_0]$

$$= a_1 x + a_1 h + a_0 - a_1 x - a_0 = a_1 h \qquad\qquad .....(1)$$

So the theorem is true for n =1

Assume now that the theorem is true for all degrees $1, 2, 3 . . . , n-1$

Consider the n-ic polynomial (i.e. nth degree polynomial)

$$f(x) = \sum_{i=0}^{n} a_i x^i \qquad\qquad .....(2)$$

Applying the properties (1) and (2) mentioned above we have

$\Delta^n f(x)\ =\ \sum_{i=0}^{n} a_i\ \Delta^n x^i$ for i<n , $\Delta^n x^i$ is the nth   difference of a polynomial of degree less than n , and   must vanish , by the induction process .

Thus we have $\Delta^n f(x) = a_n \Delta^n x^n = a_n \Delta^{n-1} (\Delta x^n)$

$$= a_n \Delta^{n-1}[\ (x + h)^n - x^n]$$

$$= a_n \Delta^{n-1}[\ nh\ x^{n-1} + g(x)]$$

Where g(x) is a polynomial of degree less than $n - 1$.

Applying induction hypothesis again, we get

$\Delta^n f(x)\ = a_n \Delta^{n-1}(\ nh\ x^{n-1}) = a_n (nh)\ (n-1\ )! h^{n-1} = a_n n! h^n$, a constant

The (n+1) th difference is $a_n n! h^n - a_n n! h^n = 0$

This proves the theorem by induction

Alternative proof: Let $f(x) = a_0 x^n + a_1 x^{n-1} + . . . + a_n$

Let h be the interval of differencing we have

$\Delta f(x) = f(x+h) - f(x) = a_0(x+h)^n + a_1(x+h)^{n-1} + \ldots + a_n - [a_0 x^n + a_1 x^{n-1} + \ldots + a_n]$

$\quad = a_0[(x+h)^n - x^n] + a_1[(x+h)^{n-1} - x^{n-1}] + \ldots + [a_n - a_n]$

$\quad = a_0 n h x^{n-1} + \text{terms Involving powers of x less than (n-1)}$

$\Rightarrow \Delta f(x)$ is a polynomial of degree (n-1)

$\Delta^2 f(x) = \Delta[\Delta f(x)] = \Delta[a_0 n h x^{n-1} + \text{terms Involving powers of x less than } (n-1)]$

$\quad = a_0 \, nh[(x+h)^{n-1} - x^{n-1}] +$

$\quad\quad\quad\quad\quad\quad \text{terms involving the powers of x lesser than } (n-2)$

$\quad = a_0 \, n(n-1) \, h^2 x^{n-2} + \text{+terms involving the powers of x lesser than } (n-2)$

$\Rightarrow \Delta^2 f(x)$ is a polynomial of degree (n−2)

Proceeding as above we get

$\Delta^n f(x) = a_0[n(n-1)\ldots 1] h^n x^{n-n} = a_0 \, n! \, h^n x^0 = a_0 \, n! \, h^n$

Which is a constant

We have $\Delta^{n+1} f(x) = \Delta[\Delta^n f(x)] = a_0 \, n! \, h^n - a_0 \, n! \, h^n = 0$

Hence, proved

Note: The converse of the theorem is also true

3.3 FORWARD DIFFERENCE OPERATOR TABLE

The forward differences of a function f(x) can be represented in the form a table known as the forward difference table as shown below

x	y	Δy	$\Delta^2 y$	$\Delta^3 y$	$\Delta^4 y$
x_0	y_0				
		Δy_0			
x_1	y_1		$\Delta^2 y_0$		
		Δy_1		$\Delta^3 y_0$	
x_2	y_2		$\Delta^2 y_1$		$\Delta^4 y_0$
		Δy_2		$\Delta^3 y_1$	
x_3	y_3		$\Delta^2 y_2$		
		Δy_3			
x_4	y_4				

The forward difference table is also called diagonal difference table.

y_0 is called the leading entry

Δy_0, $\Delta^2 y_0$, ... are called leading differences

Note: If h = 1 i.e. interval of differencing is 1 (unity), then Δ f(x) = f(x+1) − f(x)

The forward difference of a function is zero

3.4 BACKWARD DIFFERENCE OPERATOR (∇)

The backward differential operator ∇ is defined as

$$\nabla f(x) = f(x) - f(x - h)$$

Where h is the interval of differencing i.e. step length

Alternatively, if y_0 and y_1 are two consecutive values i.e. entries of y corresponding to the values (i.e. arguments) x_0 and x_1 of x then

$$\nabla y_1 = y_1 - y_0$$

From the definition we have

$$\nabla^2 f(x) = \nabla[\nabla f(x)] = \nabla[\, f(x) - f(x - h) = \nabla f(x) - \nabla f(x - h)$$
$$= f(x) - f(x - h) - [f(x - h) - f(x - 2h)\,]$$
$$= f(x) - 2f(x - h) + f(x - 2h)$$

Backward difference table

x	y	∇y	$\nabla^2 y$	$\nabla^3 y$	$\nabla^4 y$
x_0	y_0				
		∇y_1			
x_1	y_1		$\nabla^2 y_1$		
		∇y_2		$\nabla^3 y_1$	
x_2	y_2		$\nabla^2 y_2$		$\nabla^4 y_1$
		∇y_3		$\nabla^3 y_2$	
x_3	y_3		$\nabla^2 y_3$		
		∇y_4			
x_4	y_4				

Note: The table of backward differences is identical with the table of forward differences but that the entries in the table have different names

3.5 SHIFTING OPERATOR (E)

We define the shifting operator E as follows

$$Ef(x) = f(x + h)$$

Where h is the interval of differencing or step length

If y_x is a given function then the shifting operator is defined as

$$Ey_x = y_{x+h}$$

From the definitions given above we have

$$E^2f(x) = f(x+2h) , E^3f(x) = f(x+3h) , \ldots , E^nf(x) = f(x+3h)$$

$$E^2y_x = y_{x+2h} , E^3y_x = y_{x+3h} , \ldots , E^ny_x = y_{x+nh}$$

Also we have $E^2y_0 = y_2 , \ldots , E^ny_0 = y_n$

The following hold good to the operator E:

$$E(f_1(x) + f_2(x) + \ldots + f_n(x)) = E f_1(x) + Ef_2(x) + \ldots + Ef_n(x)$$

$$E(cf(x)) = c E f(x) \text{ where C is a constant}$$

$$[E^m E^n] f(x) = E^{m+n} f(x) = E^{n+m} f(x) = E^n E^m]f(x)$$

E is also called the displacement operator of translation operator

3.6 INVERSE SHIFT OPERATOR E^{-1}

The inverse shift operator E^{-1} is defined as follows

$$E^{-1}f(x) = f(x - h)$$

Where h is the step length and, $E^{-1}Ef(X) = f(x)$,

$$E^{-2}Ef(X) = E^{-1}f(X) = f(x - h).$$

$$E^{-3}Ef(X) = E^{-2}f(X) = f(x - 2h) \text{ etc}$$

3.7 AVERAGING OPERATOR(μ)

The averaging operator μ is defined by

$$\mu f(x) = \frac{1}{2} [f(x + \frac{h}{2}) + f (x - \frac{h}{2})] \text{ i.e. } \mu = \frac{1}{2} [y_{x+\frac{h}{2}} + y_{x-\frac{h}{2}}]$$

Differential operator (D)

The differential operator D is defined by

$$Df(x) = \frac{d}{dx} f(x)$$

If f(x) and g(x) are functions then we have D [f(x) + g(x)] = Df(x) + Dg(X)

3.8   UNIT OPERATOR (I)

The unit operator I is defined by I f(x) =f(x)

3.9   THE CENTRAL DIFFERENCE OPERATOR (δ)

The central difference operator δ is defined by

$$\delta f(x) = [f(x + \tfrac{h}{2}) - f (x - \tfrac{h}{2})]$$

Or $\quad \delta y_x = [y_{x+\frac{h}{2}} - y_{x-\frac{h}{2}}]$

The operator can be defined as follows:

$$y_1 - y_0 = \delta y_{1/2}$$

$$y_2 - y_1 = \delta y_{3/2}$$

$$\cdots \qquad \cdots \quad \cdots$$

$$y_n - y_{n-1} = \delta y_{n-1/2}$$

For higher order differences we have

$$\delta^{n-1} y_{r+1/2} - \delta^{n-1} y_{r-1/2} = \delta^n y_r$$

Theorem 2: If f (x) and g(x) are functions then Δ[a f(x) + b g(x)] = a Δf(x) + b g(x)

Where a and b are   constants

Proof: We have Δ[a f(x) + b g(x)] = a f(x + h) + b g(x + h) $-$[a f(x) + b g(x)]

$$= a\ f(x + h) + b\ g(x + h) - a\ f(x) - b\ g(x)]$$

$$= a\ f(x + h)) - a\ f(x) + b\ g(x + h) - b\ g(x)]$$

$$= a\ [f(x + h)) - f(x)] + b\ [g(x + h) - g(x)]$$

$$= a\ \Delta f(x) + b\ g(x)$$

The operator Δ is commutative, associative and distributive over addition

Since $\Delta^m \Delta^n = \Delta^{m+n} = \Delta^{n+m} = \Delta^n \Delta^m$

And $\quad \Delta^{(m+n)+p} = \Delta^{m+n} \Delta^p = \Delta^m \Delta^n \Delta^p = \Delta^{+m} = \Delta^n \Delta^m$

Where m, n, and p are integers

Theorem 3: If a and b are constants, then Δ [f(x).g(x)] = f(x + h) Δ g(x) + g(x) Δ f(x)

Where f(x) and g(x) are any two functions

Proof: We have Δ [f(x) .g(x)] = f(x + h) g(x + h) −f(x) g(x)

$$= f(x + h) g(x + h) - f(x + h) g(x) + f(x + h) g(x) -f(x) g(x)$$

$$= f(x + h) [g(x + h)- g(x)]+ g(x) [f(x + h) -f(x)]$$

$$= f(x + h)\Delta g(x) + g(x) \Delta f(x)$$

Hence, Δ [f(x).g(x)] = f(x + h) Δ g(x) + g(x) Δ f(x)

Theorem 4: If f (x) and g(x) are any two functions, then

$$\Delta \left[\frac{f(x)}{g(x)}\right] = \frac{g(x)\,\Delta f(x) - f(x)\,\Delta g(x)}{g(x+h)g(x)}$$

Proof :

$$\Delta \left[\frac{f(x)}{g(x)}\right] = \frac{f(x+h)}{g(x+h)} - \frac{f(x)}{g(x)} = \frac{f(x+h)g(x) - f(x)g(x+h)}{g(x+h)g(x)}$$

$$= \frac{f(x+h)g(x) - f(x)g(x) + f(x)g(x) - f(x)g(x+h)}{g(x+h)g(x)}$$

$$= \frac{[f(x+h) - f(x)]g(x) - f(x)[g(x+h) - g(x)]}{g(x+h)g(x)}$$

$$= \frac{g(x)[f(x+h) - f(x)] - f(x)[g(x+h) - g(x)]}{g(x+h)g(x)}$$

$$= \frac{g(x)\,\Delta f(x) - f(x)\,\Delta g(x)}{g(x+h)g(x)}$$

Hence proved

3.10   RELATION BETWEEN THE OPERATORS Δ AND E

From the definition we have Δ f(x) = f(x+h) – f(x)

$\Rightarrow \Delta$f(x) = E f(x) − f(x)

$\Rightarrow \Delta$f(x) = (E − 1)f(x)

$\Rightarrow \Delta$ = E – 1

Also Δ= E – 1 $\Rightarrow$ E = 1 + Δ where 1 is the identity operator

Relation between∇ and E:

From the definition we have ∇f(x) = f(x) – f (x−h)∇f(x) = f(x) – f (x − h)$\Rightarrow \nabla$f(x) = f(x) $- E^{-1}f(x)$

$\Rightarrow \nabla$f(x) = $(1-E^{-1})$f(x)

$\Rightarrow \nabla = 1-E^{-1}$

Also $\qquad \nabla = 1-E^{-1} \Rightarrow E^{-1} = 1-\nabla$

Since $(E^{-1})^{-1} = E$ we get $E = (1-\nabla)^{-1}$

3.11 RELATION BETWEEN E AND δ

From the definition we know that

$$\delta f(x) = [\,f(\,x+\tfrac{h}{2}) - f\,(x-\tfrac{h}{2})\,] \Rightarrow \delta\,f(x) = E^{\frac{1}{2}}f(x) - E^{\frac{-1}{2}}f(x)$$

$\Rightarrow \qquad \delta f(x) = (\,E^{\frac{1}{2}} - E^{\frac{-1}{2}}\,)f(x)$

$\Rightarrow \qquad \delta = E^{\frac{1}{2}} - E^{\frac{-1}{2}}$

Also $\qquad \delta = E^{\frac{1}{2}} - E^{\frac{-1}{2}} \Rightarrow \delta = E^{\frac{-1}{2}}(\,E-1) = E^{\frac{-1}{2}}\Delta$

And $\qquad \delta = E^{\frac{1}{2}}(1-E^{-1}) = E^{\frac{1}{2}}\nabla$

3.12 RELATION BETWEEN E, Δ *and* D

We know that $D\,f(x) = \dfrac{d}{dx}\,f(x)$ and by Taylor's theorem

$$f(x+h) = f(x) + \frac{h}{1!}f'(x) + \frac{h^2}{2!}f''(x) + \ldots \quad \text{to } \infty$$

$$\Rightarrow Ef(x) = f(x) + \frac{h}{1!}\,D\,f(x) + \frac{h^2}{2!}D^2 f(x) + \ldots \quad \text{to } \infty$$

$$= [1 + \frac{hD}{1!} + \frac{(hD)^2}{2!} + \ldots \quad \text{to } \infty]\,f(x)$$

$$\Rightarrow Ef(x) = e^{hD}\,f(x)$$

$$\Rightarrow E = e^{hD}$$

We have $E = e^{hD} \Rightarrow 1 + \Delta = e^{hD}$

$$\Rightarrow hD = \log E = \log(1+\Delta) = \Delta - \frac{\Delta^2}{2} + \frac{\Delta^3}{3} - \ldots$$

$$\Rightarrow D = \frac{1}{h}\,[\,\Delta - \frac{\Delta^2}{2} + \frac{\Delta^3}{3} - \ldots\,]$$

Example 1: Evaluate (a) $\Delta \log x$ (b) $\Delta\,(x^2 + e^x + 4)$

Solution:

(a) $\Delta \log x = \log(x+h) - \log x = \log \dfrac{x+h}{x} = \log\,[\,1+\dfrac{h}{x}\,]$

(b) $\Delta\,(x^2 + e^x + 4) = [\,(x+h)^2 + e^{x+h} + 4)] - (x^2 + e^x + 4)$

$$= x^2 + 2xh + h^2 + e^{x+h} + 4 - x^2 - e^x - 4$$

$$= 2xh + h^2 + e^x(e^h - 1)$$

Example 2: Form a forward difference table for the values $y_o = -5$, $y_1 = -10$, $y_2 = -9$, $y_3 = 4$

Solution: Forward difference table

X	y	Δy	Δ^2	Δ^3	Δ^4
0	−5				
	−5				
1	−10		6		
		1		6	
2	−9		12		0
		13		6	
3	4		18		
		31			
4	35				

Example 3: Evaluate (i) $\Delta \log f(x)$ (ii) $\Delta \tan^{-1} x$

Solution: (i) $\Delta \log f(x) = \log f(x + h) - \log f(x) = \log \left[\dfrac{f(x+h)}{f(x)} \right]$

$$= \log \left[\frac{f(x) + \Delta f(x)}{f(x)} \right] \quad [\because \Delta f(x) = f(x+h) - f(x) \Rightarrow f(x+h) = f(x+h)$$

$$= f(x) + \Delta f(x)]$$

$$= \log \left[1 + \frac{\Delta f(x)}{f(x)} \right]$$

(ii) $\Delta \tan^{-1} x = \tan^{-1}(x+h) - \tan^{-1} x = \tan^{-1} \dfrac{x+h-x}{1+(x+h)x} = \tan^{-1} \dfrac{h}{1+hx+x^2}$

Example 4: Find the nth difference of $\sin(ax + b)$

Solution: let $f(x) = \sin(ax + b)$, then we have

$$\Delta \sin(ax + b) = \sin[a(x+h) + b] - \sin(ax + b) = \sin(ax + ah + b) - \sin(ax + b)$$

$$= 2 \cos \left(ax + b + \frac{ah}{2} \right) \sin \frac{ah}{2} = 2 \sin \frac{ah}{2} \left[\sin ax + b + \frac{ah}{2} + \frac{\pi}{2} \right)$$

$$= 2 \sin \frac{ah}{2} \left[\sin ax + b + \frac{ah + \pi}{2} \right]$$

For the second difference we have

$$\Delta^2 \sin(ax+b) = \Delta(\Delta \sin(ax+b)) = \Delta\left[2\sin\frac{ah}{2}\sin\left(ax+b+\frac{ah+\pi}{2}\right)\right]$$

$$= 2\sin\frac{ah}{2}\ 2\sin\frac{ah}{2}\left[\sin ax+b+\frac{ah+\pi}{2}+\frac{ah+\pi}{2}\right]$$

$$= 2^2(\sin\frac{ah}{2})^2\left[\sin ax+b+2\left(\frac{ah+\pi}{2}\right)\right]$$

Similarly, proceeding as explained above we get

$$\Delta^n \sin(ax+b) = 2^n(\sin\frac{ah}{2})^n\left[\sin\left(ax+b+n\left(\frac{ah+\pi}{2}\right)\right)\right]$$

Example 5: Evaluate $\Delta^4[(1-x)(1-2x)(1-3x)(1-4x)]$

Solution: Consider $\Delta^4[(1-x)(1-2x)(1-3x)(1-4x)]$ (given)

the highest power of x in the expansion of $[(1-x)(1-2x)(1-3x)(1-4x)]$ is 4.

The coefficient of x^4 in $[(1-x)(1-2x)(1-3x)(1-4x)]$ is $(-1)(-2)(-3)(-4) = 24$

Therefore $\Delta^4[(1-x)(1-2x)(1-3x)(1-4x)] = 4!\ 1^4.\ 24 = 576$

Example 6: Find the sixth term of the sequence of numbers 8, 12, 19, 29, 42

Solution: We construct a difference table as follows

X	y	Δy	Δ^2	Δ^3	Δ^4
1	8		4		
5	12		3		
		7		0	
6	19		3		0
		10		0	
7	29		3		
		13			
8	42				

We have $y_0 = 8$, $\Delta y_0 = 4$, $\Delta^2 y_0 = 3$, $\Delta^3 y_0 = \Delta^4 y_0 = \ldots = 0$

The sixth term of the given sequence is y_5 i.e. $E^5 y_0 = (1+\Delta)^5 y_0$

$$\Rightarrow E^5 y_0 = 1 + 5\Delta y_0 + 10\,\Delta^2 y_0 + 10\,\Delta^2 y_0 + \ldots = 8 + 5.4 + 10.\ 3 + 10\ .0 + \ldots = 8 + 20$$

$+ 30 + 0 + 0 + .. = 58$

Example 7: Evaluate $\left(\frac{\Delta^2}{E}\right)x^3$

Solution: $\left(\frac{\Delta^2}{E}\right)x^3 = \left[\left(\frac{(E-1)^2}{E}\right)\right]x^3 = \left[\frac{E^2-2E+1}{E}\right]x^3 = \left(E-2+\frac{1}{E}\right)x^3$

$$= Ex^3 - 2x^3 + E^{-1}x^3 = (x+h)^3 - 2x^3 + (x-h)^3$$

$$= x^3 + 3x^2h + 3xh^2 + h^3 - 2x^3 + x^3 - 3x^2h + 3xh^2 - h^3 = 6xh^2$$

Example 8: Show that $\Delta - \nabla = \Delta.\nabla$

Solution: From the definition we have $\Delta f(x) = f(x+h) - f(x)$ and $\nabla f(x) = f(x) - f(x-h)$

Consider $\Delta.\nabla f(x) = \Delta[f(x) - f(x-h)] = \Delta f(x) - \Delta f(x-h)] = f(x+h) - f(x) - [f(x) - f(x-h)]$

$$= \Delta f(x) - \nabla f(x)$$

$$\Rightarrow \Delta - \nabla = \Delta.\nabla$$

Example 9: Show that $\left(\frac{\Delta^2}{E}\right)e^x . \frac{Ee^x}{\Delta^2 e^x} = e^x$

Solution: consider $\left(\frac{\Delta^2}{E}\right)e^x . \frac{Ee^x}{\Delta^2 e^x}$

$$We\ have\ \left(\frac{\Delta^2}{E}\right)e^x . \frac{Ee^x}{\Delta^2 e^x} = (\Delta^2 E^{-1})\,e^x . \frac{Ee^x}{\Delta^2 e^x} = \Delta^2 e^{x-h}\frac{e^{x+h}}{\Delta^2 e^x}$$

$$= e^{-h}\Delta^2 e^x \frac{e^{x+h}}{\Delta^2 e^x} = e^{-h}.e^{x+h} = e^x$$

3.13   SEPARATION OF SYMBOLS

The symbolic relation between the operators can be used to prove a number of Identities. The method used is known as the method of separation of symbols. Few examples based on this method are given below

Example 10: Use the method of separation of symbols to prove the identity

$$u_0 - u_1 + u_2 - \ldots = \frac{1}{2}u_0 - \frac{1}{4}\Delta u_0 + \frac{1}{8}\Delta^2 u_0 - \ldots$$

Solution: Consider left hand side

$$u_0 - u_1 + u_2 - \ldots = u_0 - Eu_0 + E^2 u_0 - \ldots$$

$$= (1 - E + E^2 - \ldots)u_0 = (1+E)^{-1}u_0$$

$$= (1 + 1 + \Delta)^{-1}u_0 = (2+\Delta)^{-1}u_0 =$$

$$= 2^{-1}(1+\frac{\Delta}{2})^{-1}u_0 = \frac{1}{2}(1-\frac{\Delta}{2}+\frac{\Delta^2}{2^2} - \ldots)u_0$$

$$= \frac{1}{2}u_0 - \frac{\Delta u_0}{4} + \frac{\Delta^2 u_0}{8} - \ldots)$$

Hence, proved

Example 11: Use the method of separation of symbols to prove the identity

$$u_0 + \frac{u_1}{1!}x + \frac{u_2}{2!}x^2 + \ldots = e^x \left[u_0 + \frac{\Delta u_0}{1!}x + x^2 \frac{\Delta^2 u_0}{2!} + \ldots \right]$$

Solution: Consider $u_0 + \frac{xEu_0}{1!} + \frac{x^2 E^2 u_0}{2!} + \frac{x^3 E^3 u_0}{3!} + \ldots$

$$= \left[1 + \frac{xE}{1!} + \frac{x^2 E^2}{2!} + \frac{x^3 E^3}{3!} + \ldots \right] u_0$$

$$= e^{xE} u_0 \ = e^{x(1+\Delta)} u_0$$

$$= e^x e^{x\Delta} u_0 = e^x \left[1 + \frac{x\Delta}{1!} + \frac{x^2 \Delta^2}{2!} + \ldots \right] u_0$$

$$= e^x \left[u_0 + \frac{\Delta u_0}{1!}x + x^2 \frac{\Delta^2 u_0}{2!} + \ldots \right]$$

Hence, proved

Example 12: Prove the identity $u_4 = u_3 + \Delta u_2 + \Delta^2 u_1 + \Delta^3 u_1$

Solution: Consider R.H.S

$$u_3 + \Delta u_2 + \Delta^2 u_1 + \Delta^3 u_1 = E^2 u_1 + \Delta E u_1 + \Delta^2(1 + \Delta u_1) = E^2 u_1 + \Delta E u_1 + \Delta^2(1 + \Delta) u_1$$

$$= E^2 u_1 + \Delta E u_1 + \Delta^2 E u_1 = E^2 u_1 + \Delta E u_1 + \Delta^2 E u_1$$

$$= E(E + \Delta + \Delta^2) u_1 = E (E + \Delta(1 + \Delta)) u_1 =$$

$$= E (E + \Delta E) u_1 = E^2 (1 + \Delta) u_1 = E^2 E u_1 = E^3 u_1 = u_4$$

Example 13: Prove the identity

$$u_1 x + u_2 x^2 + u_3 x^3 + \ldots = \frac{x}{1-x} u_1 + \frac{x^2}{(1-x)^2} \Delta u_1 + \frac{x^3}{(1-x)^3} \Delta^2 u_2 + \ldots$$

Solution: Consider R.H.S

$$\frac{x}{1-x} u_1 + \frac{x^2}{(1-x)^2} \Delta u_1 + \frac{x^3}{(1-x)^3} \Delta^2 u_2 + \ldots$$

$$= \frac{x}{1-x} u_1 + \frac{x^2}{(1-x)^2} (E - 1) u_1 + \frac{x^3}{(1-x)^3} (E - 1)^2 u_2 + \ldots$$

$$= \left[\frac{x}{1-x} - \frac{x^2}{(1-x)^2} + \frac{x^3}{(1-x)^3} + \ldots \right] u_1 + \left[\frac{x^2}{(1-x)^2} - \frac{2x^3}{(1-x)^3} + \ldots \right] Eu_1 + \ldots$$

$$= \frac{x}{1-x} \left(1 + \frac{x}{1-x} \right)^{-1} u_1 + \frac{x^2}{(1-x)^2} \left(1 + \frac{x}{1-x} \right)^{-2} u_2 +$$

$$\frac{x^3}{(1-x)^3} \left(1 + \frac{x}{1-x} \right)^{-3} u_3 + \ldots$$

$$= \frac{x}{1-x} \cdot \frac{1-x}{1} u_1 + \frac{x^2}{(1-x)^2} \cdot \frac{(1-x)^2}{1} u_2 + \ldots$$

$$= u_1 x + u_2 x^2 + u_3 x^3 + \ldots \quad \text{L.H.S}$$

Hence, proved

Example 14: If D, E, δ and μ are the operators with usual meaning and if hD = U, where h is the interval of differencing. Prove the following relation between the operators

(i) $E = e^U$ (ii) $\delta = 2 \sin \dfrac{hU}{2}$ (iii) $\mu = \cos \dfrac{hU}{2}$ (iv) $(E+1)\delta = 2(E-1)\mu$

Solution:

(i)   By definition we have $E = e^{hD} \Rightarrow E = E^U$ (Since U=hD)

(ii)   Consider right hand side $2 \sin \dfrac{hU}{2}$

$$2 \sin \frac{hU}{2} = 2 \left[\frac{E^{\frac{U}{2}} - E^{\frac{-U}{2}}}{2}\right] = (E^U)^{1/2} - (E^U)^{-1/2}$$

$$= (E)^{1/2} - (E)^{-1/2}\text{(since } E = E^U) = \delta$$

(iii)   Consider R.H.S i.e. $\cos \dfrac{hU}{2}$

$$\text{Cos } \frac{hU}{2} = \left[\frac{E^{\frac{U}{2}} + E^{\frac{-U}{2}}}{2}\right] = \frac{1}{2}[(E^U)^{1/2} + (E^U)^{-1/2}]$$

$$= \frac{1}{2}\left[(E)^{\frac{1}{2}} + (E)^{-\frac{1}{2}}\right]\text{(since } E = E^U) = \mu$$

(iv)   Consider L.H.S i.e. $(E + 1)\,\delta = (E + 1)(E^{1/2} - E^{-1/2})$

$$= (E^{1/2}E^{1/2} + E^{1/2}E^{-1/2}) \cdot (E^{1/2} - E^{-1/2})$$

$$= E^{1/2}(E^{1/2} + E^{-1/2}) \cdot (E^{1/2} - E^{-1/2})$$

$$= E^{1/2}(E^{1/2} - E^{-1/2})(E^{1/2} + E^{-1/2})$$

$$= E^{1/2}(E^{1/2} - E^{-1/2}) \cdot 2\left(\frac{E^{1/2} + E^{-1/2}}{2}\right)$$

$$= (E - 1) \cdot 2\,\mu$$

$$= 2\,\mu\,(E - 1)$$

Example 15: Given that

X	1	2	3	4	5
y	2	5	10	17	28

Find $\nabla^2 y_5$

Solution: we have $y_1 = 2$, $y_2 = 5$, $y_3 = 10$, $y_4 = 17$, $y_5 = 28$

$$\nabla^2 y_5 = (1 - E^{-1})^2 y_5 = (1 - E^{-1})^2 y_5 = (1 - 2E^{-1} + E^{-2})\,y_5$$

$$= y_5 - 2y_4 + y_3 = 26 - 34 + 10 = 2$$

Example 16: If Δ, ∇ and δ denote the forward, backward and the central difference operators, and E, μ are respectively the shift and averaging operators in the analysis of data with equal spacing the , prove the following

(a) $\Delta - \nabla = \delta^2$

(b) $E^{1/2} = \mu + \frac{1}{\delta}$ and $E^{-1/2} = \mu - \frac{1}{\delta}$

(c) $\Delta = \frac{1}{2}\delta^2 + \delta\sqrt{1 + \frac{\delta^2}{4}}$

(d) $\mu^2 = 1 + \frac{\delta^2}{4}$

(e) $\mu\,\delta = \frac{1}{2}\Delta E^{-1} + \frac{1}{2}\Delta$

Solution:

(a) $\Delta - \nabla = \delta E^{1/2} - \delta E^{-1/2} = \delta\,(E^{1/2} - E^{-\frac{1}{2}}) = \delta.\delta = \delta^2$

(b) $\mu + \frac{1}{2}\delta = \frac{1}{2}\,(E^{1/2} + E^{-\frac{1}{2}}) + \frac{1}{2}\,(E^{1/2} - E^{-\frac{1}{2}}) = 2.\frac{1}{2}E^{1/2} = E^{1/2}$ And $\mu - \frac{1}{2}\delta = \frac{1}{2}$

$(E^{1/2} + E^{-\frac{1}{2}}) - \frac{1}{2}\,(E^{1/2} - E^{-\frac{1}{2}}) = 2.\frac{1}{2}E^{-1/2} = E^{-1/2}$

(c) $\frac{1}{2}\delta^2 + \delta\sqrt{1 + \frac{\delta^2}{4}} = \frac{1}{2}(E^{1/2} - E^{-\frac{1}{2}})^2 + \frac{1}{2}\,(E^{1/2} - E^{-\frac{1}{2}})\sqrt{1 + \frac{1}{4}(E^{1/2} - E^{-\frac{1}{2}})^2}$

$$= \frac{1}{2}(E + E^{-1} - 2) + (E^{1/2} - E^{-\frac{1}{2}})\sqrt{\frac{E + E^{-1} + 2}{4}}$$

$$= \frac{1}{2}(E + E^{-1} - 2) + (E^{1/2} - E^{-\frac{1}{2}})\sqrt{\frac{(E^{\frac{1}{2}} + E^{\frac{-1}{2}})^2}{4}}$$

$$= \frac{1}{2}(E + E^{-1} - 2) + \frac{1}{2}(E^{1/2} - E^{-\frac{1}{2}})\,(E^{\frac{1}{2}} + E^{\frac{-1}{2}})$$

$$= \frac{1}{2}(E + E^{-1} - 2) + \frac{1}{2}(E - E^{-1}) = E - 1 = \Delta$$

(d) $\mu^2 = \frac{1}{4}\,(E^{\frac{1}{2}} + E^{\frac{-1}{2}})^2 = \frac{1}{4}(E + E^{-1} + 2) = \frac{1}{4}\,[\,(E^{\frac{1}{2}} - E^{\frac{-1}{2}})^2 + 4\,] = \frac{1}{4}[\delta^2 + 4\,] = 1 + \frac{\delta^2}{4}$

(e) $(e)\mu\,\delta = \frac{1}{2}\,(E^{\frac{1}{2}} + E^{\frac{-1}{2}})\,(E^{\frac{1}{2}} - E^{\frac{-1}{2}}) = \frac{1}{2}(E - E^{-1}) = \frac{1}{2}(E - 1 + 1 - E^{-1})$

$$= \frac{1}{2}\Delta + \frac{1}{2}\Delta E^{-1}$$

3.14 NEWTONS' BINOMIAL EXPANSION FORMULA

Let $y_0, y_1, \ldots, y_n$ denote the values of the function y = f(x) corresponding to the values

$x_0, x_1, \ldots, x_n$ of x, where $x_n = x_0 + nh$ (h = interval of differencing)and one of the values of y is missing .Since n values of the function are known we have

$$\Delta^n y_0 = 0$$

$$\Rightarrow (E - 1)^n y_0 = 0$$

$$\Rightarrow [E^n - n_{c_1} E^{n-1} + n_{c_2} E^{n-2} - \ldots + (-1)^n]\, y_0 = 0$$

$$\Rightarrow E^n y_0 - n_{c_1} E^{n-1} y_0 + n_{c_2} E^{n-2} y_0 - \ldots + (-1)^n y_0 = 0$$

$$\Rightarrow y_n - n_{c_1} y_{n-1} + n_{c_2} y_{n-2} - \ldots + (-1)^n y_0 = 0$$

The above formula is called the Binomial expansion formula and is useful in finding the missing values without constructing the difference table .

Example 17: Find the missing entry in the following table

x	1	2	3	4	5
y	2	5	7	---------	32

Solution: four values are given and one value is missing in the table

We have $x_0=1$, $x_1= 2$, $x_2= 3$, $x_3= 4$, $x_4 = 5$

$y_0 = 2$, $y_1= 5$, $y_2 = 7$, $y_3= $--, $y_4 = 32$, y_3 is missing in the table

$$\Delta^4 y_0 = (E - 1)^4 y_0 = (E^4 - 4E^3 + 6E^2 - 4E + 1)y_0 = 0$$

$$\Rightarrow y_4 - 4y_3 + 6y_2 - 4y_1 + y_0 = 0$$

$$\Rightarrow 32 - 4y_3 + 6.7 - 4.5 + 2 = 0$$

$$\Rightarrow 32 - 4y_3 + 42 - 20 + 2 = 0$$

$$\Rightarrow 56 - 4y_3 = 0 \Rightarrow 4\, y_3 = 56$$

$$\Rightarrow y_3 = 14$$

Alternative method: let α denote the missing value

The missing value can also be found by constructing the difference table

x	y	Δy	$\Delta^2 y$	$\Delta^3 y$	$\Delta^4 y$
1	2	3	-1	$\alpha - 8$	$56 - 4\alpha$
2	5	2	$\alpha - 9$	$48 - 3\alpha$	
3	7	$\alpha - 7$	$39 - 2\alpha$		
4	α	$39 - 2\alpha$			
5	32				

Since $\Delta^4 y = 0$ we have $56 - 4\alpha = 0 \Rightarrow 4\alpha = 56 \Rightarrow \alpha = 14$

3.15   FACTOR POLYNOMIAL

If the interval of differencing is h then the factorial polynomial is defined by

$x^{(n)} = x\,(x-h)(x-2h) \ldots (x-(n-1)h)$ where n is a positive integer

It can also be defined as follows:

A factorial polynomial of degree n , with unit interval denoted by $x^{(n)}$ is defined by

$$x^{(n)} = x\,(x-1)(x-2) \ldots (x-n+1)$$

From the above definition we have

$$x^{(1)} = x \ , \ x^{(2)} = x\,(x-1)\,,..$$

Note: $x^{(0)} = 1$

Differences of factor polynomial: Consider the factor polynomial

$$x^{(n)} = x\,(x-h)(x-2h)\ldots(x-(n-1)h)$$

The first difference of $x^{(n)}$ is $\Delta x^{(n)} = (x+h)^{(n)} - x^{(n)} = (x+h)\,(x)(x-h)\ldots(x-(n-2)h) - [x\,(x-h)(x-2h)\ldots(x-(n-1)h)]$

$$= (x+h)\,(x)(x-h)\ldots(x-(n-2)h)[\,x+h-x+(n-1)h\,]$$

$$= nhx^{(n-1)}$$

Similarly we get

$$\Delta^2 x^{(n)} = n(n-1)h^2 x^{(n-2)}\ldots$$

$\Delta^r x^{(n)} = n(n-1)\ldots(n-r+1)\,h^r x^{(n-r)}$ where r is a positive integer less than n.

Note: $\Delta^r x^{(n)} = 0$ If $r > n$ and $\Delta^n x^{(n)} = n!$

Example 18: Express $3x^4 + 8x^3 + 3x^2 - 27\,x + 9$ in factor polynomial and get their forward differences taking the interval of differencing h =1

Solution : Let $f(x) = 3x^4 + 8x^3 + 3x^2 - 27\,x + 9 = A\,x^{(4)} + Bx^{(3)} + C\,x^{(2)} + Dx^{(1)} + E$

which implies that $3x^4 + 8x^3 + 3x^2 - 27\,x + 9$

$$= A\,x(x-1)(x-2)(x-3) + Bx\,(x-1)(x-2) + C\,x(x-1) + Dx + E \quad \ldots\ldots(1)$$

by putting x = 0 in (1) we get E = 9

If we put x = 1 in (1) we get $D + E = -4 \Rightarrow D + 9 = -4 \Rightarrow D = -13$

If we put x = 2 in (1) we get $2C + 2D + E = 48 + 64 + 12 - 54 + 9$

$$\Rightarrow 2C - 26 + 9 = 2C = 48 + 64 + 12 - 54 + 9 + 17 = 96$$

$$\Rightarrow 2C = 96 \Rightarrow C = 48$$

If we put x = 3 in (1) we get $6B + 6C + 3D + E = 243 + 216 + 27 - 81 + 9 \Rightarrow$ B = 26

Equating the coefficients of x^3 on both sides we get A = 3

Hence, the required factorial polynomial is $3\,x^{(4)} + 26x^{(3)} + 48\,x^{(2)} - 13\,x^{(1)} + 9$

The differences are $\Delta f(x) = 12\,x^{(3)} + 78x^{(2)} + 96\,x^{(1)} - 13$

$$\Delta^2 f(x) = 36x^{(2)} + 156x^{(1)}$$

$$\Delta^2 f(x) = 72x^{(1)} + 156$$

$$\Delta^3 f(x) = 72$$

Reciprocal factorial polynomial:

If n is a positive integer , then the reciprocal factorial polynomial $x^{(-n)}$ is defined by

$$x^{(-n)} = \frac{1}{(x+h)(x+2h)...(x+nh)} \text{ where h is the interval of differencing}$$

The first difference of reciprocal factorial polynomial is

$$\Delta x^{(-n)} = \frac{-nh}{(x+h)(x+2h)...[(x+(n+1)h]} = (-n)hx^{(-(n+1))}$$

EXERCISE

1. Define the operators $\Delta, \nabla. E, \mu$ and δ

2. Given $u_0 = 4$, $u_1 = 8$, $u_2 = 21$, $u_3 = 75$, $u_4 = 32, u_5 = 16$, $u_6 = 10$, find $\Delta^6 u_0$ [**Ans:** −835]

3. Evaluate $\Delta^3 [(1-x)(1-2x)(1+3x)]$ by taking h = 2   [**Ans:**288]

4. Evaluate $\Delta^{10} [(1-ax)(1-b x^2)(1 -c x^3)(1-dx^4)]$,   [**Ans:** $abcd 2^{10} 10!$]

5. Evaluate $\Delta^3[(1+2x)(1+4x)(1+6x)]$ [**Ans:** 48 3! h^3]

6. construct a forward difference table for the data given below

x	0	1	2	3	4
y	1	3	7	13	21

7. Construct a forward difference table for f(x) = $x^3 + x^2 - 2x + 1$ for the values x = 0,1,2 ,3.4 and 5

8. Find the nth difference of a^{bx+c} [**Ans:** $a^{bx+c}(a^{bx+c})^n$]

9. Evaluate Δ^2 cos 2x   [**Ans:** $- 4 \sin^2 h \cos (2x + 2h)$]

10. Prove the following relations

(a) $\dfrac{\Delta}{\nabla} - \dfrac{\nabla}{\Delta} = \Delta + \nabla$

(b) $\nabla = 1 - e^{-hD}$

(c) $\nabla = 1 - (1 + \Delta)^{-1}$

11. Construct forward difference table for the data given below

x	0	1	2	3	4
y	1	1.5	2.2	3.1	4.6

12. Construct a forward difference table for the data given below

x	10	15	20	25	30	35
Y = f(x)	19.97	21.51	22.47	23.52	24.65	25.89

Find the values of $\Delta f(10)$, $\Delta^2 f(10)$, $\Delta^3 f(15)$ and $\Delta^4 f(15)$
[*Ans:* $\Delta f(10)$=19.97, $\Delta^2 f(10)$ = $-$ 0.38, $\Delta^3 f(15)$ = 0.19 and $\Delta^4 f(15)$ = $-$o.16]

13. If $u_0 = 1, u_1 = 5, u_2 = 8, u_3 = 3, u_4 = 7$ and $u_5 = 0$ find $\Delta^5 u_0$ [*Ans:* $-$61]

14. Prove the following (a)$y_4 = y_3 + \Delta y_2 + \Delta^2 y_1 + \Delta^3 y_1$ (b)$\Delta^2 y_0 = y_3 - 3y_2 + 3y_1 - y_0$

15. If f(x) = $2x^3 - x^2 + 3x + 1$ then evaluate $\Delta^2 f(x)$ [*Ans:* 12x + 10]

16. Find the factorial polynomial for the function $2x^3 - 3x^2 + 4x - 8$

$$[\textbf{\textit{Ans:}}\ 2x^{(3)} - 3x^{(2)} + 3x^{(1)} - 8]$$

17. Express $x^4 - 12x^3 + 24x^2 - 30x + 9$ and it's successive differences in factorial
notation   [*Ans:* f(x) = $x^{(4)} - 6x^{(3)} - 5x^{(2)} - 17 x^{(1)} + 9$]

$$\Delta f(x) = 4x^{(3)} - 18x^{(2)} - 10 x^{(1)} - 17$$
$$\Delta^2 f(x) = 12 x^{(2)} - 36 x^{(1)} - 10$$
$$\Delta^3 f(x) = 24x^{(1)} - 36$$

And $\Delta^4 f(x) = 24$]

18. Use the method of separation of symbols and prove $u_x = u_{x-1} + \Delta u_{x-2} + \Delta^2 u_{x-3} + \ldots + \Delta^n u_{x-n}$

19. Use the method of separation of symbols and prove

Following table represents the year-wise population of a district

x(year)	1881	1891	1901	1911	1921	1931
y(population)	363	393	421	--------	467	501

Find the population of the year 1911   [*Ans:* 445.2 lakhs]

20. Interpolate the missing values

x	0	1	2	3	4	5
y	0	------	8	15	------	35

[*Ans:* 3 and 24]

21. Prove the relation $\Delta f_i^2 = (f_{i+1} + f_i)\Delta f_i$

Example 19: Given that

x	1	2	3	4	5
u	2	5	10	17	26

Find $\nabla^2 u_5$

Solution: $\nabla^2 u_5 = (1 - E^{-1})^2 u_5 = (1 - 2E^{-1} + E^{-2})u_5 = u_5 - 2u_4 + u_3$

$$= 26 - 2 \times 17 + 10$$

$$= 26 - 34 + 10 = 2$$

Example 20: Find the missing entry in the following table

Given that

x	0	1	2	3	4
u	1	3	9	-	81

Find the missing entry

Solution: We have $u_0 = 1$, $u_1 = 3$, $u_2 = 9$, $u_3 = ---$, $u_4 = 81$ (given)

Four values are given. We assume that $u_x = f(x)$ is a polynomial of degree 3.

Hence $u_0 = 1$, $u_1 = 3$, $u_2 = 9$, $u_3 = --$, $u_4 = 81$

Therefore $\Delta^4 u_0 = 0$

i.e. $(E - 1)^4 u_0 = 0$

$$\Rightarrow (E^4 - 4 E^3 + 6 E^2 - 4 E + 1)u_0 = 0$$

$$\Rightarrow (E^4 u_0 - 4 E^3 u_0 + 6 E^2 u_0 - 4 E u_0 + u_0 = 0$$

$$\Rightarrow u_4 - 4u_3 + 6 u_2 - 4 u_1 + u_0 = 0$$

Substituting the given value we get in the   above equation we get

$$81 - 4u_3 + 6 \times 9 - 4 \times 3 + 1 = 0$$

$$\Rightarrow u_3 - 31 = 0 \Rightarrow u_3 = 31$$

Example 21: Find the first term of the series whose second and subsequent terms are 8, 3, 0, −1, 0...

Solution: Let u_0 denote the first term.

We have $u_1 = 8$, $u_2 = 3$, $u_3 = 0$, $u_4 = -1$, ... (given)

The difference table for the given data is

X	u	Δu	$\Delta^2 u$	$\Delta^3 u$	$\Delta^3 u$
0	u_0				
1	8	−5	2	0	0
2	3	−3	2	0	
3	0	1	2		
4	−1	1			
5	0				

Separation of symbols

The symbolic relation between the operators can be used to prove a number of identities. The method is known as the method of separation of symbols.

Example 22: Using the method of separation of symbols, show that

$$(u_1 - u_0) - (u_2 - u_1) \times + (u_3 - u_2)x^2 - \ldots \text{to}\infty = \frac{\Delta u_0}{1+x} - x\frac{\Delta^2 u_0}{(1+x)^2} + x^2\frac{\Delta^3 u_0}{(1+x)^3} - \ldots \infty$$

Solution: L.H.S $= (u_1 - u_0) - (u_2 - u_1) \times + (u_3 - u_2)x^2 - \ldots \text{to } \infty$

$$= \Delta u_0 - x\Delta u_1 + x^2\Delta u_2 - = \Delta u_0 - x\Delta E u_0 + x^2\Delta E^2 u_0 - \ldots$$

$$= \Delta (1 - x E + x^2 E^2 - \ldots)u_0 = [\Delta(1 + xE)^{-1}u_0$$

$$= \frac{\Delta u_0}{1+x} - x\frac{\Delta^2 u_0}{(1+x)^2} + x^2\frac{\Delta^3 u_0}{(1+x)^3} - \ldots \infty = \text{R.H.S}$$

$$= [\Delta(1 + xE)^{-1}u_0 = [\frac{\Delta}{1+x(1+\Delta)}]u_0$$

$$= [\frac{\Delta}{(1+x)(1+\frac{x\Delta}{1+x})}]u_0$$

$$= \frac{1}{1+x}[1 - \frac{x\Delta}{1+x} + \frac{x^2\Delta^2}{(1+x)^2} - \ldots]\Delta u_0$$

In this chapter we introduce central formulae which are used for interpolation near middle values of the given data

Consider the function y = f(x). Let y be given for (2n+1) equally spaced values of x. say

$$x_0 \pm h, x_0 \pm 2h, \ldots, x_0 \pm nh$$

Corresponding values of y be y_r (r = 0, ± 1, ± 2, $\ldots$, ± n)

Let $y = y_0$ denote the central ordinate corresponding to $x = x_0$. We can form a difference table as follows:

X	y	Δy	$\Delta^2 y$	$\Delta^3 y$	$\Delta^4 y$	$\Delta^5 y$
$x_0 - 2h$	y_{-2}					
		Δy_{-2}				
			$\Delta^2 y_{-2}$			
				$\Delta^3 y_{-2}$		
					$\Delta^4 y_{-2}$	
						$\Delta^5 y_{-2}$
$x_0 - h$	y_{-1}					
		Δy_{-1}				
			$\Delta^2 y_{-1}$			
				$\Delta^3 y_{-1}$		
					$\Delta^4 y_{-1}$	
x_0	y_0					
		Δy_0				
			$\Delta^2 y_0$			
				$\Delta^3 y_0$		
$x_0 + h$	y_1					
		Δy_1				
			$\Delta^2 y_1$			
$x_0 + 2h$	y_2					

The above table can also be written in terms of central differences as follows

X	y	Δy	$\Delta^2 y$	$\Delta^3 y$	$\Delta^4 y$	$\Delta^5 y$
$x_0 - 2h$	y_{-2}	Δy_{-2}	$\Delta^2 y_{-2}$	$\Delta^3 y_{-2}$	$\Delta^4 y_{-2}$	$\Delta^5 y_{-2}$
$x_0 - h$	y_{-1}	Δy_{-1}	$\Delta^2 y_{-1}$	$\Delta^3 y_{-1}$	$\Delta^4 y_{-1}$	
x_0	y_0	Δy_0	$\Delta^2 y_0$	$\Delta^3 y_0$		
$x_0 + h$	y_1	Δy_1	$\Delta^2 y_1$			
$x_0 + 2h$	y_2					

We now introduce the central difference operator δ to represent the successive differences operator of a function in a convenient way.

If $y_0, y_1, y_2, \ldots, y_n$ denote two successive values corresponding to the values $x_0, x_1, x_2, \ldots, x_n$

$$y_1 - y_0 = \delta y_{1/2},$$

$$y_2 - y_1 = \delta y_{3/2}$$

....

$$y_n - y_{n-1} = \delta y_{n-1/2}$$

For the higher order central differences we have

$$\delta y_{\frac{3}{2}} - \delta y_{\frac{1}{2}} = \delta^2 y_1,\ \delta y_2 - \delta y_1 = \delta^2 y_{\frac{3}{2}},\ \ldots,\ \delta^{n-1} y_{r+\frac{1}{2}} - \delta^{n-1} y_{r-\frac{1}{2}} = \delta^n y_r$$

$$= (E^{1/2} - E^{-1/2})^n y_r$$

In it's alternative notation we have $\delta f(x) = f(x+\tfrac{1}{2}h) - f(x-\tfrac{1}{2}h)$

where h is the interval of differencing

X	y	δy	$\delta^2 y$	$\delta^3 y$	$\delta^4 y$	$\delta^5 y$
x_0	y_0	$\delta y_{1/2}$	$\delta^2 y_1$	$\delta^3 y_{3/2}$	$\delta^4 y_2$	$\delta^5 y_{5/2}$
x_1	y_1	$\delta y_{3/2}$	$\delta^2 y_1$	$\delta^3 y_{5/2}$	$\delta^4 y_2$	
x_2	y_2	$\delta y_{5/2}$	$\delta^2 y_2$	$\delta^3 y_{7/2}$		
x_3	y_3	$\delta y_{7/2}$	$\delta^2 y_3$			
x_4	y_4					

In constructing the above difference table we have used the relation $\delta = \Delta E^{-1/2}$

The mean operator: Mean operator is also called the averaging operator .It is denoted by μ and is defined as follows:

$$\mu f(x) = \frac{1}{2}\left[f(x+\tfrac{1}{2}h) + f(x-\tfrac{1}{2}h)\right] \text{ or } \ \mu f(x) = \frac{1}{2}\left[E^{1/2} + E^{-1/2}\right] f(x)$$

In general we define operators are defined as follows

$$\Delta y_x = y_{x+h} - y_x \quad \text{(Forward difference operator)}$$

$$\nabla y_x = y_x - y_{x-h} \quad \text{(Back ward difference operator)}$$

$$\delta y_x = y_{x+\frac{h}{2}} - y_{x-\frac{h}{2}} \quad \text{(Central difference operator)}$$

$$\mu y_x = \frac{1}{2}[y_{x+\frac{h}{2}} + y_{x-\frac{h}{2}}] \quad \text{(Averaging operator)}$$

CHAPTER 4

Interpolation with Equal Intervals

4.1 INTRODUCTION

Interpolation is a method of constructing new data points within the range of a discrete set of known data points. It is the technique of estimating a past figure. It is defined as follows.

Definition: Interpolation is the estimation of a value within two known values in a sequence of values.

The study of interpolation is based on the assumption that there are no jumps in the values of the dependent variable for the period under consideration. It is also assumed that the rate of change of one figure to another is uniform.

Extrapolation is an estimation of a value based on extending a known sequence of values or facts beyond the area that is certainly known. In a general sense, to extrapolate is to infer something that is not explicitly stated from existing information.

4.2 INTERPOLATION WITH EQUAL INTERVALS

4.2.1 NEWTON'S FORWARD INTERPOLATION FORMULA

Let $y = f(x)$ be a function which takes the values $y_0, y_1 ,....., y_n$ corresponding to the $(n + 1)$ values $x_0, x_1, ..., x_n$ of the independent variable x. Let the values of be equally spaced i.e. $x_r = x_0 + rh$, $r = 0,1,...,n$

Where h is the interval of differencing

Let $\phi(x)$ be a polynomial of nth degree in x taking the same values of y corresponding to $x = x_0, x_1 , ..., x_n$, then $\phi(x)$ represents the continuous $y = f(x)$ such that $f(x_r) = \phi(x_r)$ for r = 0,1,2,..., n ,and at all other points.

$$f(x) = \phi(x) + R(x)$$

where R(x) is called the error term of the interpolating formula.

Let us assume that

$$f(x) \approx \phi(x) = a_0 + a_1(x-x_0) + a_2 (x-x_0)(x-x_1) + ... + a_n(x-x_0)(x-x_1)...(x-x_{n-1}) \quad .....(4.1)$$

Where $a_0, a_1 , ..., a_n$ are constants

The constants $a_0, a_1, \ldots, a_n$ can determined as follows

Putting x = x_0 in (4.1) we get $f(x_0) = y_0 = a_0 \Rightarrow \phi(x_0) \approx a_0$

Putting x = x_1 in (4.1) we get $f(x_1) = \phi(x_1) = a_0 + a_1(x_1 - x_0) = a_0 + a_1 h$ or

$$\Rightarrow y_1 = y_0 + a_1 h$$

$$\Rightarrow a_1 = \frac{y_1 - y_0}{h} = \frac{\Delta y_0}{h}$$

Putting x = x_2 in (4.1) we get $f(x_2) = a_0 + a_1(x_2 - x_0) + a_2(x_2 - x_0)(x_2 - x_1)$

Or $\qquad y_2 = a_0 + a_1(2h) + a_2 (2h)(h)$

Or $\qquad a_2 = \dfrac{y_2 - 2y_1 \div y_0}{2h^2} = \dfrac{\Delta^2 y_0}{2!h^2}$

Similarly by putting x = x_3 , x = x_4 , . . . , x = x_n in (4.1) we get

$$a_3 = \frac{\Delta^3 y_0}{3!h^3} , a_4 = \frac{\Delta^4 y_0}{4!h^4} , ..., a_n = \frac{\Delta^n y_0}{n!h^n}$$

Substituting the values of $a_0, a_1, \ldots, a_n$ in (4.1) we get

$$f(x) = \phi(x)$$

$$= y_0 + \frac{\Delta y_0}{h}(x - x_0) + \frac{\Delta^2 y_0}{2!h^2}(x - x_0)(x - x_1) + \ldots + \frac{\Delta^n y_0}{n!h^n}(x - x_0)(x - x_1)\ldots$$

$$(x - x_{n-1}) \qquad\qquad\qquad\qquad .....(4.2)$$

Writing $\qquad$ p = $\dfrac{x - x_0}{h}$ we get $(x - x_0) = ph$, $(x - x_1) = (x - x_0) + (x_0 - x_1) = (x - x_0)$

$$- (x_1 - x_0)$$

$$= ph - h = (p-1)h$$

Similarly we get $(x - x_2) = (p-2)h$, ..., $(x - x_{n-1}) = (p - n + 1)h$

Substituting these values in (4.2) we get

$$f(x) = \phi(x) = y_0 + p\frac{\Delta y_0}{h} + \frac{p(p-1)}{2!}\Delta^2 y_0 + \ldots +$$

$$\frac{p(p-1)(p-2)...(p-n+1)}{n!}\Delta^n y_0 \qquad\qquad .....(4.3)$$

The above formula is called Newton's forward interpolation formula.

Note: Newton's forward interpolation is used to interpolate the values of the function y near the beginning of a set of tabular values.

y_0 is taken at any point of, but the formula contains only those values of the function y which come after the value chosen as y_0.

The error in the Newton's formula is given by

$$R(x) = \frac{p(p-1)(p-2)...(p-n)}{(n+1)!}\Delta^{n+1} y_0$$

R(x) is also called the remainder term

The smaller the of h is taken, the more near does the formula give the actual error.

Example 1: Find a polynomial for the data given below

X:	0	1	2	3
Y:	1	3	7	13

Solution: We construct difference table for the given data

X	y	Δy	$\Delta^2 y$	$\Delta^3 y$
0	1			
		2		
1	3		2	
2	7	4		0
			2	
3	13	6		

We have h = 1 x_0= 0, y_0 = 1, Δy_0 = 2, $\Delta^2 y_0$ = 2, $\Delta^3 y_0$ = 0

From Newton's forward interpolation formula we get

$$f(x) = y_0 + p\,\frac{\Delta y_0}{h} + \frac{p(p-1)}{2!}\Delta^2 y_0 + \dots$$

$$= 1 + 2x + \frac{x(x-1)}{2!}.2 = 1 + 2x + x^2 - x$$

$$= x^2 + x + 1$$

Example 2: The following data are taken from the steam table

Temp °C   :   140   150   160   170   180

Pressure kgf/cm² :   3.685   4.854   6.302   8.076   10.225

Find the pressure at temperature t = 14 2°C

Solution: We construct difference table

t	p	Δ	Δ^2	Δ^3	Δ^4
140	3.685	1.169	0.279	0.047	0.002
150	4.854	1.448	0.326	0.049	
160	6.302	1.774	0.375		
170	8.076	2.149			
180	10.225				

We have h = 10, t_0 = 140, t_0 = 3.685, Δt_0 = 1.169, $\Delta^2 t_0$ = 0.279, $\Delta^3 t_0$ = 0.047, $\Delta^4 t_0$ = 0.047

$$\Delta^4 t_0 = 0.002,\ p = \frac{t - t_0}{h} = \frac{142 - 140}{10} = \frac{2}{10} = 0.2$$

From Newton's forward interpolation formula we get

$$f(x) = y_0 + p\frac{\Delta t_0}{h} + \frac{p(p-1)}{2!}\Delta^2 t_0 + \ldots$$

$$f(t = 142) = 3.685 + (0.2)(1.169) + \frac{(0.2)(-0.8)}{2}(0.279) +$$

$$\frac{(0.2)(-0.8)(-1.8)}{6}(0.047) + \frac{(0.2)(-0.8)(-1.8)\,(-2.8)}{24}(0.002)$$

$$= 3.685 + 0.2338 - 0.02332 + 0.002256 - 0.0000672$$

$$= 3.89766$$

Example 3: Given $\sin 45^0 = 0.7071$, $\sin 50^0 = 0.7660$, $\sin 55^0 = 0.8192$ and $\sin 60^0 = 0.8660$

Find the value of $\sin 52^0$

Solution: The forward difference table for the given values is

X	y = sin x	Δy	$\Delta^2 y$	$\Delta^3 y$
45	0.7071	0.0589	−0.0057	−0.0007
50	0.7660	0.0532	−0 0064	
55	0.8192	0.0.468		
60	0.8660			

We have $x_0 = 45^0$, h = 5, x = 52^0, p $= \frac{x - x_0}{h} = \frac{52-45}{5} = 1.4$, $y_0 = 0.7071$,

$\Delta y_0 = 0.0589$, $\Delta^2 y_0 = -\,0.0057$, $\Delta^3 y_0 = -\,0.0007$

From Newton's formula we have

$$f(x) = y_0 + p\frac{\Delta y_0}{h} + \frac{p(p-1)}{2!}\Delta^2 y_0 + \ldots$$

$$= 0.7071 + 1.4 \times 0.0589 + \frac{(1.4)(1.4-1.)}{2}(-.0057)$$

$$+ \frac{(1.4)(1.4-1)(1.4-2)}{6}(-0.0007)$$

Or $\quad f(52^0) = 0.7071 + 0.8246 - 0.0011596 + 0.0000392 = 0.788032$

$$\Rightarrow \sin 52^0 = 0.788032$$

Example 4: Consider the following data for g(x) = $\frac{\sin x}{x^2}$

x	0.1	0.2	0.3	0.4	0.5
f(x)	9.9833	4.9667	3.2836	2.4339	1.9177

Find the value of f(0.25)

Solution: The difference table for the given data is

X	f(x)		$\Delta\Delta^2\Delta^3\Delta^4$		
0.1	9.9833	−5.0166	3.3335	−2.5001	2.002
0.2	4.9667	−1.6381	0.8334	−4.999	
0.3	3.2836	−0.8497	−0.3335		
0.4	2.4339	− 0.5162			
0.5	1.9177				

We have x = 0.25, x_0 = 0.1 , p = $\frac{x-x_0}{h}$ = $\frac{0.25-1}{0.1}$ = 1.5

Applying Newtons forward interpolation formula we get

$$f(x) = f(0.25) = y_0 + p\frac{\Delta y_0}{h} + \frac{p(p-1)}{2!}\Delta^2 y_0 + \dots$$

$$= 0.9833 + (1.5)(-5.0166) + \frac{(1.5)(1.5-1)}{2}(3.3335) +$$

$$\frac{(1.5)(1.5-1)(1.5-2)}{6}(-2.5001) + \frac{(1.5)(1.5-1)(1.5-2)(1.5-3)}{24}(2.0002)$$

$$= 9.9833 - 7.5249 + 1.2\ 50063 + 0.156256 + 0.04688$$

$$= 3.911599$$

Example 5: Interpolate the figure of population for the tear 1986 from the   following data

Year	:	1970	1980	1990	2000
Population of a town	:	25,494	29,003	32,538	36,070

Solution:

Year (x)	population (y)	Δy	$\Delta^2 y\Delta^3 y$		y
1970	25,494				
		3509			
1980	29,003		16		
		3525		1	
1990	32,538		17		
		3542			
2,000	36,070				

We have x_0 = 1970, h = 10,0 Δy_o = 3509, $\Delta^2 y_0$ = 16, $\Delta^3 y_0$ = 1,

$$p = \frac{x-x_0}{h} = \frac{1986-1970}{10} = 1.6$$

$$Y(1986) = y_0 + p\frac{\Delta y_0}{h} + \frac{p(p-1)}{2!}\Delta^2 y_0 + \frac{p(p-1)(p-2)}{3!}\Delta^3 y_0 + \dots$$

$$= 25494 + 1.6 \times 3509 + \frac{1,6(1.6-1)}{2} \times 16 + \frac{1.6(1.6-1)(1.6-2)}{6} \times 1$$

$$= 25494 + 5614.4 + 7.68 - 0.064 = 31,116$$

4.2.2 NEWTON'S BACKWARD INTERPOLATION FORMULA

Let y = f(x) be a function which takes the values $y_0, y_1, \ldots, y_n$ corresponding to the (n+1) values

$x_0, x_1, \ldots, x_n$ of the independent variable x. Let the values of be equally spaced i.e.

$$x_r = x_0 + rh, \quad r = 0, 1,\ldots, n$$

Where h is the interval of differencing

Let $\phi(x)$ be a polynomial of nth degree in x taking the same values the same values as y corresponding to x = $x_0, x_1, \ldots, x_n$ i.e. $\phi(x)$ represents y = f(x) such that

$$f(x_r) = \phi(x_r), \quad r = 0.1, 2, \ldots, n$$

we may write

$$f(x)= a_0 + a_1(x - x_n) + a_2 (x - x_n)(x - x_{n-1}) + \ldots + a_n$$
$$(x - x_n)(x - x_{n-1}).\ldots(x - x_1) \qquad \ldots\ldots(4.4)$$

putting   x = x_n in (4.4) we get $f(x_n) = \phi(x_n) = a_0$ i.e. $y_n = a_0$

putting   x = x_{n-1} in (4.4) we get $f(x_{n-1}) = \phi(x_{n-1}) = a_0 + a_1 (x_{n-1} - x_n)$

i.e. $y_{n-1} = \phi(x_{n-1}) = y_n + a_1 (-h)$

or $a_1 h = y_n - y_{n-1} = \nabla y_n$

or $a_1 = \dfrac{\nabla y_n}{1!h}$

putting

get $f(x_{n-2}) = \phi(x_{n-2}) = a_0 + a_1 (x_{n-2} - x_n) + a_2 (x_{n-2} - x_n)$

or $y_{n-2} = \phi(x_{n-2}) = a_0 + a_1 (x_{n-2} - x_n) + a_2 (x_{n-2} - x_{n-1})$

or $y_{n-2} = y_n + \dfrac{y_n - y_{n-1}}{h} (-2h)+ a_2(-2h)(-h)$

or $y_{n-2} = y_n - 2y_n + 2h^2 a_2$

or $a_2 = \dfrac{y_n - y_{n-1} + y_{n-2}}{2h^2} = \dfrac{\nabla^2 y_n}{2!h^2}$

Similarly putting x = x_{n-3}, x = $x_{n-4}, \ldots$, we get

$$a_3 = \frac{\nabla^3 y_n}{3!h^3}, \quad a_4 = \frac{\nabla^4 y_n}{4!h^4}, \ldots, a_n = \frac{\nabla^n y_n}{n!h^n}$$

Substituting these values in 4.4 we get

$$y = \phi(n) = y_n + \frac{\nabla y_n}{h} (x-x_n) + \frac{\nabla^2 y_n}{2!h^2} (x-x_n)(x-x_{n-1}) + \ldots +$$
$$\frac{\nabla^n y_n}{n!h^n}(x-x_n)(x-x_{n-1})\ldots(x-x_1)\ldots \qquad \ldots\ldots(4.5)$$

writing $\quad p = \frac{x - x_n}{h}\quad$ we get $\quad x - x_n = ph$

Therefore we get $\quad x - x_{n-1} = x - x_n + x_n - x_{n-1} = ph + h = (p+1)h$

Similarly we get $\quad x - x_{n-2} = (p+2)h , \ldots , x - x_1 = (p+ n - 1)h$

Therefore substituting these in (4.5), we get

$$f(x) = \phi(x) = y_n + p \frac{\nabla y_n}{1!h} + \frac{p(p+1)}{2!} \nabla^2 y_n + \ldots +$$

$$\frac{p(p+1)(p+2)\ldots(p+n-1)}{n!} \Delta^n y_n$$

The above formula is known as Newton's backward interpolation formula

Example 6: The hourly declination of the moon on a day is given below. Find the Declination at $3^h 35^m 15^s$

Hour (x)	0	1	2	3	4
Declination(y)	$8^0 29' 7''$	$8^0 18' 19.4''$	$8^0 6' 43.5''$	$7^0 55' 6.1''$	$7^0 43' 27.2''$

Solution: We construct difference table for the given

Hour	Declination	∇y	$\nabla^2 y$	$\nabla^3 y$	$\nabla^4 y$
0	$8^0 29' 7''$	$-11' 343''$			
1	$8^0 18' 19.4''$				
2	$8^0 6' 43.5''$	$-11' 35.9''$	$-1.6''$		
3	$7^0 55' 6.1''$	$-11' 374''$	$-1.5''$	$0.1''$	$-0.1''$
4	$7^0 43' 27.2''$	$11' 38.9''$	$-1.5''$	$0.0''$	

The given values lie near the end of the table, therefore we use Newton's backward interpolation formula

When x = $3^h 35^m 15^s$ we have $p = \frac{3^h 35^m 15^s - 4^h}{1^h} = \frac{-0^h 24^m 15^m 45^s}{1^h} = -0.4125$

Substituting in Newton's backward interpolation formula we get

$$Y(3^h 35^m 15^s) = f(3^h 35^m 15^s) = y_n + p \frac{\nabla y_n}{1!h} + \frac{p(p+1)}{2!} \nabla^2 y_n + \ldots +$$

$$\frac{p(p+1)(p+2)\ldots(p+n-1)}{n!} \Delta^n y_n$$

$$= 7^0 43' 27.2'' + (-0.4125)(-11' 38.9'')$$

$$+ \frac{(-0.4125)(0.5875)}{2}(-1.5'') + \ldots$$

$$= 7^0 43' 27.2'' + 4' 48.29'' + 0.18''$$

$$= 7^0 43' 27.2'' + 4' 48''. 29 + 0.18'' = 7^0 48' 16''$$

EXERCISE

1. State and prove Newton's Forward interpolation formula

2. State and prove Newton's backward interpolation formula

3. Construct the forward difference table of the polynomial $x^3 + x^2 - 2x + 1$ by taking 1 as the interval of differencing .Also find the   value of the polynomial at x = 6

4. Construct the forward difference table for the following data and evaluate $\Delta^3 f(2)$

x	0	1	2	3	4
f(x)	1.0	1.5	2.2	3.1	4.6

5. Given that $\sin 45^0$ = 0.7071, $\sin 50^0$= 0.7660, $\sin 55^0$= 0.8192, $\sin 60^0$= 0.8660, Find $\sin 52^0$, using Newton's forward interpolation formula

6. Find the value of $e^{0.24}$ using the data given below

x	0.1	0.2	0.3	0.4	0.5
y	1.10517	1.22140	1.34986	1.49182	1.64872

7. Find the cubic polynomial which takes the following values

x	0	1	2	3
f(x)	1	2	1	10

8. Find the value of f(4.2) from the table

x	0	2	4	6
f(x)	2	10	66	218

9. Find the value of value of tan(0.12) from the table   given below

x	0.10	0.15	0.20	0.25	0.30
f(x)= tan x	0.1003	0.1511	0.2027	0.2553	0.3093

10. Ordinates f(x) of a normal curves in terms of standard deviation x are given as follows

x	1.00	1.02	1.04	1.06	1.08
f(x)	0.2420	0.2371	0.2323	0.2275	0.2227

Find the ordinate for the standard deviation x = 1.025 using Newton's forward interpolation formula

11. Find the value of $e^{0.35}$ using the data given below

x	0.0	0.1	0.2	0.3	0.4
f(x)= e^x	1.0000	1.1052	1.2214	1.3499	1.4918

12. The table gives the distances in Nautical miles of the visible horizon for the given heights in feet above the earth's surface

X = height	100	150	200	250	300	350	400
Y – distance	10.63	13.03	15.04	16.81	18.42	19.90	21.27

13. The following data are taken from steam table

Temp^{0}C	140	150	160	170	180
Pressure kg/cm^2	3.685	4.854	6.302	8.076	10.223

Find the pressure at temperature t = 142^0 and t = 175^0

14. The following temperature readings were taken on a day

time	2 a.m	6a.m	10a.m	2p.m
temp	40.2^0	42.4^0	51.0^0	72.4^0

Find the temperature at 3a.m and 4a.m

15. Using Newton's backward formula, find a polynomial of degree 3 passing through (3,6), (4,24), (5,60) and (6,120).

CHAPTER 5

Matrices, Characteristic Equation, Eigen Values and Eigen Vectors

In this section 1 we introduce, elementary row and column operations for matrices, rank of a matrix, linear dependence, liner independence of vectors, linear systems of equations, Eigen values, Eigen vectors, Cayley-Hamilton theorem, complex and unitary matrixes.

ELEMENTARY ROW AND COLUMN OPERATIONS

Let A be a given matrix. The operations which are performed on the rows of the matrix A are called elementary row operations and the operation which are performed on the columns of the matrix A are called, elementary column operations.

The following row operations can be performed on the matrix A.

(i)   Interchange of any two rows of A

(ii) Multiplication of each element of a row by a non-zero scalar say k the multiplication of a multiple of one row to another row of A

The interchange of two rows will be denoted by

$$R_i \leftrightarrow R_j$$

The multiplication of the ith row by a non–zero constant by a non-zero constant k will be denoted by

$$R_i \leftrightarrow k R_j$$

Addition of ith row, with k times the corresponding elements of kth row will be denoted by

$$R_i \leftrightarrow R_i + k R_j$$

The above operations can also be performed on the columns of A and are known as elementary column operations.

Similarly, the corresponding column operations are denoted by

$c_i \leftrightarrow c_j$ (Interchange of the columns)

$c_i \rightarrow k\, c_j$ (Multiplication of ith column by a non-zero constant k)

$c_i \rightarrow c_j + k\, c_j$ (Addition of ith column, with k times the corresponding elements of kth column)

The elementary operations are also known as the elementary transformations.

Equivalent Matrices

Two matrices A and B are said to be equivalent matrices, if one of them say B can be obtained by applying elementary operations on A. If A and B are equivalent, then we write A ~ B elementary operations are also known as elementary transformations.

If A and B are two equivalent matrixes, then there exist two non-singular matrixes P and Q such that B = PAQ.

If A and B, two equivalent matrices such that B = PAQ then A = P–1 B Q–1.

If A is a non-singular matrix, then it can be expressed as a product of elementary matrices and A can be transformed into a unit matrix I, by a sequence of elementary operations.

Elementary Matrices

A matrix obtained from a unit matrix by a single elementary transformation is called an elementary matrix.

Example 1:

$$\begin{bmatrix} 1 & 0 & 0 \\ 0 & 6 & 0 \\ 0 & 0 & 1 \end{bmatrix}$$ is the elementary matrix obtained from I3 by subjecting R2 $\rightarrow$ 6R2.

Elementary matrix is denoted by E-matrix Eij will denote the E-matrix obtained by interchanging ith and jth rows of a unit matrix I.

Ei(k) denotes the E-matrix obtained by multiplying ith row of a unit matrix I by a non-zero scalar k.

Eij (m) will denote, the elementary matrix obtained by adding to the elements of the ith row of a unit matrix, the products by any number m of the corresponding elements of the ith row.

$E_{ij}{'}$ (k) will denote the elementary matrix obtained by multiplying every element of jth column with k and adding them to the corresponding elements of ith column in I.

Properties of elementary matrices

1. All elementary matrices are non-singular

2. Every elementary matrix is a square matrix

3. $|E_i(k)| = k, (10^10)$

4. $|E_{ij}'| = -1$

5. $|E_{ij}(k)| = 1 = |E_{ij}'(k)|$

Rank of a Matrix

Let A be an m × n matrix. Then the rank of A is the order of the largest non-zero minor of A. The rank of a matrix A is denoted by $\rho(A)$.

If r is the rank of the matrix A; then

(i) There exists at least one non-zero minor of order r

(ii) Every minor of order (r + 1) or more is zero.

Properties:

(a) If A is a null matrix, then $\rho(A) = 0$

(b) If A is a non-singular matrix, the $\rho(A) > 0$,

(c) Rank of a unit matrix of order n is n

(d) If A is matrix of order m × n, them $1.0 < \rho(A) < \min(m, n)$

(e) $\rho(A) = \rho(AT)$, where AT is the transpose of A.

(f) If A is a singular matrix of order on then $\rho(A) < m$

(g) If A is a non-singular matrix of order m than $\rho(A) = m$

Example 2: Obtain the rank of the matrix $\begin{bmatrix} 5 & 1 \\ 2 & 3 \end{bmatrix}$

Solution: Let $\qquad A = \begin{bmatrix} 5 & 1 \\ 2 & 3 \end{bmatrix}$

$\therefore \qquad \det A = \begin{vmatrix} 5 & 1 \\ 2 & 3 \end{vmatrix} = 15 - 2 = 13 \neq 0$

Since det A ≠ 0; the matrix A is non-singular.

Hence, the rank of A = $\rho(A)$ = 2

Example 3: Find the rank of $\begin{bmatrix} 1 & 2 & 3 \\ 3 & 4 & 5 \\ 4 & 5 & 6 \end{bmatrix}$

Solution: Let A = $\begin{bmatrix} 1 & 2 & 3 \\ 3 & 4 & 5 \\ 4 & 5 & 6 \end{bmatrix}$

We get   det A = $\begin{vmatrix} 1 & 2 & 3 \\ 3 & 4 & 5 \\ 4 & 5 & 6 \end{vmatrix}$ = (24 – 25) –2(18 – 20) + 3 (15 – 16)

$$= -1 + 4 - 3 = 0$$

det A = 0, therefore $\rho(A) \neq 3$, i.e., $\rho(A) < 3$

Consider the minor $\begin{vmatrix} 3 & 4 \\ 4 & 5 \end{vmatrix}$ = 15 – 16 = – 1 ≠ 0

The order of non-zero minor is 2

∴ $\rho(A) = 2$

Rank and Elementary Matrices

If A is a matrix, then the rank of A remains unaltered by performing elementary operations on the matrix A. If the rank of A is r, then A can be reduced into an equivalent form having exactly r non-zero elements in its leading diagonal positions. Since, all the matrices of the same order and rank can be reduced into the same form by row and column operations, they are equivalent.

Theorem 1: Elementary row and column operations do not other the rank of a matrix.

Thus, interchange of two rows of a matrix A, do not alter the rank of A. Similarly, the inter change of any two columns of a matrix A, do not alter the rank of A. The multiplication of the elements of a matrix A, by a non-zero scalar, does not alter the rank of A. The multiplication of the elements of a row (or a column) of a matrix by a non-zero scalar does not alter the rank of a matrix.

Also, the addition to the elements of a row (or a column) of a matrix, to the corresponding elements of another row (or a column) does not alter the rank of the matrix.

Example 4: Find the rank of A, by applying elementary row operations on A, where

$$A = \begin{bmatrix} 1 & 5 & 4 & 2 \\ 3 & 15 & 12 & 6 \\ 1 & 5 & 4 & 3 \end{bmatrix}$$

Solution: We have $A = \begin{bmatrix} 1 & 5 & 4 & 2 \\ 3 & 15 & 12 & 6 \\ 1 & 5 & 4 & 3 \end{bmatrix} \rightarrow$ R2 $\rightarrow$ R2– R, R3 $\rightarrow$ R3 – R1

$$\begin{bmatrix} 1 & 5 & 4 & 2 \\ 0 & 0 & 0 & 0 \\ 0 & 0 & 0 & 1 \end{bmatrix} \rightarrow$$ R2$\leftrightarrow$ R3

$$\begin{bmatrix} 1 & 5 & 4 & 2 \\ 0 & 0 & 0 & 1 \\ 0 & 0 & 0 & 0 \end{bmatrix}$$

A non-zero minor in the matrix is $\begin{vmatrix} 4 & 2 \\ 0 & 1 \end{vmatrix}$

The order of the non-zero minor is 2.

Hence, the rank of A = $\rho(A)$ = 2

In the equivalent matrix, of A we observe that all minor of order 3 vanish and there exits a Echelon form of a matrix.

A matrix A is said to be in Echelon form if the following conditions hold:

(i) All zero rows, if any are on the bottom of the matrix

(ii) Each leading non-zero entry is to the right of the leading non-zero entries in the preceding row

Hence, a matrix A = $[a_{ij}]$ is said to be an echelon matrix, if there exist non-zero entries $a_{ij1}, a_{ij2},..., a_{ijr}$, where J1< j2< ...<jr with the property that a_{ij} = 0 for

(i) i< r, s $\leq$, 1

(ii) i$\leq$ r, j < j$_i$, and

(iii) i> r

The elements $a_{ij1}, a_{ij2},..., a_{ijr}$ are called the leading non-zero entries of the matrix A.

The following matrices are in echelon form

$$\begin{bmatrix} 1 & 5 & 9 \\ 0 & 0 & 1 \\ 0 & 0 & 0 \end{bmatrix} \qquad \begin{bmatrix} 5 & 1 & 2 & 0 & 3 \\ 0 & 0 & 1 & -2 & 0 \\ 0 & 0 & 0 & 1 & 7 \\ 0 & 0 & 0 & 0 & 0 \end{bmatrix}$$

Row Canonical form of a matrix

A matrix A is said to be in row echelon form if

(i) A is an echelon matrix

(ii) Each non-zero entry is 1

(iii) Each leading non-zero entry is the only non-zero entry in its column

Example 5: The matrix

$$A = \begin{bmatrix} 0 & 1 & 5 & 0 & 0 & 2 \\ 0 & 0 & 0 & 1 & 0 & -3 \\ 0 & 0 & 0 & 0 & 1 & 5 \end{bmatrix} \text{ is in row canonical form.}$$

The zero matrix $O_{m \times n}$ is an example of a matrix in row canonical form.

SOLVED EXAMPLES

Example 6: Find the rank of the matrix.

$$A = \begin{bmatrix} 1 & 0 & -4 & 5 \\ 2 & -1 & 3 & 0 \\ 8 & 1 & 0 & -7 \end{bmatrix}$$

Solution: A is a matrix containing 3 rows and columns. Therefore, the rank of A is less than or equal to 3 i.e., $\rho(A) < 3$.

Consider the minor $\begin{vmatrix} 1 & 0 & -4 \\ 2 & -1 & 3 \\ 8 & 1 & 0 \end{vmatrix} = 1(0-3) - 0 - 4(2+8) = -3 - 40 = -43 \neq 0$

Order the minor is 311

$$\therefore \qquad \rho(A) = 3$$

Example 7: Reduce the matrix $A = \begin{bmatrix} 5 & 3 & 14 & 4 \\ 0 & 1 & 2 & 1 \\ 1 & -1 & 2 & 0 \end{bmatrix}$ into Echelon matrix and hence find the rank.

Solution: We have $A = \begin{bmatrix} 5 & 3 & 14 & 4 \\ 0 & 1 & 2 & 1 \\ 1 & -1 & 2 & 0 \end{bmatrix} \begin{matrix} R_1 \leftrightarrow R_3 \\ \rightarrow \end{matrix}$

$$\sim \begin{bmatrix} 1 & -1 & 2 & 0 \\ 0 & 1 & 2 & 1 \\ 5 & 3 & 14 & 4 \end{bmatrix} R_3 \rightarrow R_3 - 5R$$

$$\sim \begin{bmatrix} 1 & -1 & 2 & 0 \\ 0 & 1 & 2 & 1 \\ 0 & 8 & 4 & 4 \end{bmatrix} \begin{matrix} R_3 \rightarrow R_3/4 \\ \rightarrow \end{matrix}$$

$$\sim \begin{bmatrix} 1 & -1 & 2 & 0 \\ 0 & 1 & 2 & 1 \\ 0 & 2 & 1 & 1 \end{bmatrix} \begin{matrix} R_3 \rightarrow R_3 - 2R_2 \\ \rightarrow \end{matrix}$$

$$\sim \begin{bmatrix} 1 & -1 & 2 & 0 \\ 0 & 1 & 2 & 1 \\ 0 & 0 & -3 & -1 \end{bmatrix}$$

The above matrix is in Echelon form number of non-zero rows = 3

$\therefore \qquad \rho(A) = 3$

Example 8: Find the rank of the matrix.

$$A = \begin{bmatrix} 1 & 1 & 1 \\ a & b & c \\ a^3 & b^3 & c^3 \end{bmatrix}$$

Solution: We have

$$\det A = \begin{vmatrix} 1 & 1 & 1 \\ a & b & c \\ a^3 & b^3 & c^3 \end{vmatrix} = (b-c)\,(c-a)\,(a-b)\,(a+b+c)$$

The following cases arise

(i)   When a, b, c are all different and $a + b + c \neq 0$: In this case we have det $A \neq 0$

i.e., there exists a non-zero minor of order 3.

Hence $\rho(A) = 3$

(ii)   When a, b, c are different and $a + b + c = 0$

In this case det $A = 0$

Consider the minor $\begin{vmatrix} 1 & 1 \\ b & c \end{vmatrix} = c - b \neq 0$

There exist a minor of order 2 which is non-zero

$\therefore$ $\rho(A) = 2$

(iii)   When two of the members a, b, c are equal but are different from the third

Let $a = b \neq c$.

Consider the minor $\begin{vmatrix} 1 & 1 \\ a & c \end{vmatrix} = c - a \neq 0$

The matrix A has a non-zero minor of order 2

$\therefore$ $\rho(A) = 2$

(iv)   When $a = b = c$

In this case all minors of order 2 and 3 vanish and A has a non-zero of order1

$\therefore$ Hence $\rho(A) = 1$

Example 9: If the rank of the matrix

$$A = \begin{bmatrix} 1 & -2 & 3 & 1 \\ 2 & 1 & -1 & 2 \\ 6 & -2 & \lambda & \mu \end{bmatrix} \text{ is 2, the find the values of } \lambda \text{ and } \mu.$$

Solution: We have

$$A = \begin{bmatrix} 1 & -2 & 3 & 1 \\ 2 & 1 & -1 & 2 \\ 6 & -2 & \lambda & \mu \end{bmatrix} \begin{array}{l} R_2 \rightarrow R_2 - 2R_1 \\ R_3 + R_3 - 6R_1 \\ \rightarrow \end{array}$$

$$\sim \begin{bmatrix} 1 & -2 & 3 & 1 \\ 0 & 5 & -7 & 0 \\ 0 & 10 & \lambda-18 & \mu-6 \end{bmatrix} \begin{array}{l} R_3 \to R_3 - 2R_2 \\ \\ \to \end{array}$$

$$\sim \begin{bmatrix} 1 & -2 & 3 & 1 \\ 0 & 5 & -7 & 0 \\ 0 & 10 & \lambda-4 & \mu-6 \end{bmatrix} \begin{array}{l} R_2 \to R_2 / 5 \\ \\ \to \end{array}$$

$$\sim \begin{bmatrix} 1 & -2 & 3 & 1 \\ 0 & 1 & -7/5 & 0 \\ 0 & 0 & \lambda-\mu & \mu-6 \end{bmatrix}$$

The last matrix is in echelon form

It is given that, the rank of A is 2 we must have $\lambda - 4 = 0$, $\mu - 6 = 0$

$\therefore$ $\qquad\qquad \lambda = 4$, $\mu = 6$

Example 10: Reduce the matrix

$$\begin{bmatrix} 1 & 2 & 3 & 0 \\ 2 & 4 & 3 & 2 \\ 3 & 2 & 1 & 3 \\ 6 & 8 & 7 & 5 \end{bmatrix}$$ in to Echelon form and find its rank.

Solution:

We have $\qquad A = \begin{bmatrix} 1 & 2 & 3 & 0 \\ 2 & 4 & 3 & 2 \\ 3 & 2 & 1 & 3 \\ 6 & 8 & 7 & 5 \end{bmatrix} \begin{array}{l} R_2 \to R_2 - 2R_1 \\ R_3 \to R_3 - 3R_1, R_4 \to R_4 - 2R_2 \\ \to \end{array}$

$$\sim \begin{bmatrix} 1 & 2 & 3 & 0 \\ 0 & 0 & -3 & 2 \\ 0 & -4 & -8 & 3 \\ 0 & 4 & 5 & -1 \end{bmatrix} \begin{array}{l} R_2 \leftrightarrow R_3 \\ R_4 \to R_3 + R_4 \\ \to \end{array}$$

$$\sim \begin{bmatrix} 1 & 2 & 3 & 0 \\ 0 & -4 & -8 & 2 \\ 0 & 0 & -3 & 2 \\ 0 & 0 & -3 & 2 \end{bmatrix} \quad R_4 \rightarrow R_4 - R_3 \quad \rightarrow$$

$$\sim \begin{bmatrix} 1 & 2 & 3 & 0 \\ 0 & -4 & -8 & 3 \\ 0 & 0 & -3 & 2 \\ 0 & 0 & 0 & 0 \end{bmatrix}$$

The above matrix is echelon form the number of non-zero rows = 3

Therefore $\rho(A) = 3$

Rank of a product of matrices: If A and B are matrices conformable for multiplication then the rank of a product of the matrices cannot exceed the rank of either matrix.

That is $\qquad \rho(AB) < \rho(B)$ and $\rho(AB) < \rho(A)$

Rank of a sum: If A and B are the matrices, conformable for addition them rank of $(A + B) <$ rank of B

i.e., $\rho (A + B) < \rho (A) + \rho (B)$

In other words: Rank of the sum of the matrices cannot exceed the sum of their ranks.

Example 11: If $A = \begin{bmatrix} 1 & 1 & -1 \\ 2 & -3 & 4 \\ 3 & -2 & 3 \end{bmatrix}$ and $B = \begin{bmatrix} -1 & -2 & -1 \\ 6 & 12 & 6 \\ 5 & 10 & 5 \end{bmatrix}$

Find the ranks of A, B, A + B and AB.

Solution: Rank of A:

We have $\qquad A = \begin{bmatrix} 1 & 1 & -1 \\ 2 & -3 & 4 \\ 3 & -2 & 3 \end{bmatrix} \quad R_2 \rightarrow R_2 - 2R_1, R_3 \rightarrow R_3 - 3R_1 \quad \longrightarrow$

$$\sim \begin{bmatrix} 1 & 1 & 0 \\ 0 & -5 & 6 \\ 0 & -5 & 6 \end{bmatrix} \quad R_3 \rightarrow R_3 - R_2 \quad \rightarrow$$

$$\sim \begin{bmatrix} 1 & 1 & 0 \\ 0 & -5 & 6 \\ 0 & 0 & 0 \end{bmatrix} \begin{array}{l} C_2 \to C_2 - C_1 \\ \\ \to \end{array}$$

$$\sim \begin{bmatrix} 1 & 0 & 0 \\ 0 & -5 & 6 \\ 0 & 0 & 0 \end{bmatrix} \begin{array}{l} C_2 \to C_2/_{-5}, C_3 \to C_3/_6 \\ \\ \to \end{array}$$

$$\sim \begin{bmatrix} 1 & 0 & 0 \\ 0 & -1 & 1 \\ 0 & 0 & 0 \end{bmatrix} \begin{array}{l} C_3 \to C_3 \to C_2 \\ \\ \to \end{array}$$

$$\sim \begin{bmatrix} 1 & 0 & 0 \\ 0 & 1 & 0 \\ 0 & 0 & 0 \end{bmatrix} \sim \begin{bmatrix} I_2 & 0 \\ 0 & 0 \end{bmatrix}$$

Hence, rank of $A = \rho(A) = 2$

Rank of B:

We have

$$B = \begin{bmatrix} -1 & -2 & -1 \\ 6 & 12 & 6 \\ 5 & 10 & 5 \end{bmatrix} \begin{array}{l} R_2 \to R_3 + 6R_1 \\ R_3 + R_3 + 5R_1 \\ \to \end{array}$$

$$\sim \begin{bmatrix} -1 & -2 & -1 \\ 0 & 0 & 0 \\ 0 & 0 & 0 \end{bmatrix} \begin{array}{l} C_2 \to C_2 + 2C, \\ C_3 \to C_3 + C_1 \\ \to \end{array}$$

$$\sim \begin{bmatrix} -1 & 0 & 0 \\ 0 & 0 & 0 \\ 0 & 0 & 0 \end{bmatrix} \begin{array}{l} C_2 \to C_2/(-1) \\ \\ \to \end{array}$$

$$\sim \begin{bmatrix} 1 & 0 & 0 \\ 0 & 0 & 0 \\ 0 & 0 & 0 \end{bmatrix}$$

$\therefore \qquad \rho(B) = 1$

We have,

$$A + B = \begin{bmatrix} 1 & 1 & -1 \\ 2 & -3 & 4 \\ 3 & -2 & 3 \end{bmatrix} + \begin{bmatrix} -1 & -2 & -1 \\ 6 & 12 & 6 \\ 5 & 10 & 5 \end{bmatrix} = \begin{bmatrix} 0 & -1 & -2 \\ 8 & 9 & 10 \\ 8 & 8 & 8 \end{bmatrix} \begin{matrix} R_3 \to R_3/8 \\ \to \end{matrix}$$

$$\sim \begin{bmatrix} 0 & -1 & -2 \\ 8 & 9 & 10 \\ 1 & 1 & 1 \end{bmatrix} \begin{matrix} R_1 \leftrightarrow R_3 \\ \to \end{matrix}$$

$$\sim \begin{bmatrix} 1 & 1 & 1 \\ 8 & 9 & 10 \\ 0 & -1 & -2 \end{bmatrix} \sim \begin{bmatrix} 1 & 1 & 1 \\ 0 & 1 & 2 \\ 0 & -1 & -2 \end{bmatrix} \begin{matrix} R_3 \to R_3 - R_2 \\ \to \end{matrix}$$

$$\sim \begin{bmatrix} 1 & 1 & 1 \\ 0 & 1 & 2 \\ 0 & 0 & 0 \end{bmatrix} \begin{matrix} C_2 \to C_2 - C_1 \\ C_3 \to C_3 - C_1 \\ \to \end{matrix}$$

$$\sim \begin{bmatrix} 1 & 0 & 0 \\ 0 & 1 & 2 \\ 0 & 0 & 0 \end{bmatrix} \begin{matrix} C_3 \to C_3 - 2C_2 \\ \to \end{matrix}$$

$$\sim \begin{bmatrix} 1 & 0 & 0 \\ 0 & 1 & 0 \\ 0 & 0 & 0 \end{bmatrix} \sim \begin{bmatrix} I_2 & 0 \\ 0 & 0 \end{bmatrix}$$

i.e., rank of $(A + B) = \rho(A + B) = 2$

We have

$$AB = \begin{bmatrix} 1 & 1 & -1 \\ 2 & -3 & 4 \\ 3 & -2 & 3 \end{bmatrix} \begin{bmatrix} -1 & -2 & -1 \\ 6 & 12 & 6 \\ 5 & 10 & 5 \end{bmatrix} = \begin{bmatrix} 0 & 0 & 0 \\ 0 & 0 & 0 \\ 0 & 0 & 0 \end{bmatrix}$$

$\therefore \qquad \rho(AB) = 0$

Theorem 2: Every m × n matrix of rank r, can be reduced to the form

$\begin{bmatrix} I_2 & 0 \\ 0 & 0 \end{bmatrix}$ by a finite chain of elementary operations, where I_r is the r-rowed unit matrix.

Corollary 1: The rank of an m × n matrix A is r if and only it. It can be reduced to the form $\begin{bmatrix} I_r & 0 \\ 0 & 0 \end{bmatrix}$ by a finite number of elementary operations.

Corollary 2: If A is an m × n matrix of rank r, there exists non-zero matrices P and Q such that

Such that $PAQ = \begin{bmatrix} I_r & 0 \\ 0 & 0 \end{bmatrix}$

Example 12: Find the rank of

$$A = \begin{bmatrix} 4 & 2 & 1 & 3 \\ 6 & 3 & 4 & 7 \\ 2 & 1 & 0 & 1 \end{bmatrix}$$ by applying column operations on A.

Solution: We have

$$A = \begin{bmatrix} 4 & 2 & 1 & 3 \\ 6 & 3 & 4 & 7 \\ 2 & 1 & 0 & 1 \end{bmatrix} \begin{array}{c} C_4 \to C_4 - C_2 - C_3 \\ \to \end{array}$$

$$\sim \begin{bmatrix} 4 & 2 & 1 & 0 \\ 6 & 3 & 4 & 0 \\ 2 & 1 & 0 & 0 \end{bmatrix} \begin{array}{c} C_2 \to C_2 - 2C_3, C_1 \to C_1 - 4C_3, \\ \to \end{array}$$

$$\sim \begin{bmatrix} 0 & 0 & 1 & 0 \\ -10 & -5 & 4 & 0 \\ 2 & 1 & 0 & 0 \end{bmatrix}$$ Each minor of order in the above matrix is zero.

There exists of minor $\begin{vmatrix} 0 & 1 \\ -5 & 4 \end{vmatrix} = 0 + 5 = 5$, which is non-zero

Therefore, the rank of A = 2.

Example 13: Reduce the matrix

$$A = \begin{bmatrix} 1 & -1 & 2 & -3 \\ 4 & 1 & 0 & 7 \\ 0 & 3 & 0 & 4 \\ 0 & 1 & 0 & 2 \end{bmatrix}$$ to normal form and find its rank.

Solution: We have

`

$$A = \begin{bmatrix} 1 & -1 & 2 & -3 \\ 4 & 1 & 0 & 7 \\ 0 & 3 & 0 & 4 \\ 0 & 1 & 0 & 2 \end{bmatrix} \begin{array}{l} C_2 \to C_2 + C_1, C_3 \to C_3 - 2C_1 \\ C_4 \to C_4 + 3C_1 \\ \to \end{array}$$

$$\sim \begin{bmatrix} 1 & 0 & 0 & 0 \\ 4 & 5 & -8 & 14 \\ 0 & 3 & 0 & 4 \\ 0 & 1 & 0 & 2 \end{bmatrix} \begin{array}{l} R_2 \to R_2 - 4R_1 \\ \to \end{array}$$

$$\sim \begin{bmatrix} 1 & 0 & 0 & 0 \\ 0 & 5 & -8 & 14 \\ 0 & 3 & 0 & 4 \\ 0 & 1 & 0 & 2 \end{bmatrix} \begin{array}{l} R_2 \leftrightarrow R_4 \\ \to \end{array}$$

$$\sim \begin{bmatrix} 1 & 0 & 0 & 0 \\ 0 & 1 & 0 & 2 \\ 0 & 3 & 0 & 4 \\ 0 & 5 & -8 & 14 \end{bmatrix} \begin{array}{l} C_4 \to C_4 - 2C_2 \\ \to \end{array}$$

$$\sim \begin{bmatrix} 1 & 0 & 0 & 0 \\ 0 & 1 & 0 & 0 \\ 0 & 3 & 0 & -2 \\ 0 & 5 & -8 & 4 \end{bmatrix} \begin{array}{l} R_3 \to R_3 - 3R_2 \\ R_4 + R_4 - 5R_2 \\ \to \end{array}$$

$$\sim \begin{bmatrix} 1 & 0 & 0 & 0 \\ 0 & 1 & 0 & 0 \\ 0 & 0 & 0 & -2 \\ 0 & 0 & -8 & 4 \end{bmatrix} \begin{array}{l} R_4 \to R_4 + 2R_3 \\ \to \end{array}$$

$$\sim \begin{bmatrix} 1 & 0 & 0 & 0 \\ 0 & 1 & 0 & 0 \\ 0 & 0 & 0 & -2 \\ 0 & 0 & -8 & 0 \end{bmatrix} \begin{array}{l} R_3 \leftrightarrow R \\ \to \end{array}$$`

$$\sim \begin{bmatrix} 1 & 0 & 0 & 0 \\ 0 & 1 & 0 & 0 \\ 0 & 0 & 1 & 0 \\ 0 & 0 & 0 & -2 \end{bmatrix} \begin{array}{l} C_3 \to C_3/-8, C_4 \to C_4/-2 \\ \to \end{array}$$

$$\sim \begin{bmatrix} 1 & 0 & 0 & 0 \\ 0 & 1 & 0 & 0 \\ 0 & 0 & 1 & 0 \\ 0 & 0 & 0 & 1 \end{bmatrix} \sim [I4] \text{ which is the required normal form.}$$

Hence, the rank of A = 4.

Example 14: Find two non-singular matrices P and Q such that P A Q is in the normal form, where

$$A = \begin{bmatrix} 1 & 1 & 1 \\ 1 & -1 & -1 \\ 3 & 1 & 1 \end{bmatrix}$$

Also, find the rank of A.

Solution: A is a matrix of order 3 × 3. We write A = I3 AI3 as shown below.

$$\begin{bmatrix} 1 & 1 & 1 \\ 1 & -1 & -1 \\ 3 & 1 & 1 \end{bmatrix} = \begin{bmatrix} 1 & 0 & 0 \\ 0 & 1 & 0 \\ 0 & 0 & 1 \end{bmatrix} A \begin{bmatrix} 1 & 0 & 0 \\ 0 & 1 & 0 \\ 0 & 0 & 1 \end{bmatrix}$$

We now perform elementary operations on the matrix A, until it is reduced to the normal for every elementary operation is applied to the prefatory I3 of the product on the right hand slide of the above equation and every elementary column operation is applied to the post factor I3 of the product on the right hand side.

Applying   R2 → R2 → R1,   R3 → R3 − 3R1 we get

$$\begin{bmatrix} 1 & 1 & 1 \\ 0 & -2 & -2 \\ 0 & -2 & -2 \end{bmatrix} = \begin{bmatrix} 1 & 0 & 0 \\ -1 & 1 & 0 \\ 3 & 0 & 1 \end{bmatrix} A \begin{bmatrix} 1 & 0 & 0 \\ 0 & 1 & 0 \\ 0 & 0 & 1 \end{bmatrix}$$

Applying   C2 → C2 → C1,   C3 → C3 − C1   we get

$$\begin{bmatrix} 1 & 0 & 0 \\ 0 & -2 & -2 \\ 0 & -2 & -2 \end{bmatrix} = \begin{bmatrix} 1 & 0 & 0 \\ -1 & 1 & 0 \\ -3 & 0 & 1 \end{bmatrix} A \begin{bmatrix} 1 & -1 & -1 \\ 0 & 1 & 0 \\ 0 & 0 & 1 \end{bmatrix}$$

Performing $R2 \rightarrow -\dfrac{1}{2} R2$, we get

$$\begin{bmatrix} 1 & 0 & 0 \\ 0 & 1 & 1 \\ 0 & -2 & -2 \end{bmatrix} = \begin{bmatrix} 1 & 0 & 0 \\ \dfrac{1}{2} & -\dfrac{1}{2} & 0 \\ -3 & 0 & 1 \end{bmatrix} A \begin{bmatrix} 1 & -1 & -1 \\ 0 & 1 & 0 \\ 0 & 0 & 1 \end{bmatrix}$$

Performing $R3 \rightarrow R3 + 2R2$ we get

$$\begin{bmatrix} 1 & 0 & 0 \\ 0 & 1 & 1 \\ 0 & 0 & 0 \end{bmatrix} = \begin{bmatrix} 1 & 0 & 0 \\ \dfrac{1}{2} & -\dfrac{1}{2} & 0 \\ -2 & -1 & 1 \end{bmatrix} A \begin{bmatrix} 1 & -1 & -1 \\ 0 & 1 & 0 \\ 0 & 0 & 1 \end{bmatrix}$$

Performing $c3 \rightarrow c3 - c2$, we get

$$\begin{bmatrix} 1 & 0 & 0 \\ 0 & 1 & 0 \\ 0 & 0 & 0 \end{bmatrix} = \begin{bmatrix} 1 & 0 & 0 \\ \dfrac{1}{2} & -\dfrac{1}{2} & 0 \\ -2 & -1 & 1 \end{bmatrix} A \begin{bmatrix} 1 & -1 & 0 \\ 0 & 1 & -1 \\ 0 & 0 & 1 \end{bmatrix}$$

which is of the form $PAQ = \begin{bmatrix} I_2 & 0 \\ 0 & 0 \end{bmatrix}$

where $P = \begin{bmatrix} 1 & 0 & 0 \\ \dfrac{1}{2} & -\dfrac{1}{2} & 0 \\ -2 & -1 & 1 \end{bmatrix}, Q_2 = \begin{bmatrix} 1 & -1 & 0 \\ 0 & 1 & -1 \\ 0 & 0 & 1 \end{bmatrix}$

A is reduced to the form $\begin{bmatrix} I_2 & 0 \\ 0 & 0 \end{bmatrix}$

Therefore, rank of A = $\rho(A) = 2$

EXERCISE 1

I. **Find the rank of each of the following matrices**

1. $\begin{bmatrix} 3 & -1 & 2 \\ -6 & 2 & -4 \\ -3 & 1 & -2 \end{bmatrix}$ *[Ans. 1]*

2. $\begin{bmatrix} 2 & 1 & 3 \\ 4 & 7 & 13 \\ 4 & -3 & -1 \end{bmatrix}$ [*Ans.* 2]

3. $\begin{bmatrix} 1 & 0 & 2 & 3 \\ 2 & 1 & 0 & 1 \\ 4 & 1 & 4 & 7 \end{bmatrix}$ [*Ans.* 2]

4. $\begin{bmatrix} 2 & 3 & 7 \\ 3 & -2 & 4 \\ 1 & -3 & -1 \end{bmatrix}$ [*Ans.* 2]

II. Find the ranks the following matrices

(a) $\begin{bmatrix} 1 & 1 & 1 & -1 \\ 1 & 2 & 3 & 4 \\ 3 & 4 & 5 & 2 \end{bmatrix}$ (b) $\begin{bmatrix} 8 & 0 & 0 & 1 \\ 1 & 0 & 8 & 1 \\ 0 & 0 & 1 & 8 \\ 0 & 8 & 1 & 8 \end{bmatrix}$ [*Ans.*(a)2 (b)4]

III. Define (a) Rank of a matrix (b) Normal form of a matrix

Reduce the matrix to its normal form and find its rank

(a) $\begin{bmatrix} 9 & 7 & 3 & 6 \\ 5 & -1 & 4 & 1 \\ 6 & 8 & 2 & 4 \end{bmatrix}$ (b) $\begin{bmatrix} 0 & 1 & -3 & -1 \\ 1 & 0 & 1 & 1 \\ 3 & 1 & 0 & 2 \\ 1 & 1 & -2 & 0 \end{bmatrix}$ [*Ans.* (a) 3 (b)2]

VECTOR

Definition: Any ordered n-type of numbers is called an n-vector.

Example 15: $\alpha = (x_1, x_2, x_3)$, $\beta = (-1, 0, 2)$, $\gamma = (1, 0, 2, 0, -3)$ α are vectors.

A vector may be written either as a row or as a column vector.

If $\alpha = (x_1, x_2, x_3, x_4)$ then α is called n-vector and $\beta = (x_1, x_2, ..., x_n)$ is called a n-vector.

If α, and β are n-vectors we can define $\alpha + \beta$ i.e., addition of the vector a and b where $\alpha + \beta$ is also a n-vector.

The vector a is called the scalar multiple of the vector a by the scalar a.

Properties of addition and scalar multiplication of vectors.

If α, β, γ are vectors and a, b are scalars then

(i) $\alpha + \beta = \beta + \alpha$

(ii) $\alpha + (\beta + \gamma) = (\alpha + \beta) + \gamma$

(iii) $a(\alpha + b) = a\,\alpha + a\beta$

(iv) $(a + b)\,\alpha = a\,\alpha + b\,\alpha$

(v) $a(b\alpha) = (ab)\alpha$

Linear dependence of vectors: A set of n vectors α_1, α_2,....,α_n are said to be linearly dependent if there exist scalars α_1, α_2, ...,α_n not all zero such that $a_1x_1 + a_2x_2 + \dots + a_nx_n = 0$

Implies $a_1 = a_2 = \dots = a_n = 0$

where $\bar{0}$ is a zero vector.

Example 16: Show that the vectors $\alpha_1 = (1, 3, -4)$, $\alpha_2 = (2, 6, -8)$ are linearly dependent.

Solution: Let there exist scalars $a1$, $a2$ such that $a1\alpha1 + a2\alpha2 = \bar{0}$

i.e., $a_1(1, 3, -4) + a_2(2, 6, -8) = (0, 0, 0)$

or $(a_1 + 2a_2, 3a_1 + 6a_2, -4a_1 - 8a_2) = (0, 0, 0)$

$\therefore$ $a_1 + 2a_2 = 0,\ 3a_1 + 6a_2 = 0,\ -4a_1 - 8a_2 = 0$

or $\dfrac{a_1}{2} = \dfrac{a_2}{-1}$

$\therefore$ there exist numbers 2, -1 such that

$$2(1, 3, -4) + (-1)(2, 6, -8) = (0, 0, 0)$$

Therefore the vectors $\alpha1$, $\alpha2$ are linearly dependent

Example 17: Show that the vectors are linearly dependent $\alpha = (1, 1, 1)$ $\beta = (2, -1, 3)$, $\gamma = (1, -5, 3)$

Solution: Let there exist scalars a, b, c, such that $a\alpha + b\beta + c\gamma = \bar{0}$

i.e., $(1, 1, 1) + b(2, -1, 3) + c(1, -5,3) = (0, 0, 0)$

or $(a + 2b + c, a - b - 5c, a + 3b + 3c) = (0, 0, 0)$

we get $a + b + c = 0,\ a - b - 5c = 0,\ a + 3b + 3c = 0$

Solving the above equation, we obtain $a = 3$, $b = -2$, $c = 1$

Such that $3(1, 1, 1) + (-2)(2, -1, 3) + 1 - (1, -5, 3) = (0, 0, 0)$

i.e., $\qquad 3\alpha - 2\beta + \gamma = \overline{0}$

Hence $\qquad \alpha, \beta, \gamma$ are linearly dependent

Linear independence of vectors: Vectors $\alpha_1, \alpha_2, \ldots \alpha_n$ are said to be linearly independent if every relation of the form $a_1x_1 + a_2x_2 + \ldots . + a_nx_n = 0$

Implies $\qquad a_1 = a_2 = \ldots . = a_n = 0$

Example 18: Show that the vectors $(1, -2, -3)$, $(2, 3, -1)$ and $(3, 2, 1)$ are linearly independent.

Solution: Let $\alpha 1 = (1, -2, -3)$, $\alpha 2 = (2, 3, -1)$ and $\alpha 3 = (3, 2, 1)$. Let there exist scalars a1, a2, a3 such that $\quad a1\alpha 1 + a2\alpha 2 + a3\alpha 3 = \overline{0}$

i.e., $\qquad a1(1, -2, -3) + a2(2, 3, -1) + a3 (3, 2, 1) = (0, 0, 0)$

or $\qquad (a1 + 2a2 + 3a3, -2a, + 3a2 + 2a3, -3a1 -a2 + a3) = (0, 0, 0)$

$\therefore \qquad a1 + 2a2 + 3a3 = 0 \qquad\qquad\qquad\qquad\qquad .....(5.1)$

$\qquad\qquad -2a1 + 3a2 + 2a3 = 0 \qquad\qquad\qquad\qquad .....(5.2)$

and $\qquad -3a1 + a2 + a3 = 0 \qquad\qquad\qquad\qquad\qquad .....(5.3)$

Solving (5.1), (5.2), (5.3) we get $a1 = a2 = a3 = 0$.

Hence, the given vectors are linearly independent.

EXERCISE 2

Determine whether or not the following vectors are linearly dependent.

1. $(3, 1, -4), (2, 2, -3)$ $\qquad\qquad\qquad$ [**Ans:** Linearly independent]

2. $(3, -2, -1), (1, -6, -5), (1, 2, 3)$ $\qquad\qquad$ [**Ans:** Linearly dependent]

3. $(1, 0, 1), (1, 1, 1) (0, 1, 1)$ $\qquad\qquad\qquad$ [**Ans:** Linearly independent]

4. $(1, 0, 0, 1), (0, 1, 2, 1) (1, 2, 3, 4)$ $\qquad$ [**Ans:** Linearly independent]

5. $(1, 2) (1, -1), (2, 5)$ $\qquad\qquad\qquad\qquad$ [**Ans.** Linearly dependent]

Systems of linear equations: In this section we investigate the systems linear equations, and the methods which are used to find their solution.

An equation of the form

$$a_1x_1 + a_2x_2 + \cdots . a_nx_n = b \qquad\qquad\qquad .....(5.4)$$

$$a_1x_1 + a_2x_2 + \cdots . a_nx_n = 0$$

where $x_1, x_2, .., x_n$ are unknown and $a_1, a_2,..,a_n,$ b are constants is called a linear equation.

The constants a_i's are called the coefficients of x_i and b is called the constant of the equation. A set of n values for all the unknowns x_i, (i = 1, 2,.., n) is an n-type $(k_1, k_2, .., k_n)$ of constants is called the solution of the equation if

$$a_1k_1 + a_2k_2 + ... + a_nk_n = b$$

That is if $(k_1, k_2, .., k_n)$ satisfies the equation 5.4. If b = 0, (5.4) is called a homogeneous equation.

Systems of Homogenous Equations: A system of equations given by

$$a_{11}x_1 + a_{12}x_2 + ... + a_{1n}x_n = b_1$$
$$a_{21}x_1 + a_2x_2 + ... + a_{2n}x_n = b_2$$
$$... \quad ... \quad \quad ... \quad ...$$
$$... \quad ... \quad \quad ... \quad ...$$
$$a_{n1}x_1 + a_{n2}x_2 + ... + a_{nn}x_n = b_n$$

is called a system of non-homogenous equation in n unknown $x_1, x_2, ..., x_n$

The above system can be written as

$$A x = 0$$

where

$$A = \begin{bmatrix} a_{11} & \cdots & a_{1n} \\ \vdots & \ddots & \vdots \\ a_{m1} & \cdots & a_{mn} \end{bmatrix},$$

$$x = \begin{bmatrix} x_1 \\ x_2 \\ \vdots \\ x_m \end{bmatrix}, \quad 0 = \begin{bmatrix} 0 \\ 0 \\ \vdots \\ 0 \end{bmatrix}$$

A is an m × n, matrix,

x is an n × n matrix,

and   0 is an m × 1 matrix

A is called the coefficient matrix

$x_1 = 0, x_2 = 0, .., x_n = 0$ is a trivial solution of the equation.

If $x_1, x_2,$ are any two solutions of A X = 0

Then $k_1 x_1 + k_2 x_2$ where k_1, k_2 are constant is also a solution of A X = 0

Method of solving a homogenous system of linear equation:

Consider the homogenous system

$$a_{11} x_1 + a_{12} x_2 + \dots + a_{1n} x_n = 0$$
$$a_{21} x_1 + a_2 x_2 + \dots + a_{2n} x_n = 0$$

$$\dots \quad \dots \quad \quad \dots \quad \quad \dots$$

$$\vdots$$

$$a_{m1} n_1 + a_{m2} x_2 + \dots + a_{mn} x_n = 0$$

where $\qquad x_1, x_2, \dots x_n$ are unknowns

The homogenous system (1) always has a solution the n-type (0, 0, …,0) is always a solution of (1). It is called the trivial solution or zero solution. A homogenous system with fewer equations and more unknowns gas a non-zero solution A is to find the solution of (1). We reduce the coefficient matrix A to Echelon form by applying elementary row operations. If A is a matrix of the type m × n and r is the rank of A, the process of reducing A to Echelon form involves eliminating of m − r equations provided r < m. The given system of m equations will thus be replaced by an equivalent system of r equations. Solving these equations we get the solution. There are two possibilities. If r = n, the system has only are solution i.e., the trivial solution. If r < n, the system has a non-zero solution.

In this case there will be infinity of solutions. If the rank of A is r, the number of independent solutions of the system is n − r.

Example 19: Solve the system

$$x + y - z = 0$$

$$2x + 4y - z = 0$$

$$3x + 2y + 2z = 0$$

Solution: The given system can be expressed as

A x = 0, where

$$A = \begin{bmatrix} 1 & 1 & -1 \\ 2 & 4 & -1 \\ 3 & 2 & 2 \end{bmatrix}, X = \begin{bmatrix} x \\ y \\ z \end{bmatrix}, 0 = \begin{bmatrix} 0 \\ 0 \\ 0 \end{bmatrix}$$

Consider $\qquad A = \begin{bmatrix} 1 & 1 & -1 \\ 2 & 4 & 1 \\ 2 & 2 & 2 \end{bmatrix}$ applying $\begin{array}{l} R_2 \rightarrow R_2 - 2R_1 \\ R_3 \rightarrow R_3 - 3R_1 \end{array}$

We get
$$\sim \begin{bmatrix} 1 & 1 & -1 \\ 0 & 2 & -1 \\ 0 & -1 & 5 \end{bmatrix} \text{ applying } R2 \leftrightarrow R3$$

We get
$$A \sim \begin{bmatrix} 1 & 1 & -1 \\ 0 & -1 & 5 \\ 0 & 2 & 1 \end{bmatrix} \text{ now applying } R3 \leftrightarrow R3 + 2R2$$

$$A \sim \begin{bmatrix} 1 & 1 & -1 \\ 0 & -1 & 5 \\ 0 & 0 & 11 \end{bmatrix}$$

The echelon form of A has three non-aero rows.

Therefore the echelon form of the given system

$$A x = \begin{bmatrix} 1 & 1 & -1 \\ 0 & -1 & 5 \\ 0 & 0 & 11 \end{bmatrix} \begin{bmatrix} x \\ y \\ x \end{bmatrix} = \begin{bmatrix} 0 \\ 0 \\ 0 \end{bmatrix}$$

i.e.,
$$x + y - z = 0$$
$$-y + 5z = 0$$
$$11\,z = 0$$

Since, the echelon form has three equations in three unknowns, the given system has only the zero solution (0, 0, 0)

Example 20: Solve completely the system of equations

$$x + 3y - 2z = 0$$
$$2x - y + 4z = 0$$
$$x - 11y + 14z = 0$$

Solution: The given system is a homogenous system equivalent to $A \times = 0$

where
$$A = \begin{bmatrix} 1 & 3 & -2 \\ 2 & -1 & 4 \\ 1 & -4 & 14 \end{bmatrix}, X = \begin{bmatrix} x \\ y \\ z \end{bmatrix}, 0 = \begin{bmatrix} 0 \\ 0 \\ 0 \end{bmatrix}$$

Consider
$$A = \begin{bmatrix} 1 & 3 & -2 \\ 2 & -1 & 4 \\ 1 & -4 & 14 \end{bmatrix} \begin{matrix} R_2 \rightarrow R_2 - 2R_1 \\ R_3 \rightarrow R_3 - R_1 \\ \rightarrow \end{matrix}$$

$$\sim \begin{bmatrix} 1 & 3 & -2 \\ 0 & -7 & 8 \\ 0 & -14 & 16 \end{bmatrix} \begin{matrix} R_3 \to R_3 - 2R_2 \\ \\ \to \end{matrix}$$

$$\sim \begin{bmatrix} 1 & 3 & -2 \\ 0 & -7 & 8 \\ 0 & 0 & 0 \end{bmatrix}$$

The echelon form of A is $\begin{bmatrix} 1 & 3 & -2 \\ 0 & -7 & 8 \\ 0 & 0 & 8 \end{bmatrix}$

The rank of A is 2. The system has $n - r = 3 - 2 = 1$ independent solution.

The system is reduced to the form

$$A X = \begin{bmatrix} 1 & 3 & -2 \\ 0 & -7 & 8 \\ 0 & 0 & 0 \end{bmatrix} \begin{bmatrix} x \\ y \\ z \end{bmatrix} = \begin{bmatrix} 0 \\ 0 \\ 0 \end{bmatrix}$$

i.e., $x + 3y - 2z = 0$ .....(5.5)

$-7y + 8z = 0$ .....(5.6)

Choosing $z = k$, we get

$$y = \frac{8}{7}\, z = \frac{8}{7}\, k \text{ (form (5.6)}$$

Substituting these values in 5.5

$$x + 3y - 2z = 0$$

We get $x = -\dfrac{10}{7} k,\ y = \dfrac{8}{7} k,\ z = k$

is the general solution of the system, where k is an arbitrary parameter.

System linear non-homogenous equations: A system of a linear equations of the form

$$a_1,\, x_1 + a_{12}\, x_2 + \dots + a_{1n} x_n = b_1$$

$$a_{21},\, x_1 + a_{22}\, x_2 + \dots + a_{2n} x_n = b_2$$

$$\vdots \quad \vdots \qquad\qquad \vdots$$

$$a_{m1},\, x_1 + a_{m2}\, x_2 + \dots + a_{mn} x_n = b_n$$

is called a system of non-homogenous equations. If at least one of b_i's $\neq 0$ ($i = 1, 2, \dots, n$)

It can be expressed as

$$A X = B$$

Where

$$A = \begin{bmatrix} a_{11} & a_{12} & \cdots & a_{1n} \\ a_{21} & a_{22} & \cdots & a_{2n} \\ .. & \cdots & \cdots & \cdots \\ a_{m1} & a_{m2} & \cdots & a_{mn} \end{bmatrix}, \quad X = \begin{bmatrix} x_1 \\ x_2 \\ \vdots \\ x_m \end{bmatrix}, B = \begin{bmatrix} b_1 \\ b_2 \\ M \\ b_n \end{bmatrix}$$

The matrix

$$[A \vdots B]. = \begin{bmatrix} a_{11} & \cdots & a_{1n}b_1 \\ \vdots & \ddots & \vdots \\ a_{m1} & \cdots & a_{mn}b_n \end{bmatrix},$$

Is called augmented matrix. If $\rho(A) = \rho[A:B] = r <$ the number of unknowns (n) then there will be $n - r$ linearly independent solutions. The set of $n - r$ solutions is called the fundamental set of solutions.

Consistency: A system of linear equations A X = B is said to be consistent if it has one or more solution. If a system has no solution then it is said to be inconsistent.

If a system of linear equation is consistent it may have one or more solutions.

Condition for consistency: The system of equations A X = B is consistent if and only if the coefficient matrix A and the augmented matrix [A:B] are of the same rank.

Example 21: Show that the system

$$x + y + z = 6$$

$$x + 2y + 3z = 14$$

and

$$x + 4y + 7z = 30$$

is consistent and find its solution.

Solution: The given system can be expressed as

$$A X = B \text{ where } A = \begin{bmatrix} 1 & 1 & 1 \\ 1 & 2 & 3 \\ 1 & 4 & 7 \end{bmatrix}, X = \begin{bmatrix} x \\ y \\ z \end{bmatrix}, B = \begin{bmatrix} 6 \\ 14 \\ 30 \end{bmatrix}$$

The augmented matrix is

$$[A : B] \begin{bmatrix} 1 & 1 & 1 & \vdots & 6 \\ 1 & 2 & 3 & \vdots & 14 \\ 1 & 4 & 7 & \vdots & 30 \end{bmatrix}$$

The above matrix can be reduced to Echelon form as follows. Applying R2 → R2 – R1, R3 → R3 – R1

We get

$$[A : D] \sim \begin{bmatrix} 1 & 1 & 1 & \vdots & 6 \\ 0 & 1 & 2 & \vdots & 8 \\ 0 & 3 & 6 & \vdots & 24 \end{bmatrix} \begin{matrix} R_3 \to R_3 - 3R_3 \\ \to \end{matrix}$$

$$\sim \begin{bmatrix} 1 & 1 & 1 & \vdots & 6 \\ 0 & 1 & 2 & \vdots & 8 \\ 0 & 0 & 60 & \vdots & 0 \end{bmatrix}$$

which is Echelon form

∴ Rank of [A : B] = Number of non-zero rows in the Echelon form = 2

By elementary operations, we get

$$A \sim \begin{bmatrix} 1 & 1 & 1 \\ 0 & 1 & 2 \\ 0 & 0 & 0 \end{bmatrix}$$

∴ $\rho(A)$ = rank of A = 2

Since rank of A = rank of (A : B) = 2

The given system of equation is consistent we have r = 2, n = number of unknown = 3

∴   r = 2 < n

Hence the given has an infinite number of

Solution: The equivalent form of the given system is

$$\begin{bmatrix} 1 & 1 & 1 \\ 0 & 1 & 2 \\ 0 & 0 & 0 \end{bmatrix} \begin{bmatrix} x \\ y \\ z \end{bmatrix} = \begin{bmatrix} 6 \\ 8 \\ 10 \end{bmatrix}$$

We have $x + y + z = 6$

               $y + 23 = 8$

Taking z = k, we obtain $y = 8 - 2k$

and $x = k - 2$

The general solution of the system is

               $X = k - 2, y = 8 k$, and $z = k$

where k is an arbitrary constant.

Example 22: Solve $x + y + z = 9$

$$2x + 5y + 8z = 5z$$

$$2x + y - z = 0$$

Solution: The matrix form of the given system of equation is

$$A\,X = B \text{ i.e.,} \begin{bmatrix} 1 & 1 & 1 \\ 2 & 5 & 7 \\ 2 & 1 & -1 \end{bmatrix} \begin{bmatrix} x \\ y \\ z \end{bmatrix} = \begin{bmatrix} 9 \\ 52 \\ 0 \end{bmatrix}$$

The augmented matrix is

$$[A : B] = \begin{bmatrix} 1 & 1 & 1 & \vdots & 9 \\ 2 & 5 & 7 & \vdots & 52 \\ 2 & 1 & -1 & \vdots & 0 \end{bmatrix} \begin{matrix} R_2 \to R_2 - 2R_1 \\ R_3 \to R_3 - 2R_1 \\ \to \end{matrix}$$

$$\sim \begin{bmatrix} 1 & 1 & 1 & \vdots & 9 \\ 0 & 3 & 5 & \vdots & 34 \\ 0 & -1 & -3 & \vdots & -18 \end{bmatrix} \begin{matrix} \\ R_2 \leftrightarrow R_3 \\ \to \end{matrix}$$

$$\sim \begin{bmatrix} 1 & 1 & 1 & \vdots & 9 \\ 0 & -1 & -3 & \vdots & -18 \\ 0 & 3 & 5 & \vdots & 34 \end{bmatrix} \begin{matrix} \\ R_3 \to R_3 + 3R_2 \\ \to \end{matrix}$$

$$\sim \begin{bmatrix} 1 & 1 & 1 & \vdots & 9 \\ 0 & -1 & -3 & \vdots & -18 \\ 0 & 0 & 4 & \vdots & -20 \end{bmatrix}$$

The above matrix is in Echelon form

Therefore, rank of $[A: B] = 3$ and rank of $A = 3$

The equivalent form of given system is

$$\begin{bmatrix} 1 & 1 & 1 \\ 0 & -1 & -3 \\ 0 & 0 & 4 \end{bmatrix} \begin{bmatrix} x \\ y \\ z \end{bmatrix} = \begin{bmatrix} 9 \\ -18 \\ -20 \end{bmatrix}$$

i.e., $x + y + z = 9$ ......(5.7)

$$-y - 3z = -18 \qquad ......(5.8)$$

$$-4z = -20 \qquad ......(5.9)$$

Solving (5.7), (5.8), (5.9), we get (by back substitution)

$$z = 5, y = 3 \text{ and } x = 1$$

The solution is $x = 1, y = 3, z = 5$

Example 23: Show that the system of equations

$$x + y + 4z = 6$$

$$x + 2y - 2z = 6$$

$$k x + b + z = 6$$

is consistent when $k \neq \dfrac{7}{10}$, and in consistent where $k = \dfrac{7}{10}$.

Solution: The equivalent matrix form of given equations is $A X = B$

i.e.,
$$\begin{bmatrix} 1 & 1 & 4 \\ 1 & 2 & -2 \\ k & 1 & 1 \end{bmatrix} \begin{bmatrix} x \\ y \\ z \end{bmatrix} = \begin{bmatrix} 6 \\ 6 \\ 6 \end{bmatrix}$$

The augmented matrix is

$$[A : B] = \begin{bmatrix} 1 & 1 & 4 & \vdots & 6 \\ 1 & 2 & -2 & \vdots & 6 \\ k & 1 & 1 & \vdots & 6 \end{bmatrix} \begin{matrix} R_1 \to R_2 - R_1 \\ R_3 \to R_3 - R_1 \\ \to \end{matrix}$$

$$\sim \begin{bmatrix} 1 & 1 & 4 & \vdots & 6 \\ 0 & 1 & -6 & \vdots & 0 \\ 0 & 1-k & 1-4k & \vdots & 6-6k \end{bmatrix} \begin{matrix} R_1 \to R_2 - R_1 \\ R_3 \to R_3 - R_1 \\ \to \end{matrix}$$

We have
$$A = \begin{bmatrix} 1 & 1 & 4 \\ 0 & 1 & -6 \\ 0 & 1-k & 1-4k \end{bmatrix}$$

A is non-singular if and only is

$$1 - 4k + 6(1 - k) \neq 0$$

i.e.,
$$1 - 4k + 6 - 6k \neq 0$$

or
$$10 k \neq 7$$

or
$$k \neq \dfrac{7}{10}$$

When $k \neq \dfrac{7}{10}$ the given system is consistent (the system has unique solution)

When k = $\neq \dfrac{7}{10}$, we have

$$[A:B] = \begin{bmatrix} 1 & 1 & 4 & \vdots & 6 \\ 0 & 1 & -6 & \vdots & 0 \\ 0 & \dfrac{3}{10} & -\dfrac{18}{10} & \vdots & \dfrac{18}{10} \end{bmatrix} \quad R_3 \rightarrow R_3 - \dfrac{3}{10}R_2 \rightarrow$$

$$\sim \begin{bmatrix} 1 & 1 & 4 & \vdots & 6 \\ 0 & 1 & -6 & \vdots & 0 \\ 0 & 0 & 0 & \vdots & \dfrac{18}{10} \end{bmatrix}$$

The given system of equations reduce to the form

$$\begin{bmatrix} 1 & 1 & 4 \\ 0 & 1 & -6 \\ 0 & 0 & 0 \end{bmatrix}\begin{bmatrix} x \\ y \\ z \end{bmatrix} = \begin{bmatrix} 6 \\ 8 \\ \dfrac{18}{10} \end{bmatrix}$$

which is not consistent

Example 24: Investigate for what values of λ and μ the equations

$$x + y + z = 6$$

$$x + 2y + 3z = 10$$

$$x + 2y + \lambda z = \lambda$$

(i) no solution (ii) a unique solution (iii) on infinite number of solution

Solution: The given system of equations can be written as

$$A X = B \text{ i.e.,} \begin{bmatrix} 1 & 1 & 1 \\ 1 & 2 & 3 \\ 1 & 2 & \lambda \end{bmatrix}\begin{bmatrix} x \\ y \\ z \end{bmatrix} = \begin{bmatrix} 6 \\ 10 \\ \mu \end{bmatrix}$$

The augmented matrix is

$$[A:B] = \begin{bmatrix} 1 & 1 & 1 & \vdots & 6 \\ 1 & 2 & 3 & \vdots & 10 \\ 1 & 2 & \lambda & \vdots & \mu \end{bmatrix} \begin{matrix} R_2 \rightarrow R_2 - R_1 \\ R_3 \rightarrow R_3 - R_1 \\ \rightarrow \end{matrix}$$

$$\sim \begin{bmatrix} 1 & 1 & 1 & \vdots & 6 \\ 0 & 1 & 2 & \vdots & 4 \\ 0 & 1 & \lambda-1 & \vdots & \mu-6 \end{bmatrix} \begin{matrix} \\ R_3 \to R_3 - R_2 \\ \to \end{matrix}$$

$$\sim \begin{bmatrix} 1 & 1 & 1 & \vdots & 6 \\ 0 & 1 & 2 & \vdots & 4 \\ 0 & 0 & \lambda-3 & \vdots & \mu-10 \end{bmatrix}$$

We have the following possibilities.

If $\lambda \neq 3$, we have rank of (A : B) = Rank of A in this case the given system of equations in consistent. The number of unknown is 3 since $\rho(A) = 3$ = number of unknown, the given system has a unique solution for any value of μ.

If $\lambda = 3$ and $\mu \neq 10$, we have

$\rho(A : B) = 3$ and $\rho(A) = 2$ $\rho(A : B) \neq \rho(A)$

In this case the given system has no solution. Hence, the given system is consistent when $\lambda = 3$ and $\mu \neq 10$.

When $\lambda = 3$ and $\mu = 10$, we have

$$\rho[A : B] = \rho(A)$$

The given system is consistent in this case since rank (A) = 2 which is less than number of unknown, the given system has infinite number of solutions.

EXERCISE 3

1. Solve $x + y + z = 0$, $2x + 5y + 7z = 0$, $2x - 5y + 3z = 0$ [**Ans:** $x = 0$, $y = 0$, $z = 0$]

2. Solve $x + 2y + 3z = 0$, $3x + 4y + 4z = 0$, $7x + 10y + 12z = 0$ [**Ans:** $x = 0$, $y = 0$, $z = 0$]

3. Solve $3x + 4y - z - 6\omega = 0$

   i. $2x + 3y + 2z - 3\omega = 0$

   ii. $x + 3y + 12z + 3\omega = 0$

   [**Ans:** $x = 11k1 + 6k2$, $y = - 8k, - 3k2$ $z = 5$, $\omega = k2$]

4. Solve $x + y + z \neq 0$, $2x - y - 3z = 0$, $3x - 5y + 4z = 0$ $x + 17y + 4z = 0$

   [**Ans:** $x = y = z = 0$]

5. Solve $x + y + z = - 3$, $3x + y - 2z = -2$, $2x + 4y + 7z = 7$ [**Ans:** No solution]

6. Show that the system is consistent and find the solution $2x - y + 3z = 8$, $- x + 2y + z = 4$, $3x + y - 4z = 0$.   [**Ans:** $x = y = z = 2$]

7. For what values of λ, will the following equations find to have unique solution

$$3x - y + \lambda z = 1, \ 2x + y + z = 2, \ x + 2y - \lambda z = -1 \qquad \textbf{[Ans: } \lambda = {}^{-7}\!/_2 \textbf{]}$$

8. Solve completely the equations $3x - 2y - w = 2, 2y + 2z + w = 1,$ $x - 2y - 3z + 2w = 3, \ y + 2z + w = 1$ **[Ans: $x = 1, y = 0, z = 0, w = 1$]**

9. Show that the three equations $-2x + y + z = a, \ x - 2y + z = b, \ x + y - 2z = c$ have no solutions unless $a + b + c = 0$, in which case they have infinitely many solutions. Find these solutions when $a = 1, b = 1, c = -2$.

$$\textbf{[Ans: } x = k - 1, y = k - 1, z = k \ (k \ \text{arbitrary})\textbf{]}$$

10. Show that the equations $x - 3y - 8z = -10, \ 3x + y - 4z = 0, \ 2x + 5y + 6z = 13$ are consistent and solve the equations.

$$\textbf{[Ans: } x = -1 + 2k, y = 3 - 2k, z = k \ (k \ \text{arbitrary})\textbf{]}$$

11. Solve $x + 2y - 5z + 9 = 0, 3x - y + 2z = 5, 2x + 3y - z = 3, 4x - 5y + z = -3$

$$\textbf{[Ans: } x = {}^1\!/_2, y = {}^3\!/_2, z = {}^5\!/_2 \textbf{]}$$

12. Solve $2x - y + 3z = 9, x + y + z = 6, x - y + z = 2$.   [Ans. $x = 1, y = 2, z = 3$]

Eigen values and Eigen vectors

If $A_0, A_1, A_2, \ldots\ldots, A_{n-1}, A_n$ are all square matrices of the same order and λ is an indeterminate, then the expression of the form $A_0 + A_1\lambda + A_2\lambda^2 + \ldots\ldots + A_{n-1}\lambda^{n-1} + A_n\lambda^n$ is called a matrix polynomial of degree n, provided A_n is not a zero matrix.

Every square matrix whose elements are ordinary polynomials in λ, can essentially be expressed as a matrix polynomial in λ of degree n, where n is the highest power of λ, occurring in any element of the matrix.

Let A be an n-square matrix and λ be a scalar. Then λ is called an eigen value of A if

$$AX = \lambda X$$

$$\ldots..(5.10)$$

where X is a column vector.

Every vector X, satisfying the relation (5.10) is called an eigen vector of A.

The terms characteristic value and characteristic vectors are used instead of the terms eigen value and eigen vector respectively.

It is obvious that the zero vector $X = 0$ is a solution of (5.10) for any value of λ.

If I denotes the unit vector of order, then the equation (5.10) can be written as

$$AX - \lambda IX = 0$$

or $\qquad (A - \lambda I) X = 0$ $\qquad\qquad\qquad\qquad$ .....(5.11)

Definition: Let $\quad A = \begin{bmatrix} a_{11} & a_{12} & \cdots & a_{1n} \\ a_{21} & a_{22} & \cdots & a_{2n} \\ .. & \cdots & \cdots & \cdots \\ a_{n1} & a_{n2} & \cdots & a_{nn} \end{bmatrix}$

be an n-rowed square matrix X and λ be an indeterminate. The matrix $\left[A - \lambda I\right]$, is called the characteristic matrix of A, where I is the unit matrix of order n.

The determinant $\left|A - \lambda I\right| = 0$

i.e., $\qquad \begin{vmatrix} a_{11} - \lambda & a_{12} & \cdots & a_{1n} \\ a_{21} & a_{22} - \lambda & \cdots & a_{2n} \\ .. & \cdots & \cdots & \cdots \\ a_{n1} & a_{n2} & \cdots & a_{nn} - \lambda \end{vmatrix} = 0$

is called, the characteristic equation of A and the roots of this equation are called characteristic roots or characteristic values or eigen values of A. They are also know as proper values of A.

Corresponding to each eigen value λ, there exists a non-zero vector X such that $(A - \lambda I) X = 0$

i.e., $\qquad\qquad$ AX = λX

Definition: Let A be a square matrix of order n. If λ is an eigen value of A, then there exists a non-zero vector X such that

$\qquad\qquad$ AX = λX

which is called a characteristic or eigen vector of A corresponding to a the eigen value λ.

If A is a square matrix of order n, then the following are equivalent,

A scalar λ is an eigen value of A.

The matrix $A - \lambda I$ is singular.

Example 25: Determine the eigen values of A $= \begin{bmatrix} 2 & 5 & 6 \\ 0 & 3 & 0 \\ 0 & 1 & 1 \end{bmatrix}$

Solution: We have $A = \begin{bmatrix} 2 & 5 & 6 \\ 0 & 3 & 0 \\ 0 & 1 & 1 \end{bmatrix}$

The characteristic equation of A is $|A - \lambda I| = 0$

i.e.,
$$\begin{vmatrix} 2-\lambda & 5 & 6 \\ 0 & 3-\lambda & 0 \\ 0 & 1 & 1-\lambda \end{vmatrix} = 0$$

or $\quad (2-\lambda)\big|(3-\lambda)(1-\lambda) - 0\big| - 5\big|0\big| + 6\big|0\big| = 0$

or $\quad (2-\lambda)(3-\lambda)(1-\lambda) = 0$

or $\lambda = 1,2,3$ are the roots of the characteristic equation.

Therefore the eigen values are 1, 2, 3.

Example 26: Determine the eigen values of $A = \begin{bmatrix} 5 & 4 \\ 1 & 2 \end{bmatrix}$

Solution: The characteristic equation of A is $|A - \lambda I| = 0$

i.e.,
$$\begin{vmatrix} 5-\lambda & 4 \\ 1 & 2-\lambda \end{vmatrix} = 0$$

or $\quad (5-\lambda)(2-\lambda) - 4 = 0$

or $\quad \lambda^2 - 7\lambda + 10 - 4 = 0$

or $\quad \lambda^2 - 7\lambda + 6 = 0 \Rightarrow (\lambda - 1)(\lambda - 6) = 0$

The eigen values are $\lambda = 1,6$

Example 27: Find the eigen values and the eigen vectors of

Solution: The characteristic equation of A is $|A - \lambda I| = 0$

i.e.,
$$\begin{vmatrix} 4-\lambda & 2 \\ 3 & -1-\lambda \end{vmatrix} = 0$$

or $\quad (4-\lambda)(-1-\lambda) - 6 = 0$

or $\quad -4 - 4\lambda + \lambda + \lambda^2 - 6 = 0$

or $\quad \lambda^2 - 3\lambda - 10 = 0 \Rightarrow (\lambda - 5)(\lambda + 2) = 0$

$\therefore \lambda = -2, 5$ are the eigen values of A

When $\lambda = -2$: Let $\begin{bmatrix} x \\ y \end{bmatrix}$ be the corresponding eigen vector of A.

We have $\qquad (A - \lambda I)X = 0$

i.e., $\qquad \begin{bmatrix} 4+2 & 2 \\ 3 & 1 \end{bmatrix} \begin{bmatrix} x \\ y \end{bmatrix} = \begin{bmatrix} 0 \\ 0 \end{bmatrix}$

i.e., $\qquad 6x + 2y = 0$

$\qquad\qquad 3x + y = 0$

the above system has only one independent solution given by $\dfrac{x}{-1} = \dfrac{y}{3}$

The eigen vector corresponding to $\lambda = -2$ is $\begin{bmatrix} -1 \\ 3 \end{bmatrix}$

When $\lambda = 5$: Let $\begin{bmatrix} x' \\ y' \end{bmatrix}$ be the eigen vector corresponding to $\lambda = 5$. The homogeneous

system corresponding to $\lambda = 5$ is $(A - \lambda I)X = 0$

i.e., $\qquad \begin{bmatrix} 4-5 & 2 \\ 3 & -1-5 \end{bmatrix} \begin{bmatrix} x' \\ y' \end{bmatrix} = \begin{bmatrix} 0 \\ 0 \end{bmatrix}$

or $\qquad \begin{bmatrix} -1 & 2 \\ 3 & -6 \end{bmatrix} \begin{bmatrix} x' \\ y' \end{bmatrix} = \begin{bmatrix} 0 \\ 0 \end{bmatrix}$

or $\qquad -x' + 2y' = 0$

$\qquad\qquad 3x' - 6y' = 0$

or $\qquad -x' + 2y' = 0$

The system has only one independent solution.

The solution is $\dfrac{x'}{2} = \dfrac{y'}{1}$

The corresponding eigen vector is $\begin{bmatrix} 2 \\ 1 \end{bmatrix}$

Example 28: Determine the eigen values and eigen vectors of the matrix

$$A = \begin{bmatrix} 6 & -2 & 2 \\ -2 & 3 & -1 \\ 2 & -1 & 3 \end{bmatrix}$$

Solution: The characteristic equation of the matrix A is $|A - \lambda I| = 0$

i.e.,
$$\begin{vmatrix} 6-\lambda & -2 & 2 \\ -2 & 3-\lambda & -1 \\ 2 & -1 & 3-\lambda \end{vmatrix} = 0$$

or
$$(2-\lambda)(\lambda-2)(\lambda-8) = 0$$

The eigen values of A are $\lambda = 2,2,8$

The eigen vector corresponding to $\lambda = 2$ is given by

$$(A - 2I)X = 0$$

i.e.,
$$\begin{bmatrix} 6-2 & -2 & 2 \\ -2 & 3-2 & -1 \\ 2 & -1 & 3-2 \end{bmatrix} \begin{bmatrix} x \\ y \\ z \end{bmatrix} = \begin{bmatrix} 0 \\ 0 \\ 0 \end{bmatrix}$$

or
$$4x - 2y + 2z = 0$$

$$-2x + y - z = 0$$

$$2x - y + z = 0$$

The above system is represented by the single equation $2x - y + z = 0$

Obviously $X_1 = \begin{bmatrix} 1 \\ 2 \\ 0 \end{bmatrix}$ and $X_2 = \begin{bmatrix} -1 \\ 0 \\ 2 \end{bmatrix}$ are two linearly independent solutions of the system.

The eigen vector corresponding to $\lambda = 8$ is given by $(A - 8I)X = 0$

i.e.,
$$\begin{bmatrix} 6-8 & -2 & 2 \\ -2 & 3-8 & -1 \\ 2 & -1 & 3-8 \end{bmatrix} \begin{bmatrix} x \\ y \\ z \end{bmatrix} = \begin{bmatrix} 0 \\ 0 \\ 0 \end{bmatrix}$$

or
$$-2x - 2y + 2z = 0$$

$$-2x - 5y - z = 0$$

$$2x - y - 5z = 0$$

Adding last two equations of the system we get $6y - 6z = 0$ or $\dfrac{y}{-1} = \dfrac{z}{1}$

Substituting in the first equation we get $x = 2$

Therefore, the eigen vector corresponding to $\lambda = 8$ is $\begin{bmatrix} 2 \\ -1 \\ 1 \end{bmatrix}$

Example 29: Solve the system

$$x - 2y + z = 7$$

$$2x - y + 4z = 17$$

$$3x - 2y + 2z = 14$$

Solution: The Augmented matrix of the given system is

$$[A \vdots B] = \begin{bmatrix} 1 & -2 & 1 & \vdots & 7 \\ 2 & -1 & 4 & \vdots & 17 \\ 3 & -2 & 2 & \vdots & 14 \end{bmatrix} \rightarrow \begin{matrix} R_2 \rightarrow R_2 - 2R_1 \\ R_3 \rightarrow R_3 - 3R_1 \end{matrix}$$

$$\sim \begin{bmatrix} 1 & -2 & 1 & \vdots & 7 \\ 0 & 3 & 2 & \vdots & 3 \\ 0 & 4 & -1 & \vdots & -7 \end{bmatrix} \rightarrow R_3 \rightarrow -4R_2 + 3R_3$$

$$\sim \begin{bmatrix} 1 & -2 & 1 & \vdots & 7 \\ 0 & 3 & 2 & \vdots & 3 \\ 0 & 0 & -11 & \vdots & -33 \end{bmatrix}$$

We have rank $[A \vdots B]$ = rank of A = 3

The given system is consistent.

The reduced system is

$$x - 2y + z = 7$$

$$3x + 2y = 3$$

$$-11z = -33$$

The solution of the system is $z = 3, y = -1, x = 2$

Example 30: If A is non-singular, then show that the eigen values of A^{-1} are reciprocals of the eigen values of A.

Solution: Let λ be the eigen value of A, and let X be the corresponding eigen vector of A

Then, we have $AX = \lambda X$

or $X = A^{-1}(\lambda X)$

or $X = \lambda(A^{-1}X)$

or $\dfrac{1}{\lambda}X = A^{-1}X$

i.e., $A^{-1}X = \dfrac{1}{\lambda}X$

$\therefore \dfrac{1}{\lambda}$ is an eigen value of A^{-1} and X is corresponding eigen vector

Conversely, let us suppose that λ' is an eigen value of A^{-1}

Since A^{-1} is non-singular and $(A^{-1})^{-1} = A$, it follows that $\dfrac{1}{\lambda}$ is an eigen value of

$(A^{-1})^{-1} = A$

Therefore, each eigen value of A^{-1} is equal to the reciprocal of some eigen value of A

Example 31: Find the characteristic roots and characteristic vectors of the matrix

$$A = \begin{bmatrix} 8 & -6 & 2 \\ -6 & 7 & -4 \\ 2 & -4 & 3 \end{bmatrix}$$

Solution: The characteristic matrix of A is given by $|A - \lambda I| = 0$

i.e., $$\begin{vmatrix} 8-\lambda & -6 & 2 \\ -6 & 7-\lambda & -4 \\ 2 & -4 & 3-\lambda \end{vmatrix} = 0$$

i.e., $(8-\lambda)\left[21 - 10\lambda + \lambda^2 - 16\right] + 6\left[-18 + 6\lambda + 8\right] + 2\left[24 - 14 + 2\lambda\right] = 0$

or $-\lambda^3 + 18\lambda^2 - 45\lambda = 0$

or $-\lambda(\lambda - 3)(\lambda - 15) = 0$

$\therefore \lambda = 0, 3, 15$ are the characteristic roots

Case 1: When $\lambda = 0$

The characteristic vector corresponding to $\lambda = 0$ is $(A - \lambda I)X = 0$

i.e.,
$$\begin{bmatrix} 8 & 0 & -6 & 2 \\ -6 & 7-0 & -4 \\ 2 & -4 & 3-0 \end{bmatrix} \begin{bmatrix} x \\ y \\ z \end{bmatrix} = \begin{bmatrix} 0 \\ 0 \\ 0 \end{bmatrix}$$

or　　　　$8x - 6y + 2z = 0$

$$-6x + 7y - 4z = 0$$

$$2x - 4y + 3z = 0$$

from the last two equations we get $\dfrac{x}{5} = \dfrac{y}{10} = \dfrac{z}{10} = k$

or　　　　$x = k, y = 2k, z = 2k$

The characteristic vector corresponding to $\lambda = 0$ is $X_1 = k \begin{bmatrix} 1 \\ 2 \\ 2 \end{bmatrix}$

Where k is a non-zero parameter

Case 2: When $\lambda = 3$

The corresponding Characteristic matrix is given by $(A - \lambda I)X = 0$

i.e.,
$$\begin{bmatrix} 5 & -6 & 2 \\ -6 & 4 & -4 \\ 2 & -4 & 0 \end{bmatrix} \begin{bmatrix} x \\ y \\ z \end{bmatrix} = \begin{bmatrix} 0 \\ 0 \\ 0 \end{bmatrix}$$

i.e.,　　　　$5x - 6y + 2z = 0$

$$-6x + 4y - 4z = 0$$

$$2x - 4y = 0$$

solving we get　　　$x = 2k, y = k, z = -2k$

The corresponding characteristic vector is $X_2 = k \begin{bmatrix} 2 \\ 1 \\ -2 \end{bmatrix}, k \neq 0$

Case 3: When $\lambda = 15$

The characteristic matrix is $(A - \lambda I)X = 0$

i.e.,
$$\begin{bmatrix} 8-15 & -6 & 2 \\ -6 & 7-15 & -4 \\ 2 & -4 & 3-15 \end{bmatrix}\begin{bmatrix} x \\ y \\ z \end{bmatrix} = \begin{bmatrix} 0 \\ 0 \\ 0 \end{bmatrix}$$

i.e.,
$$-7x - 6y + 2z = 0$$

$$-6x - 8y - 4z = 0$$

$$2x - 4y - 12z = 0$$

Solving these equations we get $x = 2k, y = -2k, z = k$

The corresponding characteristic vector is $X_3 = k\begin{bmatrix} 2 \\ -2 \\ 1 \end{bmatrix}, k \neq 0$

The eigen values of A are $\lambda = 0, 3, 15$ and the eigen vectors of A are

$$\begin{bmatrix} 1 \\ 2 \\ 2 \end{bmatrix}, \begin{bmatrix} 2 \\ 1 \\ -2 \end{bmatrix}, \begin{bmatrix} 2 \\ -2 \\ 1 \end{bmatrix}$$

Example 32: Show that the square matrices A and A' (i.e., transpose of A) have the same characteristic values.

Solution: Let λ be a scalar. Then $(A - \lambda I)' = A' - \lambda I'$

$$= A' - \lambda I (\because I' = I)$$

We have
$$\left|(A - \lambda I)'\right| = \left|A' - \lambda I\right|$$

$\therefore$
$$\left|(A - \lambda I)\right| = \left|A' - \lambda I\right| \qquad \text{(since } \det A = \det A')$$

$\therefore$
$$\left|A - \lambda I\right| = 0 \qquad \text{If and only if } \left|A' - \lambda I\right| = 0$$

Hence λ is characteristic value of A if and only if λ is characteristic value of A'

Example 33: If λ is characteristic root of a square matrix, then show that $k + \lambda$ is a characteristic root of $A + kI$

Solution: Let λ be a characteristic root of and X be the corresponding characteristic vector of A

Then,

$$AX = \lambda X \text{ and } (A + kI)X = AX + k(IX)$$

$$= \lambda X + kX$$

$$= (\lambda + k)X$$

Since X is non-zero, $k + \lambda$ is a characteristic root of $A + kI$.

CAYLEY-HAMILTON THEOREM

Every square matrix satisfies its characteristic equation.

Proof: Let A be a square matrix of order n

The characteristic equation of A is given by $|A - \lambda I| = 0$

Since, the elements of $A - \lambda I$ are almost of the first degree in λ, the elements of $\text{Adj}(A - \lambda I)$ are ordinary polynomials in λ of degree $n - 1$ or less. Therefore can be written as a matrix polynomials in λ, which is given by

$$\text{Adj}(A - \lambda I) = B_0 \lambda^{n-1} + B_1 \lambda^{n-2} + \ldots\ldots + B_{n-2}\lambda + B_{n-1}$$

Where $B_0, B_1, \ldots, B_{n-1}$ are the matrices of type $n \times n$

Since $\quad AAdjA = |A|I_n$, we have

$$(A - \lambda I)\text{Adj}(A - \lambda I) = |(A - \lambda I)|I$$

$$\Rightarrow (A - \lambda I)\left(B_0 \lambda^{n-1} + B_1 \lambda^{n-2} + \ldots\ldots + B_{n-2}\lambda\right) = (-1)^n\left[\lambda^n + a_1\lambda^{n-1} + \ldots + a_n\right]$$

Comparing coefficients of like powers of λ on both sides, we get

$$-IB_0 = (-1)^n I$$

$$AB_0 - IB_1 = (-1)^n a_1 I$$

$$AB_1 - IB_2 = (-1)^n a_2 I$$

.

.

.

$$AB_{n-1} = (-1)^n a_n I$$

Pre-multiplying, these successively by $A^n, A^{n-1}, ..., I$ and adding we get

$$0 = (-1)^n \left[A^n + a_1 A^{n-1} + ... + a_n I \right]$$

Thus, $A^n + a_1 A^{n-1} + ... + a_n I = 0$

Hence proved.

Example 34: Show that $A = \begin{bmatrix} 1 & 0 & 2 \\ 0 & 2 & 1 \\ 2 & 0 & 3 \end{bmatrix}$ satisfies its own characteristic equation and

ind A^{-1}

Solution: The characteristic equation of A is $|A - \lambda I| = 0$

i.e.,
$$\begin{vmatrix} 1-\lambda & 0 & 2 \\ 0 & 2-\lambda & 1 \\ 2 & 0 & 3-\lambda \end{vmatrix} = 0$$

or
$$(1-\lambda)\left[(2-\lambda)(3-\lambda) - 2\right] - 0 + 2\left[0 - 2(2-\lambda)\right] = 0$$

or
$$-\lambda^3 + 6\lambda^2 - 7\lambda - 2 = 0$$

or
$$\lambda^3 - 6\lambda^2 + 7\lambda + 2 = 0$$

now consider $A = \begin{bmatrix} 1 & 0 & 2 \\ 0 & 2 & 1 \\ 2 & 0 & 3 \end{bmatrix}$

we have $A^2 = \begin{bmatrix} 1 & 0 & 2 \\ 0 & 2 & 1 \\ 2 & 0 & 3 \end{bmatrix}\begin{bmatrix} 1 & 0 & 2 \\ 0 & 2 & 1 \\ 2 & 0 & 3 \end{bmatrix}$

$$= \begin{bmatrix} 5 & 0 & 8 \\ 2 & 4 & 5 \\ 8 & 0 & 13 \end{bmatrix}$$

and
$$A^3 = A.A^2 = \begin{bmatrix} 1 & 0 & 2 \\ 0 & 2 & 1 \\ 2 & 0 & 3 \end{bmatrix} . \begin{bmatrix} 5 & 0 & 8 \\ 2 & 4 & 5 \\ 8 & 0 & 13 \end{bmatrix}$$

$$= \begin{bmatrix} 21 & 0 & 34 \\ 12 & 8 & 23 \\ 34 & 0 & 55 \end{bmatrix}$$

Consider $A^3 - 6A^2 + 7A + 2I$

$$= \begin{bmatrix} 21 & 0 & 34 \\ 12 & 8 & 23 \\ 34 & 0 & 55 \end{bmatrix} - 6\begin{bmatrix} 5 & 0 & 8 \\ 2 & 4 & 5 \\ 8 & 0 & 13 \end{bmatrix} + 7\begin{bmatrix} 1 & 0 & 2 \\ 0 & 2 & 1 \\ 2 & 0 & 3 \end{bmatrix} + 2\begin{bmatrix} 1 & 0 & 0 \\ 0 & 1 & 0 \\ 0 & 0 & 1 \end{bmatrix}$$

$$= \begin{bmatrix} 21-30+7+2 & 0-0+0+0 & 34-48+14+0 \\ 12-12+0+0 & 8-24+14+2 & 23-30+7+0 \\ 34-48+14+0 & 0+0+0+0 & 55-78+21+2 \end{bmatrix}$$

$$= \begin{bmatrix} 0 & 0 & 0 \\ 0 & 0 & 0 \\ 0 & 0 & 0 \end{bmatrix} = 0$$

$\therefore$ $A^3 - 6A^2 + 7A + 2I = 0$

Hence, A satisfies its own characteristic equation.

i.e., Cayley Hamilton theorem is verified.

To find A^{-1} :

Now consider $A^3 - 6A^2 + 7A + 2I = 0$

Multiplying the above equation by A^{-1} we get

$$A^2 - 6A + 7I + 2A^{-1} = 0$$

or
$$2A^{-1} = -\left[A^2 - 6A + 7I\right]$$

$$A^{-1} = -\frac{1}{2}\left[A^2 - 6A + 7I\right]$$

$$= -\frac{1}{2}A^2 + 3A - \frac{7}{2}I$$

$$= -\frac{1}{2}\begin{bmatrix} 5 & 0 & 8 \\ 2 & 4 & 5 \\ 8 & 0 & 13 \end{bmatrix} + 3\begin{bmatrix} 1 & 0 & 2 \\ 0 & 2 & 1 \\ 2 & 0 & 3 \end{bmatrix} - \frac{7}{2}\begin{bmatrix} 1 & 0 & 0 \\ 0 & 1 & 0 \\ 0 & 0 & 1 \end{bmatrix}$$

$$= \begin{bmatrix} -3 & 0 & 2 \\ -1 & \dfrac{1}{2} & \dfrac{1}{2} \\ 2 & 0 & -1 \end{bmatrix}$$

Example 35: Find the characteristic roots and the characteristic vectors of the matrix

$$A = \begin{bmatrix} 1 & 0 & 1 \\ 1 & 4 & 3 \\ 0 & 2 & 0 \end{bmatrix}$$

Solution: The characteristic equation of A is $|A - \lambda I| = 0$

i.e.,
$$\begin{vmatrix} 1-\lambda & 0 & 1 \\ 1 & 4-\lambda & 3 \\ 0 & 2 & 0-\lambda \end{vmatrix} = 0$$

or
$$(1-\lambda)\left[(4-\lambda)(0-\lambda)-6\right]+(2-0)=0$$

or
$$(1-\lambda)(\lambda^2 - 4\lambda - 6)+2=0$$

or
$$\lambda^3 - 5\lambda^2 - 2\lambda + 4 = 0$$

or
$$(\lambda+1)(\lambda^2 - 6\lambda + 4) = 0$$

or
$$\lambda = -1, \lambda = 3 \pm \sqrt{5}$$

Hence, the characteristic roots are $-1, 3-\sqrt{5}, 3+\sqrt{5}$

Characteristic vectors

Let $X = \begin{bmatrix} x_1 \\ x_2 \\ x_3 \end{bmatrix}$ be a characteristic vector of A, then $(A - \lambda I)X = 0$

When $\lambda = -1$, we get $\begin{bmatrix} 2 & 0 & 1 \\ 1 & 5 & 3 \\ 0 & 2 & 1 \end{bmatrix}\begin{bmatrix} x_1 \\ x_2 \\ x_3 \end{bmatrix} = \begin{bmatrix} 0 \\ 0 \\ 0 \end{bmatrix}$

or
$$2x_1 + x_3 = 0$$

$$x_1 + 5x_2 + 3x_3 = 0$$

$$2x_2 + x_3 = 0$$

Solving the above system of equations we get

$$x_1 = -1, x_2 = -1, x_3 = 2$$

$$\therefore X_1 = \begin{bmatrix} -1 \\ -1 \\ 2 \end{bmatrix} \text{ is the eigen vector when } \lambda = -1$$

When $\lambda = 3 + \sqrt{5}$, we get

$$\begin{bmatrix} -2-\sqrt{5} & 0 & 0 \\ 1 & 1-\sqrt{5} & 3 \\ 0 & 2 & -3-\sqrt{5} \end{bmatrix} \begin{bmatrix} x_1 \\ x_2 \\ x_3 \end{bmatrix} = \begin{bmatrix} 0 \\ 0 \\ 0 \end{bmatrix}$$

or
$$\left(-2-\sqrt{5}\right)x_1 + x_3 = 0$$

$$x_1 + \left(1-\sqrt{5}\right)x_2 + 3x_3 = 0$$

$$2x_2 - \left(-3-\sqrt{5}\right)x_3 = 0$$

Solving the above system and by putting $x_3 = k$ we get

$$x_1 = \left(-2+\sqrt{5}\right)k, x_2 = \frac{\left(3+\sqrt{5}\right)}{2}k, x_3 = k$$

The eigen vector corresponding to $\lambda = 3 + \sqrt{5}$ is $X_2 = \begin{bmatrix} -2+\sqrt{5} \\ \dfrac{\left(3+\sqrt{5}\right)}{2} \\ 1 \end{bmatrix}$

When $\lambda = 3 - \sqrt{5}$ we get $\begin{bmatrix} -2+\sqrt{5} & 0 & 0 \\ 1 & 1+\sqrt{5} & 3 \\ 0 & 2 & -3+\sqrt{5} \end{bmatrix} \begin{bmatrix} x_1 \\ x_2 \\ x_3 \end{bmatrix} = \begin{bmatrix} 0 \\ 0 \\ 0 \end{bmatrix}$

i.e.,
$$\left(-2+\sqrt{5}\right)x_1 + x_3 = 0$$

$$x_1 + \left(1+\sqrt{5}\right)x_2 + 3x_3 = 0$$

$$2x_2 + \left(3-\sqrt{5}\right)x_3 = 0$$

Putting $x_3 = k$, in the above system of equations,

We get $\qquad x_1 = \left(-2-\sqrt{5}\right)k, x_2 = \dfrac{\left(-3+\sqrt{5}\right)}{2}k, x_3 = k$

The eigen vector corresponding to $\lambda = 3-\sqrt{5}$ is $X_3 = \begin{bmatrix} -2-\sqrt{5} \\ \dfrac{\left(-3+\sqrt{5}\right)}{2} \\ 1 \end{bmatrix}$

Example 36: Find the eigen values of the matrix $A = \begin{bmatrix} 1 & 4 \\ 2 & 3 \end{bmatrix}$ and verify Cayley-Hamilton

theorem for this matrix. Express $A^5 - 4A^4 - 7A^3 + 11A^2 - A - 10I$ as a linear polynomial in A

Solution: The characteristic equation of A is $\left|A - \lambda I\right| = 0$

$$\left|A - \lambda I\right| = \begin{vmatrix} 1-\lambda & 4 \\ 2 & 3-\lambda \end{vmatrix} = 0$$

i.e., $\qquad \left(1-\lambda\right)\left(3-\lambda\right) - 8 = 0$

or $\qquad \lambda^2 - 4\lambda - 5 = 0$

or $\qquad \left(\lambda+1\right)\left(\lambda-5\right) = 0$

$\qquad \therefore \lambda = -1, 5$ are the eigen values of

Consider $\qquad A^2 - 4A - 5I$

i.e., $\qquad A^2 - 4A - 5I = A.A - 4A - 5I$

$$= \begin{bmatrix} 1 & 4 \\ 2 & 3 \end{bmatrix} \cdot \begin{bmatrix} 1 & 4 \\ 2 & 3 \end{bmatrix} - 4\begin{bmatrix} 1 & 4 \\ 2 & 3 \end{bmatrix} - 5\begin{bmatrix} 1 & 0 \\ 0 & 1 \end{bmatrix}$$

$$= \begin{bmatrix} 9 & 16 \\ 8 & 17 \end{bmatrix} - \begin{bmatrix} 4 & 16 \\ 8 & 12 \end{bmatrix} - \begin{bmatrix} 5 & 0 \\ 0 & 5 \end{bmatrix} = \begin{bmatrix} 0 & 0 \\ 0 & 0 \end{bmatrix} = 0$$

$$\therefore A^2 - 4A - 5I = 0$$

Thus, Cayley-Hamilton theorem is verified.

Consider
$$A^5 - 4A^4 \quad 7A^3 + 11A^2 - A - 10I$$
$$= \left(A^2 - 4A - 5I\right)\left(A^3 - 2A + 3I\right) + A + 5I$$
$$= 0 + A + 5I = A + 5I$$

A + 5I is the required linear polynomial in A.

COMPLEX MATRICES

Definition:

Let A be matrix. If the entries of A are complex numbers, then A is called as complex matrix

Example 37: If $A = \begin{bmatrix} 2-i & 0 \\ 1 & 1+i \end{bmatrix}$, then A is a complex matrix.

If A is a complex matrix, then the matrix $\overline{A}$, is the matrix obtained by taking conjugate of each entry in A i.e., if $A = \left[a_{ij}\right]$ then $\overline{A} = \left[\overline{a_{ij}}\right]$ where $\overline{A} = \left[\overline{a_{ij}}\right]$ is the conjugate of a_{ij}

Example 38: If $A = \begin{bmatrix} 5-2i & i \\ 2-3i & 5+3i \end{bmatrix}$ then $\overline{A}$ is the matrix $\begin{bmatrix} 5+2i & -i \\ 2+3i & 5-3i \end{bmatrix}$

If A is a complex matrix, then the two operations of transpose and conjugation commute for the matrix A.

We have $\left(\overline{A}\right)' = \left(\overline{A'}\right)$

If $\overline{A}$ and $\overline{B}$ are the conjugates of matrices A and B then

(i) $\overline{\left(\overline{A}\right)} = A$

(ii) $\overline{\left(A+B\right)} = \overline{A} + \overline{B}$

(iii) $\overline{\left(kA\right)} = \overline{k}\overline{A}$, where k is complex number

(iv) $\overline{\left(AB\right)} = \overline{A}\,\overline{B}$, A and B are conformable for multiplication

TRANSPOSED CONJUGATE OF A MATRIX

Let A be any matrix. The transpose of the conjugate of the matrix A is called the transposed conjugate of A and is denoted by A^θ or by A^*

We have $\left(\overline{A}\right)' = \left(\overline{A'}\right) = A^\theta$

Example 39: If $A = \begin{bmatrix} 2+5i & 3-2i & 4-3i \\ 5+i & 2+7i & 5-3i \\ 7 & 6+i & 5 \end{bmatrix}$

Then $A^\theta = \begin{bmatrix} 2-5i & 5-i & 7 \\ 3+2i & 2-7i & 6-i \\ 4+3i & 5+3i & 5 \end{bmatrix}$

If A^θ and B^θ are the transposed conjugations of A and B respectively then we have

(i) $\left(A^\theta\right)^\theta = A$

(ii) $\left(A+B\right)^\theta = A^\theta + B^\theta$, where A and B are of the same size.

(iii) $\left(kA\right)^\theta = \overline{k}A^\theta$, k being any complex number

(iv) $\left(AB\right)^\theta = B^\theta A^\theta$, A and B being conformable for multiplication.

UNITARY MATRIX

A square matrix A is said to be unitary if $A^\theta = A^{-1}$ i.e.., if $AA^\theta = I$

NORMAL MATRIX

A square matrix A is said to be normal if $AA^\theta = A^\theta A$

Example 40: Show that $A = \dfrac{1}{2}\begin{bmatrix} 1 & -i & -1+i \\ i & 1 & 1+i \\ 1+i & -1+i & 0 \end{bmatrix}$ is unitary.

Solution: We have $A = \dfrac{1}{2}\begin{bmatrix} 1 & -i & -1+i \\ i & 1 & 1+i \\ 1+i & -1+i & 0 \end{bmatrix}$

$$\therefore A^{\theta} = \frac{1}{2}\begin{bmatrix} 1 & -i & 1-i \\ i & 1 & -1-i \\ -1-i & 1-i & 0 \end{bmatrix}$$

And

$$AA^{\theta} = \frac{1}{2}\begin{bmatrix} 1 & -i & -1+i \\ i & 1 & 1+i \\ 1+i & -1+i & 0 \end{bmatrix} \cdot \frac{1}{2}\begin{bmatrix} 1 & -i & 1-i \\ i & 1 & -1-i \\ -1-i & 1-i & 0 \end{bmatrix}$$

$$= \frac{1}{4}\begin{bmatrix} 1+1+2 & -i-i+2i & 1-i+i-1+0 \\ i+i-2i & 1+1+2 & i+1-1-i \\ 1+i-i-1 & -i+1-1+i+0 & 2+2+0 \end{bmatrix}$$

$$= \frac{1}{4}\begin{bmatrix} 4 & 0 & 0 \\ 0 & 4 & 0 \\ 0 & 0 & 4 \end{bmatrix} = \begin{bmatrix} 1 & 0 & 0 \\ 0 & 1 & 0 \\ 0 & 0 & 1 \end{bmatrix} = I$$

Therefore, A is unitary.

SYMMETRIC MATRIX

A Square matrix $A = \left[a_{ij}\right]_{n\times n}$ is said to be symmetric if its $(i,j)^{th}$ element is the same as its $(j,i)^{th}$ element i.e., $a_{ij} = a_{ji}$ for all i,j

Example 41: If $A = \begin{bmatrix} a & h & g \\ h & b & f \\ g & f & c \end{bmatrix}$ then A is symmetric

$$A = \begin{bmatrix} 5 & 7 \\ 7 & 3 \end{bmatrix}$$ is also symmetric matrix.

Theorem 3: A necessary and sufficient condition for a matrix A to be symmetric is that A and A' are equal.

Proof: Let $A = \left[a_{ij}\right]$ be an n-rowed symmetric matrix. Then $a_{ij} = a_{ji}$

A' Will also be an n-rowed square matrix.

We have $(i,j)^{th}$ Element of A'

$$= (j,i)^{th} \text{ element of } A = a_{ij} = a_{ji}$$

$$= (i,j)^{th}, \text{ element of } A$$

Thus $\qquad A' = A$

Conversely, Let $A' = A$ then A must be a square matrix.

And $(i,j)^{th}$ element of $A' = (i,j)^{th}$ element of A'

$$= (j,i)^{th} \text{ element of } A' = A$$

Hence A is symmetric

SKEW-SYMMETRIC MATRIX

A square matrix $A = \begin{bmatrix} a_{ij} \end{bmatrix}$ is said to be skew symmetric if the $(i,j)^{th}$ element of A is the negation of $(j,i)^{th}$ element of A i.e., if $a_{ij} = -a_{ji}$ for all i,j

If A is a Skew- Symmetric matrix then, we have $a_{ij} = -a_{ji}$

i.e., $\qquad\qquad a_{ii} = -a_{ii}$ for all i

or $\qquad\qquad 2a_{ii} = 0$

or $\qquad\qquad a_{ii} = 0$ for all i

Hence the diagonal elements of a skew-symmetric matrix are all zero.

Example 42: The matrices $A = \begin{bmatrix} 0 & -2i & -4 \\ 2i & 0 & 6 \\ 4 & -6 & 0 \end{bmatrix}$ and $A = \begin{bmatrix} 0 & -h & -g \\ h & 0 & -f \\ g & f & 0 \end{bmatrix}$ are skew-symmetric matrices.

Theorem 4: A necessary and sufficient condition for a matrix to be a skew-symmetric matrix is that $A' = -A$

Hermitian matrix

Let $A = \begin{bmatrix} a_{ij} \end{bmatrix}_{n \times n}$ be a square complex matrix. Then A is said to be Hermitian, if $(i,j)^{th}$ element of A is equal to the $(j,i)^{th}$ conjugate complex of the element of A i.e., $a_{ij} = \overline{a_{ji}}$ if for all i, j.

Example 43: $A = \begin{bmatrix} 3 & 1+2i \\ 1-2i & 5 \end{bmatrix}$ and $A = \begin{bmatrix} 2 & 2-3i & 3+4i \\ 2+3i & 0 & 5-4i \\ 3-4i & 5+4i & 1 \end{bmatrix}$ are Hermitian matrices.

If A is a Hermitian matrix, then we have $a_{ii} = -\overline{a_{ii}}$ for all i

i.e., a_{ii} is real for all. Hence, each diagonal element of a Hermitian matrix is a real number.

SKEW-HERMITIAN MATRIX

A Square matrix $A = \begin{bmatrix} a_{ij} \end{bmatrix}_{n \times n}$ is said to be a Skew-Hermitian matrix if the $(i,j)^{th}$ element of A is equal to the negative of the conjugate complex of $(j,i)^{th}$ element of i.e., if $a_{ij} = -\overline{a_{ji}}$ for all i,j.

If A is a Skew-Hermitian matrix, then $a_{ii} = -\overline{a_{ii}}$ for all i

i.e., $a_{ii} + \overline{a_{ii}} = 2a_{ii} = 0$

or $a_{ii} = 0$ or a_{ii} is pure imaginary i.e., the diagonal element of a Skew-Hermitian matrix must be pure imaginary numbers or zero.

Example 44: $\begin{bmatrix} 0 & -5-i \\ 5-i & 0 \end{bmatrix}$ is a Skew-Hermitian matrix

SOLVED EXAMPLES

Example 45: If A is a symmetric matrix, then show that kA is also a symmetric matrix.

Solution: Let A be a symmetric matrix.

Then we have $A' = A$

and $(kA)' = kA'$

$$= kA \text{ (since } A' = A \text{)}$$

Therefore kA is also a symmetric matrix.

Example 46: If A and B are symmetric matrices, then show that $A + B$ is also symmetric.

Solution: Let A and B be symmetric matrices. Then $A' = A$ and $B' = B$

We have $(A+B)' = A' + B' = A + B$

i.e., $\qquad (A+B)' = A+B$

Therefore $A+B$ is symmetric

Example 47: If A and B are Skew-Symmetric then show that $A+B$ is also skew-symmetric.

Solution: Let A and B be skew-symmetric matrices.

Then we have $A' = -A$ and $B' = -B$

Now $\qquad (A+B)' = A'+B' = -A-B = -(A+B)$

Therefore $A+B$ is also a skew-symmetric matrix.

Example 48: If A and B are Hermitian matrices of the same order, then show that $AB+BA$ is hermitian and $AB-BA$ is Skew-Hermitian matrix.

Solution: Let A and B be two Hermitian matrices of the same order. Then we have

$$A^\theta = A, B^\theta = B$$

Now $\qquad (AB+BA)^\theta = (AB)^\theta + (BA)^\theta = B^\theta A^\theta + A^\theta B^\theta$

$$= BA + AB = AB + BA$$

$AB + BA$ is Hermitian

Now $\qquad (AB-BA)^\theta = (AB)^\theta - (BA)^\theta$

$$= B^\theta A^\theta - A^\theta B^\theta$$

$$= BA - AB$$

$$= -(AB-BA)$$

$AB-BA$, is Skew-Hermitian.

Example 49: If A is a square matrix, then show that $A+A'$ is symmetric and $A-A'$ is Skew-symmetric.

Solution: We have $(A+A')' = A' + (A')'$

$$= A' + A$$

$$= A + A'$$

Therefore $A+A'$ is symmetric

Now $\qquad (A-A')' = A' - (A')'$

$$= A' - A$$

$$= -(A-A')$$

Hence $A-A'$ is skew-symmetric

Example 50: Express $A = \begin{bmatrix} 2 & 3 \\ 7 & 8 \end{bmatrix}$ as the sum of a symmetric and skew-symmetric matrices.

Solution: We have $A = \begin{bmatrix} 2 & 3 \\ 7 & 8 \end{bmatrix}$

$\therefore \qquad A' = \begin{bmatrix} 2 & 7 \\ 3 & 8 \end{bmatrix}$

Now $\qquad A + A' = \begin{bmatrix} 2 & 3 \\ 7 & 8 \end{bmatrix} + \begin{bmatrix} 2 & 7 \\ 3 & 8 \end{bmatrix} = \begin{bmatrix} 4 & 10 \\ 10 & 16 \end{bmatrix}$

$A - A' = \begin{bmatrix} 2 & 3 \\ 7 & 8 \end{bmatrix} - \begin{bmatrix} 2 & 7 \\ 3 & 8 \end{bmatrix} = \begin{bmatrix} 0 & -4 \\ 4 & 0 \end{bmatrix}$

Example 51: Show that every square matrix is uniquely expressive as the sum of a symmetric and skew-symmetric matrix.

Solution: Let A be any square matrix. We can write

$$A = \frac{1}{2}(A + A') + \frac{1}{2}(A - A')$$

Let $\qquad \frac{1}{2}(A + A') = P$ and $\frac{1}{2}(A - A') = Q$

Then $\qquad P' = \left(\frac{1}{2}(A + A') \right)' = \frac{1}{2}\left(A' + (A')' \right)$

$$= \frac{1}{2}(A' + A) = \frac{1}{2}(A + A')$$

P is symmetric

And $\qquad Q' = \left(\frac{1}{2}(A - A') \right)' = \frac{1}{2}\left(A' - (A')' \right)$

$$= \frac{1}{2}(A' - A) = -\frac{1}{2}(A + A')$$

Hence, the matrix Q is skew-symmetric.

$\therefore A = P + Q$, is a sum of symmetric and a skew-symmetric matrices.

Clearly $\frac{1}{2}(A + A')$ is symmetric.

And $\dfrac{1}{2}(A - A')$ is skew-symmetric

We have $\qquad A = \dfrac{1}{2}(A + A') + \dfrac{1}{2}(A - A')$

EXERCISE 4

1. State and prove Cayley-Hamilton theorem.

2. Show that $\begin{bmatrix} 0 & c & -b \\ -c & 0 & a \\ b & -a & 0 \end{bmatrix}$ satisfies Cayley-Hamilton theorem.

3. Define
    (i) Symmetric matrix
    (ii) Skew-symmetric matrix
    (iii) Hermitian matrix
    (iv) Skew-Hermitian matrix
    (v) Unitary marix

4. If $A = \begin{bmatrix} 3 & 1 \\ -1 & 2 \end{bmatrix}$, express $2A^5 - 3A^4 + A^2 - 4I$ as a linear polynomial in A, using Cayley-Hamilton theorem.

5. Verify Cayley-Hamilton theorem for the matrix $A = \begin{bmatrix} 0 & 0 & 1 \\ 3 & 1 & 0 \\ -2 & 1 & 4 \end{bmatrix}$

6. Verify that the matrix $A = \begin{bmatrix} 1 & 2 & 0 \\ 2 & -1 & 0 \\ 0 & 0 & -1 \end{bmatrix}$ satisfies its own characteristic equation.

7. Show that the matrix $A = \begin{bmatrix} 1 & 2 & 1 \\ 0 & 1 & -1 \\ 3 & -1 & 1 \end{bmatrix}$ satisfies its own characteristic equation.

8. Show that $A = \begin{bmatrix} 0 & 0 & 1 \\ 3 & 1 & 0 \\ -2 & 1 & 4 \end{bmatrix}$ satisfies its own characteristic equation and find A^{-1}

9. Find the eigen values of the matrix A, where $A = \begin{bmatrix} -2 & 2 & -3 \\ 2 & 1 & -6 \\ -1 & -2 & 0 \end{bmatrix}$. Also find the eigen vectors.

10. If $A = \begin{bmatrix} 1 & 2 \\ -1 & 3 \end{bmatrix}$ then express $A^6 - 4A^5 + 8A^4 - 12A^3 + 14A^2$ as a linear polynomial in A.

ANSWERS

4. $138A - 403\,I$

8. $A^1 = \dfrac{1}{5} \begin{bmatrix} 4 & 1 & -1 \\ -12 & 2 & 3 \\ 5 & 0 & 0 \end{bmatrix}$

9. $5, -3, -3,\ \begin{bmatrix} 1 \\ 2 \\ -1 \end{bmatrix}, \begin{bmatrix} -2 \\ 1 \\ 0 \end{bmatrix}$

10. $-4A + 5I$

CHAPTER 6

Central Difference Interpolation Formulae

In this chapter we introduce central formulae which are used for interpolation near middle values of the given data.

Consider the function y = f(x). Let y be given for (2n+1) equally spaced values of x. say

$$x_0 \pm h, x_0 \pm 2h, \ldots, x_0 \pm nh$$

Corresponding values of y be y_r (r = 0, ± 1, ± 2, $\ldots$, $\pm n$)

Let y = y_0 denote the central ordinate corresponding to x = x_0. We can form a difference table as follows:

X	y	Δy	$\Delta^2 y$	$\Delta^3 y$	$\Delta^4 y$	$\Delta^5 y$
$x_0 - 2h$	y_{-2}					
		Δy_{-2}				
			$\Delta^2 y_{-2}$			
				$\Delta^3 y_{-2}$		
					$\Delta^4 y_{-2}$	
						$\Delta^5 y_{-2}$
$x_0 - h$	y_{-1}					
		Δy_{-1}				
			$\Delta^2 y_{-1}$			
				$\Delta^3 y_{-1}$		
					$\Delta^4 y_{-1}$	
x_0	y_0					
		Δy_0				
			$\Delta^2 y_0$			

Contd...

$$\Delta^3 y_0$$
$$x_0 + hy_1$$
$$\Delta y_1$$
$$\Delta^2 y_1$$
$$x_0 + 2hy_2$$
$$x_0 - 2hy_{-2}\Delta y_{-2}\Delta^2 y_{-2}\Delta^3 y_{-2}\Delta^4 y_{-2}\Delta^5 y_{-2}$$
$$x_0 - hy_{-1}\Delta y_{-1}\Delta^2 y_{-1}\Delta^3 y_{-1}\Delta^4 y_{-1}$$
$$x_0 y_0 \Delta y_0 \Delta^2 y_0 \Delta^3 y_0$$
$$x_0 + hy_1 \qquad \Delta y_1 \Delta^2 y_1$$
$$x_0 + 2hy_2$$

The above table can also be written in terms of central differences as follows

We now introduce the central difference operator δ to represent the successive differences operator of a function in a convenient way.

If $y_0, y_1, y_2, \ldots, y_n$ denote two successive values corresponding to the values x_0 , x_1, $x_2, \ldots, x_n$

$$y_1 - y_0 = \delta y_{1/2} ,$$

$$y_2 - y_1 = \delta y_{3/2}$$

....

$$y_n - y_{n-1} = \delta y_{n-1/2}$$

For the higher order central differences we have

$$\delta y_{\frac{3}{2}} - \delta y_{\frac{1}{2}} = \delta^2 y_1, \ \delta y_2 - \delta y_1 = \delta^2 y_{\frac{3}{2}} , \ \ldots, \ \delta^{n-1} y_{r+\frac{1}{2}} - \delta^{n-1} y_{r-\frac{1}{2}} = \delta^n y_r$$

$$= (E^{1/2} - E^{-1/2})^n y_r$$

In it's alternative notation we have $\delta\, f(x) = f(x+\tfrac{1}{2}h) - f(x-\tfrac{1}{2}h)$

Where h is the interval of differencing

X	y	δy	$\delta^2 y$	$\delta^3 y$	$\delta^4 y$	$\delta^5 y$
x_0	y_0	$\delta y_{1/2}$	$\delta^2 y_1$	$\delta^3 y_{3/2}$	$\delta^4 y_2$	$\delta^5 y_{5/2}$
x_1	y_1	$\delta y_{3/2}$	$\delta^2 y_1$	$\delta^3 y_{5/2}$	$\delta^4 y_2$	
x_2	y_2	$\delta y_{5/2}$	$\delta^2 y_2$	$\delta^3 y_{7/2}$		
x_3	y_3	$\delta y_{7/2}$	$\delta^2 y_3$			
x_4	y_4					

In constructing the above difference table we have used the relation $\delta = \Delta E^{-1/2}$

The mean operator: Mean operator is also called the averaging operator. It is denoted by μ and is defined as follows:

$$\mu \, f(x) = \frac{1}{2}\left[\, f(x+\tfrac{1}{2}h) + f(x-\tfrac{1}{2}h)\right] \text{ or } \mu \, f(x) = \frac{1}{2}\left[E^{1/2} + E^{-1/2}\right] f(x)$$

In general we define operators are defined as follows

$$\Delta y_x = y_{x+h} - y_x \quad \text{(forward difference operator)}$$

$$\nabla y_x = y_x - y_{x-h} \quad \text{(forward difference operator)}$$

$$\delta y_x = y_{x+\frac{h}{2}} - y_{x-\frac{h}{2}} \quad \text{(central difference operator)}$$

$$\mu y_x = \frac{1}{2}[y_{x+\frac{h}{2}} + y_{x-\frac{h}{2}}] \quad \text{(averaging operator)}$$

and $E y_x = y_{x+h}$ [shift operator]

Gauss forward interpolation formula

The Newton's forward interpolation formula is

$$Y = f(x) = y_0 + u\,\Delta y_0 + \frac{u(u-1)}{2!}\Delta^2 y_0 + \frac{u(u-1)(u-2)}{3!}\Delta^3 y_0 + \dots \qquad \dots(6.1)$$

Where u = $\frac{x(x-x_0)}{h}$ and $x = x_0$ is the origin.

From the   central difference table we have $\Delta^2 y_0 = \Delta^2 y_{-1} + \Delta^3 y_{-1}$

$$\Delta^3 y_0 = \Delta^3 y_{-1} + \Delta^4 y_{-1}$$

$$\Delta^4 y_0 = \Delta^4 y_{-1} + \Delta^5 y_{-1}$$

$$\Delta^3 y_{-1} = \Delta^3 y_{-2} + \Delta^4 y_{-2}$$

$$\Delta^4 y_{-1} = \Delta^4 y_{-2} + \Delta^5 y_{-2}$$

$$\dots$$

Substituting the above values in (6.1) we get

$$Y = f(x) = y_0 + u\,\Delta y_0 + \frac{u(u-1)}{2!}[\Delta^2 y_{-1} + \Delta^3 y_{-1}]$$

$$+ \frac{u(u-1)(u-2)}{3!}[\Delta^3 y_{-1} + \Delta^4 y_{-2}] + \dots$$

The above formula may be written as

$$Y = f(x) = y_0 + u\,\Delta y_0 + \frac{u(u-1)}{2}\Delta^2 y_{-1} + \frac{(u+1)u(u-1)}{6}\Delta^3 y_{-1}$$

$$+ \frac{(u+1)u(u-1)(u-2)}{24}\Delta^4 y_{-2} + \dots \qquad \dots(6.2)$$

The equation (6.2) is called Gauss's forward interpolation formula

Gauss's backward interpolation formula: $y = f(x) = y_0 + u\,\Delta y_0 + \frac{u(u-1)}{2}\Delta^2 y_{-1} + \frac{(u+1)u(u-1)}{6}\Delta^3 y_{-1} + \ldots$

Substituting

$$\Delta^2 y_0 = \Delta^2 y_{-1} + \Delta^3 y_{-1}$$

$$\Delta^3 y_0 = \Delta^3 y_{-1} + \Delta^4 y_{-1}$$

$$\Delta^4 y_0 = \Delta^4 y_{-1} + \Delta^5 y_{-1}$$

$$\Delta^3 y_{-1} = \Delta^3 y_{-2} + \Delta^4 y_{-2}$$

$$\Delta^4 y_{-1} = \Delta^4 y_{-2} + \Delta^5 y_{-2}$$

...

In Newton's forward interpolation formula we get

$$Y = f(x) = y_0 + \frac{u}{1!}(\Delta y_{-1} + \Delta^2 y_{-1}) + \frac{u(u-1)}{2!}(\Delta^2 y_{-1} + \Delta^3 y_{-1}) + \ldots$$

Or

$$y_u = y_0 + \frac{u}{1!}\Delta y_{-1} + \frac{u(u+1)}{2!}\Delta^2 y_{-1} + \frac{(u+1)\,u(u-1)}{3!}\Delta^3 y_{-2} + \frac{(u+2)(u+1)\,u(u-1)}{4!}\Delta^4 y_{-2} + \ldots$$

The above formula is known as Gauss's backward interpolation formula

Remarks: The Gauss's forward interpolation formula employs odd differences above the central line through y_0 and even differences on the central line, whereas Gauss's backward difference formula employs odd differences below the central line through y_0 and even differences on the central line.

Gauss's forward interpolation formula is used to interpolate the values of the function for the values of u such that $0<u<1$ and Gauss's backward interpolation formula is used to interpolate the value of the function for a negative value of u which lies between $-1<u<0$

Bessel's formula

Changing the origin in the Gauss's backward interpolation formula from 0 to 1, we get

$$y_u = y_1 + (u-1)\Delta y_0 + \frac{u(u-1)}{2!}\Delta^3 y_0 + \frac{u(u-1)(u-2)}{2!}\Delta^3 y_{-1} + \ldots$$

Taking mean of the above formula and the Gauss's forward interpolation formula we get

$$y_u = \frac{1}{2}[y_0 + y_1] + \left(u - \frac{1}{2}\right)\Delta y_0 + \frac{u(u-1)}{2!}\frac{1}{2}[\Delta^2 y_{-1} + \Delta^2 y_0] + \frac{\left(u-\frac{1}{2}\right)u(u-1)}{3!}\Delta^3 y_{-1} + \ldots$$

Which is known as Bessel's formula

Bessel's formula involves odd differences below the central line and means of the even differences on and below the line

Stirling's formula

Gauss's forward interpolation formula is

$$y_u = f(x) = y_0 + u\,\Delta y_0 + \frac{u(u-1)}{2}\Delta^2 y_{-1} + \frac{(u+1)u(u-1)}{6}\Delta^3 y_{-1}$$

$$+ \frac{(u+1)u(u-1)(u-?)}{24}\Delta^4 y_{-2} + \frac{(u+2)(u+1)u(u-1)(u-2)}{120}\Delta^5 y_{-2} + \ldots \qquad \ldots(6.3)$$

Gauss's backward interpolation formula is

$$y_u = y_0 + \frac{u}{1!}\Delta y_{-1} + \frac{u(u+1)}{2!}\Delta^2 y_{-1} +$$

$$\frac{(u+1)\,u(u-1)}{3!}\Delta^3 y_{-2} + \frac{(u+2)(u+1)\,u(u-1)}{4!}\Delta^4 y_{-2} + \ldots$$

Taking the mean of the two Gauss's formulae, we get

$$y_u = y_0 + u\left[\frac{\Delta y_0 + \Delta y_{-1}}{2}\right] + \frac{u^2}{2}\Delta^2 y_{-1} + \frac{(u+1)u(u-1)}{3!}\frac{\Delta^3 y_{-1} + \Delta^3 y_{-2}}{2} + \frac{u^2(u^2-1)}{4!}\Delta^4 y_{-2} + \ldots$$

Which is known as Stirling's formula $-0.25 \le u \le 0.25$. Therefore we have to choose x_0 such that $u \in [-0.25,\ 0.25]$

Laplace-Everett's formula

Eliminating odd differences in Gauss's forward interpolation formula by using the relations

$$\Delta y_0 = y_1 - y_0$$

$$\Delta^3 y_{-1} = \Delta^2 y_0 - \Delta^2 y_{-1}$$

$$\Delta^5 y_{-2} = \Delta^4 y_{-1} - \Delta^4 y_{-2}$$

We get

$$Y = f(x) = y_0 + \frac{u}{1!}(y_1 - y_0) + \frac{u(u-1)}{2!}\Delta^2 y_{-1} +$$

$$\frac{(u+1)u(u-1)}{3!}(\Delta^2 y_0 - \Delta^2 y_{-1}) + \frac{(u+1)u(u-1)(u-2)}{4!}\Delta^4 y_{-2} +$$

$$+ \frac{(u+2)(u+1)u(u-1)(u-2)}{5!}(\Delta^4 y_{-1} - \Delta^4 y_{-2}) + \ldots \qquad \ldots(6.4)$$

Writing $v = 1 - u$ i.e. $u = 1 - v$ and changing the terms of (1) with negative sign, we get

$$Y = vy_0 + \frac{v(v^2 - 1^2)}{3!}\Delta^2 y_{-1} + \frac{v(v^2-1^2)(v^2-2^2)}{5!}\Delta^4 y_{-2} + \ldots +$$

$$uy_1 + \frac{u(u^2-1^2)}{3!}\Delta^2 y_0 + \frac{u(u^2-1^2)(u^2-2^2)}{5!}\Delta^2 y_{-1} + \ldots$$

Example 1: If $u_{20} = 51203$, $u_{30} = 43931$, $u_{40} = 34563$, $u_{50} = 24348$

Find the value of u_{35} by applying Gauss's forward interpolation formula

Solution: Here we have $x_0 = 30$, $x_0 - h = 20$, $x_0 + h = 40$, $x_0 + 2h = 50$, $x = 35$

$$y_0 = 43831,\ y_{-1} = 51203,\ y_1 = 34563,\ y_2 = 24348$$

Where h = 10, therefore $u = \dfrac{x-x_0}{h} = \dfrac{35-30}{10} = \dfrac{5}{10} = 0.5$

x	t	Δ	Δ²	Δ³
20	51203			
		−7272		
30	43931		−2096	
		−9468		1249
40	43931		−847	
		−10215		
50	24348			

From the table we have $\Delta y_{-1} = -7272,\ \Delta y_0 = -9468,\ \Delta^2 y_{-1} = -2096,\ \Delta^3 y_{-1} = 1249$

From Gauss's forward interpolation formula we have

$$Y = f(x) = y_0 + u\,\Delta y_0 + \frac{u(u-1)}{2}\Delta^2 y_{-1} + \frac{(u+1)u(u-1)}{6}\Delta^3 y_{-1} +$$

$$\frac{(u+1)u(u-1)(u-2)}{24}\Delta^4 y_{-2} + \dots$$

Substituting the values of $y_0, \Delta y_{-1}, \Delta y_0, \Delta^2 y_{-1}, \Delta^3 y_{-1}$ in the above formula we get

$$\Rightarrow \quad f(0.5) = u_{35} = 4392 + (0.5)(-9468)$$

$$+ \frac{(0.5)(-0.5)}{2}(-2096) + \frac{(1.5)(0.5)(-0.5)}{6}(1249)$$

$$= 39430.9 \ (\text{approx})$$

Example 2: Use Gauss's forward interpolation formula to find y for x = 30 given that

x	21	25	29	33	37
y	18.4708	17.8144	17.1070	16.3432	15.5154

Solution: Taking x_0 (origin) = 29

We have $x_0 - h = 25,\ x_0 - 2h = 21,\ x_0 + h = 33,\ x_0 + 2h = 37.$

The corresponding values of y are

$$y_0 = 17.1070,\ y_{-1} = 17.8144,\ y_{-2} = 18.4708,\ y_1 = 16.3432,$$

$$y_2 = 15.5154,\ u = \frac{30-29}{4} = 0.25$$

Since the value of y lies between 0 and 1, Gauss's forward interpolation is a suitable interpolation formula for finding the value of y at x = 30

x	y	Δ	Δ^2	Δ^3	Δ^4
21	10.4708	−0.6564	−0.0510	−0.0054	−0.0022
25	17.8144	−0.7074	−0.0564	−0.0076	
29	17.1070	−0.7638	−0.0640		
33	16.3432	−0.8278			
37	15.5154				

X	y = f(x)	Δ	Δ^2	Δ^3
25	0.2707	0.0320	0.0039	0.0010
30	0.3027	0.0359	0.0049	
35	0.3386	0.0408		
40	0.3794			

Gauss's forward interpolation is

$$y_u = f(x) = y_0 + u\,\Delta y_0 + \frac{u(u-1)}{2}\Delta^2 y_{-1} + \frac{(u+1)u(u-1)}{6}\Delta^3 y_{-1} +$$

$$\frac{(u+1)u(u-1)(u-2)}{24}\Delta^4 y_{-2} + \frac{(u+2)(u+1)u(u-1)(u-2)}{120}\Delta^5 y_{-2} + \ldots$$

$$\Rightarrow y_{0.25} = 17.1070 + (0.25)(-0.7638) + \frac{(0.25)(-0.75)}{2}(-0.0564) + \frac{(1.25)(0.25)(-0.75)}{6}$$

$$(-0.0076) + \frac{(1.25)(0.25)(-0.75)\,(-1.75)}{24}(-0.0022) = 16.9216$$

Example 3: Find the value of f(2.5) using the following table by applying Gauss's forward interpolation formula

x	1	2	3	4
f(x)	1	8	27	64

Solution: Taking $x_0 = 2$ (i.e. origin) we get the central difference table for the given data is as follows

X	f(x)	Δ	Δ^2	Δ^3
1	1	7	12	6
2	8	19	18	
3	27	37		
4	64			

Here we have $x_0= 2$, $x = 2$, $h = 1$ $u = \dfrac{x-x_0}{h} = \dfrac{2.5-2}{1} = 0.5$

Using Gauss forward interpolation formula we get,

$$y_u = f(2.5) = y_0 + u\,\Delta y_0 + \frac{u(u-1)}{2}\Delta^2 y_{-1} + \frac{(u+1)u(u-1)}{6}\Delta^3 y_{-1} + \dots$$

$$= 8+(0.5)\,(19) + \frac{(0.5)(0.5-1)}{2}(12) + \frac{(0.5+1)0.5(0.1-1)}{6}(6) = 15.625$$

Example 4: Use Gauss's backward interpolation formula to find f(32), given that f(25) = 0.2707, f(30) = 0.3027, f(35) = 0.3386 and f(40) = 0.3794

Solution: Taking the x_0 (origin) = 35 we have $x_{-1}= 30$, $x_{-2} = 25$, $x_1 = 40$, $x = 32$

$$y_u = y_0 + \frac{u}{1!}\Delta y_{-1} + \frac{u(u+1)}{2!}\Delta^2 y_{-1} + \frac{(u+1)\,u(u-1)}{3!}\Delta^3 y_{-2} + \frac{(u+2)(u+1)\,u(u-1)}{4!}\Delta^4 y_{-2} + .$$

$$y_u = 0.3386 + (-0.6)(0.0359) + \frac{(-0.6)(-0.6+1)}{2}(0.0049) + \frac{(-0.6)(-0.6+1)\,((-0.6-1)}{6}(0.0010) =$$
0.3165

Example 5: Apply Gauss's backward interpolation formula and find the population of a town in 1946, with the help of the following data

year (x):	1931	1941	1951	1961	1971
Population (y) in Thousands :	15	20	27	30	32

Solution: We have x = 1946, h = 10

Taking the origin at 1951 we get $u = \dfrac{1946-1951}{10} = -0.5$

The difference table is

with the help of the following data

year (x)	Y	Δ	Δ^2	Δ^3	Δ^4
−2	15	5	2	3	−7
−1	20	7	5	−4	
0	27	12	1		
1	39	13			
2	52				

Substituting in Gauss's backward interpolation formula we get

$$y_u = y_0\frac{u}{1!}\Delta y_{-1} + \frac{u(u+1)}{2!}\Delta^2 y_{-1} + \frac{(u+1)\,u(u-1)}{3!}\Delta^3 y_{-2} +$$
$$\frac{(u+2)(u+1)\,u(u-1)}{4!}\Delta^4 y_{-2} + \dots \; y_{-0.5} = 27 + (-0.5)(7) +$$
$$\frac{(0.5)(-0.5+1)}{2}(5) + \frac{(0.5)(-0.5+1)(-0.5-1)}{6}(3) +$$
$$\frac{(0.5)(-0.5+1)(-0.5-1)(-0.5+2)}{6}(-7) + \dots \; = 22.8984$$

Example 6: Find by Gauss backward interpolation formula to find $\tan 50^0 42'$ given

x^0	50	51	52	53	54
$Tan x^0$	1.1918	1.2349	1.2799	1.3270	1.3764

Solution: We have h = 1 and $x - 50^0 42'$, taking the origin at $x_0 = 51^0$ we get u = $\frac{x-x_0}{h}$

$= \frac{50^0 42' - 51}{1^0} = -0.3$. The difference table is given below

X	y	Δ	Δ^2	Δ^3	Δ^4
50	1.1918	0.0431	0.0019	0.0002	0
51	1.2349	0.0451	0.0021	0.0002	
52	1.2799	0.0471	0.0023		
53	1.3270	0.0494			
54	1.3764				

From table we have $\Delta y_0 = 1.2349$, $\Delta y_{-1} = 0.0431$, $\Delta^2 y_{-1} = 0.0019$, $\Delta^3 y_{-2} = 0.0002$, $\Delta^4 y_{-2} = 0$

$$y_u = y_0 + \frac{u}{1!}\Delta y_{-1} + \frac{u(u+1)}{2!}\Delta^2 y_{-1} + \frac{(u+1)\,u(u-1)}{3!}\Delta^3 y_{-2} +$$
$$\frac{(u+2)(u+1)\,u(u-1)}{4!}\Delta^4 y_{-2} + \ldots$$

$= 1.2349 + (-0.03)(0.0431) + \frac{(-0.3+1)(-0.3)}{2!}(0.0019) = 1.2349 - 0.01293$

$\qquad - 0.0001995$

$\Rightarrow \qquad \tan 50^0 42' = 1.2218$

Example 7: Using Bessel's formula find the value of f(x) at x = 35, given

x	10	20	30	40	50
Y = f(x)	17.36	34.20	50.00	64.28	75.60

Solution: Here we have h = 10, x = 35, taking the origin at x = 3 i.e. $x_0 = 30$

$\qquad\qquad y_0 = 50.00$, $y_1 = 64.28$,

we get $\qquad$ u = $\frac{35-30}{10} = 0.5$,

The difference table is

X	y = f(x)	Δ	Δ^2	Δ^3	Δ^4
10	17.36	16.84	−1.04	−0.48	0.04
20	34.20	15.80	−1.52	−0.44	
30	50.00	14.28	−1.96		
40	64.28	12.32			
50	76 .60				

From the difference table we have

$$\Delta^2 y_0 = -1.96,\ \Delta^2 y_{-1} = -1.52,\ \Delta^4 y_{-2} = 0.04$$

Bessel's formula is

$$y_u = \frac{1}{2}[y_0 + y_1] + (u - \frac{1}{2})\Delta y_0 + \frac{u(u-1)}{2!}\frac{1}{2}[\Delta^2 y_{-1} + \Delta^2 y_0] + \frac{(u-\frac{1}{2})u(u-1)}{3!}\Delta^3 y_{-1} + \dots$$

$$\Rightarrow\quad y_{35} = \frac{1}{2}[50.00 + 64.28] - \frac{1}{8}[\frac{-1.52 - 1.96}{2}] + \frac{3}{128}(0.04) = 57.14 +$$

$$0.3125 + 0.0009375 = 57.45$$

Example 8: Use Bessel's interpolation formula to estimate $\sqrt[3]{46.24}$ given that $y = \sqrt[3]{x}$ for

X	41	45	49	53
Y	3.4482	3.5569	3.6593	3.7563

Solution: Here we have h = 4. Take origin $(x_0) = 45$, x = 46.24 we get $u = \frac{x - x_0}{h} = \frac{46.24 - 45}{4} = 0.31$

The difference table for the given data is

X	Y	Δ	Δ^2	Δ^3
41	3.4482	0.1087	−0.0063	0.0009
45	3.5569	0.1024	−0.0054	
49	3.6593	0.0970		
53	3.7563			

$$y_u = \frac{1}{2}[y_0 + y_1] + (u - \frac{1}{2})\Delta y_0 + \frac{u(u-1)}{2!}\frac{1}{2}[\Delta^2 y_{-1} + \Delta^2 y_0] + \frac{(u-\frac{1}{2})u(u-1)}{3!}\Delta^3 y_{-1}$$

$$\Rightarrow\quad y = \frac{1}{2}[3.5569 + 3.6593] + [0.031 - 0.5][0.1024] + \frac{(0.31)(-0.69)}{2}[\frac{-0.0063 - 0.0054}{2}] + \frac{(0.031 - 0.5)(0.31)(0.31 - 1)}{6}(0.0009)$$

$$= 3.6081 - 0.001945 + 0.0006256 + 0.0000061 = 3.5893$$

Example 9: Apply Everette's formula to obtain the value of y_{25}, given $y_{20} = 2854$,

$$y_{24} = 3162,\ y_{28} = 3544\ \text{and}\ y_{32} = 3992$$

Solution: Here we have h = 4, x = 25. Taking origin at $x_0 = 24$, we get $\frac{25 - 24}{4} = 0.25$

The difference table is

X	y	Δ	Δ²	Δ³
20	2854	308	74	− 28
24	3162	382	66	
28	3544	448		
32	3992			

From the table we have $y_0 = 3162, y_1 = 3544, \Delta^2 y_{-1} = 74, \Delta^2 y_0 = 66$

The Laplace- Everette's formula is

$$Y = f(x) = y_0 + \frac{u}{1!}(y_1 - y_0) + \frac{u(u-1)}{2!}\Delta^2 y_{-1} +$$

$$\frac{(u+1)u(u-1)}{3!}(\Delta^2 y_0 - \Delta^2 y_{-1}) + \frac{(u+1)u(u-1)(u-2)}{4!}\Delta^4 y_{-2} + \dots$$

$$= 3162 + (0.25)(3544 - 3162) + \frac{(0.25)(0.25-1)}{2}(74) + \frac{(0.25+1)(0.25)(0.25-1)}{3!}$$

$$(66 - 74) + \dots = 3251$$

Example 10: Assuming that 3rd and 5th terms are constant show that

$$y_{x+\frac{1}{2}} = \frac{1}{2}[y_x + y_{x+1}] - \frac{1}{16}[\Delta^2 y_{x-1} + \Delta^2 y_x]$$

Solution: From Bessel's formula we have

$$y_u = \frac{1}{2}[y_0 + y_1] + (u - \frac{1}{2})\Delta y_0 + \frac{u(u-1)}{2!}\frac{1}{2}[\Delta^2 y_{-1} + \Delta^2 y_0] +$$

$$\frac{(u-\frac{1}{2})u(u-1)}{3!}\Delta^3 y_{-1} + \dots (1)$$

Substituting $u = \frac{1}{2}$ in (1) we get

$$y_{\frac{1}{2}} = \frac{1}{2}[y_0 + y_1] + + \frac{\frac{1}{2}(\frac{1}{2}-1)}{2!}\frac{1}{2}[\Delta^2 y_{-1} + \Delta^2 y_0]$$

Shifting the origin to x we get

$$y_{x+\frac{1}{2}} = \frac{1}{2}[y_x + y_{x+1}] - \frac{1}{16}[\Delta^2 y_{x-1} + \Delta^2 y_x]$$

Example 11: Few values of sin h x are given below. Find the value of x when sin h x = 62

X	4.80	4.81	4.82	4.83	4.84
Y = sinh x	60.7511	61.3617	61.9785	62.5015	63.2307

Solution: We have h = 0.01. Clearly the value of x lies between x = 4.82 and x = 4.83

Taking the origin $x_0 = 4.82$ we get the following difference table

The difference table

X	y	Δ	Δ²	Δ³
4.80	60.7511	0.6106	0.0062	0
4.81	61.3617	0.6168	0.0062	0
4.82	61.9785	0.6230	0.0062	
4.83	62.6015	0.6292		
4.84	62.2307			

From the difference table we have

$h = 1$, $y_0 = 61.9785$, $\Delta y_0 = 0.6230$, $\Delta y_{-1} = 0.6168$, $\Delta^2 y_{-1} = 0.0062$

from Stirling's formula we have

$$y_u = y_0 + u\left[\frac{\Delta y_0 + \Delta y_{-1}}{2}\right] + \frac{u^2}{2}\Delta^2 y_{-1} + \frac{(u+1)u(u-1)}{3!}\frac{\Delta^3 y_{-1} + \Delta^3 y_{-2}}{2} + \frac{u^2(u^2-1)}{4!}\Delta^4 y_{-2} + \dots$$

Substituting the values of y_0, Δy_0, Δy_{-1}, $\Delta^2 y_{-1}$ we get

$$62 = 61.9785 + u\frac{(0.6239 + 0.6168)}{2} + \frac{u^2}{2}(0.0062)$$

Simplifying the above equation we get $31u^2 + 6199u - 215 = 0$

Solving the equation and neglecting the negative root we get u = 0.0347 (positive root)

Now consider x = x_0+uh

Substituting the values of x_0, h and u   we get x = 4.82 +0.0347 ×(0.01) =4.8203(approx.)

EXERCISE

1.   Use Gauss backward interpolation formula to find the value of tan $50^0 42'$, given

x^0	50	51	52	53	54
tan x^0	1.1918	1.2349	1.2799	1.3270	1.3764

[**Ans:** 1.2218]

2.   Use Gauss forward interpolation formula to find the value of $e^{-1.7425}$

x	1.72	1.73	1.74	1.75	1.76
$y = e^{-x}$	0.17907	0.17728	0.17552	0.17377	0.17204

[**Ans:** 0.1751]

3. Given the table

X	310	320	330	340	350	360
Y = log x	2.4914	2.5052	2.5185	2.5315	2.5441	2.5563

Find the value of log 337 by Gauss forward interpolation formula   [**Ans:** 2.5283]

4. Use Gauss forward interpolation formula to find f(32) from the following table

x	25	30	35	40
f(x)	0.2707	0.3027	0.3386	0.3794

[**Ans:** 0.3165]

5. Apply Gauss forward interpolation formula and find the value of f(x) at x =3.5, given

x	2	3	4	5
f(x)	2.626	3.454	4.784	6.986

[**Ans:** 4.034]

6. Find the value of f(x) at x = 3.75 by applying Gauss 's forward interpolation formula from the following data

x	2.5	3.0	3.5	4.0	4.5	5.0
f(x)	24.145	22.043	20.225	18.644	17.262	16.047

[**Ans:** 19.407]

7. Find the value of cos $51^0 42'$ by Gauss's backward interpolation formula given that

x	50^0	51^0	52^0	53^0	54^0
Cos x	0.6428	0.6293	0.6157	0.6018	0.5878

[**Ans:** 0.6198]

8. Given that f(1) = 1.0000, f(1.10) = 1.049, f(1.20) = 1.096, f(1.30) = 1.040. Use Everette's formula find f(1.15)   [**Ans:** 1.0728]

9. Using Bessel's formula find f(25) given that f(20) = 24, f(24) = 32, f(28) = 35, f(32) = 40

[**Ans :** 32.95]

10. Use Gauss's backward formula to find the population in the year 1936, given the following table

year	1901	1911	1921	1931	1941	1951
Population ('000s)	12	15	20	27	39	52

[**Ans:** 32.95]

11. Compute the value of $\sin 25^0 40' 30''$, by using (i) Bessel's formula (ii) Everett's formula

x	$25^0 40' 0''$	$25^0 40' 20''$	$25^0 40' 40''$	$25^0 41' 0''$	$25^0 41' 20''$
$Y = \sin \theta$	0.43313479	0.43322218	0.4330956	0.43339695	0.43358433

[Ans: 0.43326587, 0.43322218]

12. Find the value of y_{15} if $y_{10} = 2854$, if $y_{14} = 2854$, $y_{18} = 3544$, $y_{22} = 3992$ **[Ans:** 3251]

13. Find u_{35} if $u_{20} = 51203$, $u_{30} = 43931$, $u_{40} = 34563$, $u_{50} = 24348$, by Gauss forward interpolation **[Ans:** 323437]

14. Use Everette's formula and find f(1.15) if f(1) = 1.00, f(1.10) = 1.049, f(1.20) = 1.096, f(1.30) = 1.40 **[Ans :** 1.0728]

CHAPTER 7

Numerical Differentiation

7.1 INTRODUCTION

In this chapter we introduce Numerical differentiation. In the case of differentiation, we first write the interpolating formula on the interval (x_0, x_n) and the differentiate the polynomial term by term to get an approximated polynomial to the derivative of the function. When the tabular points are equidistant, one uses either the Newton's Forward or Newton's Backward Formula or Sterling's Formula; otherwise Lagrange's formula is used. Newton's Forward or Backward formula is used depending upon the location of the point at which the derivative is to be computed. In case the given point is near the midpoint of the interval, Sterling's formula can be used. We illustrate the process by taking (i) Newton's Forward formula, and (ii) Sterling's formula.

Let f be a given function that is only let f be a given function that is only known at a number of isolated points. The problem of numerical differentiation is to compute an approximation to the derivative f by suitable combinations of the known values of f.

Example 1: From the table of values below compute $\dfrac{dy}{dx}$ and $\dfrac{d^2y}{dx^2}$ for x = 1

x	1	2	3	4	5	6
y	1	8	27	64	125	216

Solution: The difference table is as follows:

x	y	Δy	$\Delta^2 y$	$\Delta^3 y$	$\Delta^4 y$
1	1				
		7			
2	8		12		
		19		6	
3	27		18		0
		37		6	
4	64		24		0
		61		6	
5	125		30		
		91			
6	216				

We have x 0 = 1, h = 1, x = 1 is at the beginning of the table.

We use Newton's forward formula

$$\left(\frac{dy}{dx}\right)_{x=x_0} = \frac{1}{h}\left[\Delta y_0 - \frac{1}{2}\Delta^2 y_0 + \frac{1}{3}\Delta^3 y_0 - \frac{1}{4}\Delta^4 y_0 + \ldots\right]$$

so that

$$\left(\frac{dy}{dx}\right)_{x=1} = \frac{1}{h}\left[7 - \frac{1}{2}12 + \frac{1}{3}6 - 0 + \ldots\right] = 7 - 6 + 2 = 3$$

and

$$\left(\frac{d^2y}{dx^2}\right)_{x=x_0} = \frac{1}{h^2}\left[\Delta^2 y_0 - \Delta^3 y_0 + \frac{11}{12}\Delta^4 y_0 - \ldots\right]$$

or

$$\left(\frac{d^2y}{dx^2}\right)_{x=1} = \frac{1}{1^2}\left[12 - 6\right] = 6$$

Hence

$$\left(\frac{dy}{dx}\right)_{x=1} = 3, \quad \left(\frac{d^2y}{dx^2}\right)_{x=1} = 6$$

Example 2: From the following table of values of x and v find $\dfrac{dy}{dx}$ and $\dfrac{d^2y}{dx^2}$ for x = 1.05.

X	1.00	1.05	1.10	1.15	1.20	1.25	1.30
y	1.00000	1.02470	1.04881	1.07238	1.09544	1.11803	1.14017

Solution: The difference table is as follows:

X	y	Δ	Δ^2	Δ^3	Δ^4	Δ^5
1.00	1.00000					
		0.02470				
1.05	1.02470		− 0.00059			
		0.002411		− 0.00002		
1.10	1.04881		− 0.00054		0.00003	
		0.02357		− 0.00001		− 0.00006
1.15	1.07238		− 0.00051		− 0.00003	
		0.02306		− 0.00002		
1.20	1.09544		− 0.00047			
		0.02259				
1.25	1.11803		− 0.00045			
		0.02214				
1.30	1.14017					

Taking x 0 = 1.05, h = 0.05, we have

$$\Delta\,y0 = 0.02411,$$

$$\Delta^2 y_0 = 0.00054,$$

$$\Delta^3 y_0 = 0.00003,$$

$$\Delta^4 y_0 = -0.00001,$$

$$\Delta^5 y_0 = -0.00003,$$

From Newton's formula, we have

$$\left(\frac{dy}{dx}\right)_{x=x_0} = \frac{1}{h}\left[\Delta y_0 - \frac{1}{2}\Delta^2 y_0 - \frac{1}{3}\Delta^3 y_0 - \frac{1}{4}\Delta^4 y_0 + \ldots\right]$$

i.e.,

$$\left(\frac{dy}{dx}\right)_{x=1.05} = \frac{1}{0.05}\left[0.02411 - \frac{0.00054}{2} + \frac{1}{3}(0.00003)\right]$$

$$+ \frac{1}{0.05}\left[-\frac{1}{4}(0.00001) + \frac{1}{5}(0.00003)\right]$$

Hence

$$\left(\frac{dy}{dx}\right)_{x=1.05} = 0.48763$$

and

$$\left(\frac{d^2 y}{dx^2}\right)_{x=x_0} = \frac{1}{h^2}\left[\Delta^2 y_0 - \Delta^3 y_0 + \frac{11}{12}\Delta^4 y_0 - \frac{5}{6}\Delta^5 y_0 \ldots\right]$$

or

$$\left(\frac{d^2 y}{dx^2}\right)_{x=1.05} = \frac{1}{(0.05)^2}$$

$$\left[-0.00054 - 0.0003 + \frac{11}{12}(0.00001) - \frac{5}{6}(-0.00003)\right]$$

or

$$\left(\frac{d^2 y}{dx^2}\right)_{x=1.05} = -02144.$$

Example 3: A rod is rotating in a plane about one a/its ends. If the following table gives the angle θ radians through which the rod has turned for different values of time t seconds, find its angular velocity at t = 7 sees.

t seconds	0.0	0.2	0.4	0.6	0.8	1.0
θ radians	0.0	0.12	0.48	0.10	2.0	3.20

Solution: The difference table is given below:

t	θ	$\Delta\theta$	$\Delta^2\theta$	$\Delta^3\theta$	$\Delta^4\theta$
0.0	0.0				
		0.12			
0.2	0.12		0.24		
		0.36		0.02	
0.4	0.48		0.26		0
		0.62		0.02	
0.6	1.10		0.28		0
		0.90		0.02	
0.8	2.0		0.30		
		1.20			
1.0	3.20				

Here $\qquad$ $x_n = t_n = 1.0$, $h = 0.2$, $x = t = 0.7$

$$u = \frac{x - x_n}{h} = \frac{0.7 - 1.0}{0.2} = -15.$$

From the Newton's backward interpolation formula, we have

$$\left(\frac{d\theta}{dt}\right)_{n=0.7} = \frac{1}{h}\left[\nabla\theta_0 + \frac{2u+1}{2}\nabla^2\theta_0 + \frac{3u^2+6u+2}{6}\nabla^2\theta_n\right]$$

$$= \frac{1}{0.2}\left[1.20 - 0.30 + \frac{3(-15)^2 - 6(-15) + 2}{6}(0.02)\right]$$

$$= 5(1.20 - 0.30 - 0.0008) = 4.496 \text{ radian/sec}$$

$\therefore \qquad \dfrac{d\theta}{dt} = 4.496$ radian/sec

and $\qquad \left(\dfrac{d^2\theta}{dt^2}\right)_{t=0.7} = \dfrac{1}{h^2}\left[\nabla^2\theta_0 + (u+1)\nabla^2\theta_0\right]$

$$= \frac{1}{(0.2)^2}[0.30 - 0.5 \times 0.02]$$

$$= 25 \times 0.29 = 7.25 \text{ radian/sec2.}$$

Hence, Angular velocity = 4.496 radian/sec and

Angular acceleration = 7.25 radian/sec2.

Example 4: Find $\dfrac{dy}{dx}$ at $x = 0.6$ of the function $y - f(x)$, tabulated below:

x	0.4	0.5	0.6	0.7	0.8
y	1.5836494	1.7974426	2.0442376	2.3275054	2.6510818

Solution: The difference table is as follows:

x	y	Δy	$\Delta^2 y$	Δ^3	Δ^4
0.4	1.5836494				
		0.2137932			
0.5	17974426		0.0330018		
		0.2467950		0.0034710	
0.6	2.0442376		0.0364728		0.0003648
		0.2832678		0.0038358	
0.7	2.3275054		0.0403084		
		0.3235764			
0.8	2.6510818				

Substituting these values in the Stirling's formula, i.e., in

$$\left(\frac{dy}{dx}\right)_{x=x_0} = \frac{1}{h}\left[\frac{\Delta y_0 + \Delta y_{-1}}{2} - \frac{1}{6}\frac{\Delta^3 y_{-1} + \Delta^3 y_{-2}}{2} + \ldots\right]$$

we get or $\left(\dfrac{dy}{dx}\right)_{x=0.6} = \dfrac{1}{0.1}\left[\dfrac{1}{2}(0.2832678 + 0.2467950) - \dfrac{1}{2}(0.0038358 + 0.0034710)\right]$

$$= 10(0.2650314 - 0.0006089) = 2.644225$$

$$\left(\frac{dy}{dx}\right)_{x=0.6} = 2.644225.$$

Example 5: From the following table find x correct to two decimal places, for which y is maximum and find this value of y

x	1.2	1.3	1.4	1.5	1.6
y	0.9320	0.9636	0.9855	0.9975	0.996

Solution: The forward difference table is as follows:

x	y	Δy	$\Delta^2 y$	$\Delta^3 y$
1.2	0.9320			
		0.0316		
1.3	0.9636		-0.0097	
		0.0219		-0.0002
1.4	0.9855		-0.0099	
		0.0120		-0.0002
1.5	0.9975		-0.0099	
		0.0021		
1.6	0.9996			

we have x_0, = 1.2. For maximum value of y we take $\dfrac{dy}{dx} = 0$.

Differentiating Newton's forward interpolation formula w.r.t. *u* and neglecting terms of second differences, we get

$$0 = 0.0316 + \frac{2u-1}{2}\,(\text{-}0.0097)$$

$$\Rightarrow \quad 0 = 00712 - (2u - 1)\,(0.0097)$$

or $\qquad (2u - 1)\,(0.0097) = 0.0712$

$$\Rightarrow \quad u - 38.$$

Substituting in $x = x_0 + uh$,

we get $\qquad x = 1.2(3.8\,)\,(0.1) = 1.58.$

The value 1.58 is closer to x = 1.6, hence we use Newton's backward difference formula

$$F(1.58) = 0.9996 - (0.2)\,(0.0021) + \frac{(-0.2)(-0.2+1)}{2}\,(\text{-}0.0099)$$

$$= 0.9996 - 0.0004 + 0.0008 = 1.000$$

The maximum value occurs at x = $-$ 1.58 and the maximum value is 1.000.

Example 6: Find the maximum and the minimum values of the function y = f(x) from the following data:

x	0	1	2	3	4	5
f(x)	0	0.25	0	2.25	16.00	56.25

Solution: The forward difference table is as follows:

x	y	Δ	Δ^2	Δ^3	Δ^4	Δ^5
$0 = x_0$	$0 = y_0$					
		$0.25 = \Delta y_0$				
1	0.25		$-0.50 = \Delta^2 y_0$			
		-0.25		$3.00 = \Delta^3 y_0$		
2	0		2.50		$6 = \Delta^4 y_0$	
		2.25		9.00		$0 = \Delta^5 y_0$
3	2.25		11.50		6	
		13.75		15.00		
4	16.00		26.50			
		40.25				
5	56.25					

We have $x_0 = 0$, $h = 1$, differentiating Newton-Gregory forward interpolation formula, we get

$$f'(x) = \frac{1}{h}\left[\Delta y_0 + \frac{2u-1}{2!}\Delta^2 y_0 + \frac{3u^2 - 6u + 2}{3!}\Delta^3 y_0 + \frac{4u^3 - 18u^2 + 22u - 6}{4!}\Delta^4 y_0 + \dots \right]$$

or

$$f'(x) = \frac{dy}{dx} = 0.25 = \frac{2u-1}{2}(-0.50) + \frac{1}{6}(3u2 - 6u + 2)(3.00)$$

$$+ \frac{1}{24}(4u3 - 18u2 + 22u - 6)(6.00)$$

or

$$\frac{dy}{dx} = u3 - 3u2 + 24.$$

At a maximum point or at a minimum point we have $\dfrac{dy}{dx} = 0$

or

$$u_3 - 3u_2 + 2u = 0$$

or

$$u(u-1)(u-2) = 0 \Rightarrow u = 0, u = 1, u = 2$$

$$\frac{d^2 y}{dx^2} = fn(x) = \frac{1}{h^2}\left[\Delta^2 y_0 + (u-1)\Delta^3 y_0 + \frac{6u^2 - 18u + 1}{12}\Delta^4 y_0 + \dots \right]$$

$$= \frac{1}{1^2}\left[-0.5 + (u-1)(3.00) + \frac{6u^2 - 18u + 1}{12} \times 6 \right]$$

Clearly $\qquad f^*(0) = \left(\dfrac{d^2y}{dx^2}\right) x_{=0} > 0$

and $\qquad f^*(2) = \left(\dfrac{d^2y}{dx^2}\right) x_{=2} > 0$

hence $f(x)$ is minimum at $x = 0$ and $x = 2$. The minimum values are $f(0) = 0$, $f(2) = 0$.

Since

$$f^*(1) = \left(\dfrac{d^2y}{dx^2}\right) x_{=1} < 0$$

$f(x)$ has a maximum at $x = 1$. The maximum value is $f(1) = 0.25$.

Example 7: Compute from following table the value of the derivative of $y = f(x)$ at $x = 1.7489$

X	1.73	1.74	1.75	1.76	1.77
Y	1.772844100	1.155204006	1.737739435	1.720448638	1.703329888

Solution: We have

$u = (1.7489 - 1.75)/0.01 = -0.11$ $x_0 = 1.75$, $h = 0.01$,

$\Delta y_0 = -.0017290797$, $\Delta^2 y_0 = .0000172047$, $\Delta^3 y_0 = -.0000001712$,

$\Delta y_{-1} = -.0017464571$, $\Delta^2 y_{-1} = .0000173774$, $\Delta^3 y_{-1} = -.0000001727$,

$\Delta^3 y_{-2} = -.0000001749$, $\Delta^4 y_{-2} = -.0000000022$

Using Newton's Forward difference formula, we get

$$f'(1.4978) \approx \dfrac{1}{0.01}\left[-0.0017290797 + (2 \times -0.11 - 1) \times \dfrac{0.0000172047}{2}\right.$$

$$\left. + (3 \times (-0.11)^2 - 6 \times -0.11 + 2) \times \dfrac{-0.0000001712}{3!}\right] = -0.173965150143$$

EXERCISE

1. Compute $f'(4)$ from the   following table

x	1	2	4	8	10
F(x)	0	1	5	21	27

[*Ans:* 2.833]

2. From the following table find x, correct to four decimal places, for which y is minimum and find this value of y

x	0.60	0.65	0.70	0.75
y	0.6221	0.6155	0.6138	0.6170

[**Ans :** 0.6137]

3. Using the following data find x for which y is minimum and find this value of y.

x	0	2	4	6
y	3	3	11	27

[**Ans:** 2.25]

4. Find the first and second derivatives of the function tabulated below at the point x = 1.5

x	1.5	2.0	2.5	3.0	3.5	4.0
Y = f(x)	3.375	7.0	13.625	24.0	38.875	59.0

[**Ans:** 4.75,9]

5. Find $\dfrac{dy}{dx}$ at x = 1.

x	1	2	3	4	5	6
Y = f(x)	1	8	27	64	125	216

also find $\dfrac{d^2y}{dx^2}$ at x = 1.

6. Find $\dfrac{dy}{dx}$ at x = 3.0 of the function tabulated below:

x	3.0	3.2	3.4	3.6	3.8	4.0
y	-14.000	-10.032	-5.296	0.256	6.672	14.000

7. Find f'(0.4) from the following table :

x	0.1	0.2	0.3	0.4
f(x)	1.10517	1.22140	1.34986	1.49182

8. Find f'(0.96) and f" (0.96) from the following table:

x	0.96	0.98	1.00	1.02	1.04
f(x)	0.7825	0.7739	0.7651	0.7563	0.7473

9. Compute the values of $\dfrac{dy}{dx}$ and $\dfrac{d^2y}{dx^2}$ at $x = 0$ from the following table:

x	0	2	4	6	8	10
f(x)	0	12	248	1284	4080	9980

10. The elevations above a datum line of seven points of roads 300 units apart are 135, 149, 157, 183, 201, 205, 193 unit. Find the gradient of the road at the middle point.

11. From the following table, calculate $\dfrac{dy}{dx}$ and $\dfrac{d^2y}{dx^2}$ at $x = 1.35$

x	1.1	1.2	1.3	1.4	1.5	1.6
y	-1.62628	0.15584	2.45256	5.39168	9.125001	3.83072

12. In a machine a slider moves along a fixed straight rod. Its distance x units along the rod is given below for various values of the time t seconds.

    Find (a) the velocity of the slider, and (b) its acceleration when $t = 0.3$ sec.

t (time in sec)	0	0.1	0.2	0.3	0.4	0.5	0.6
x (distance in units)	3.013	3.162	3.207	3.364	3.395	3.381	3.324

13. Using Bessel's formula find $f''(x)$ at $x = 0.04$ from the following table:

x	0.01	0.02	0.03	0.04	0.05	0.06
f(x)	0.1023	0.1047	0.1071	0.1096	0.1122	0.1148

14. From the following table, find die value of x for which y is minimum and find this value of y.

f	0.60	0.65	0.70	0.75
f(x)	0.0221	0.6155	0.6138	0.6170

15. From the following data, evaluate $\dfrac{dy}{dx}$ at $x = 0.00$.

X	0.00	0.05	0.10	0.15	0.20	0.25
Y	0.0000	0.10017	0.20134	0.30452	0.41075	0.52110

16. A rod is rotating in a place the following table gives die angle θ (radians) through with the rod has turned for various values of the time t seconds find (i) the angular velocity of the rod, (ii) its angular acceleration when $t = 0.6$ sec.

T	0	0.2	0.4	0.6	0.8	1.0	1.2
θ	0	0.122	0.493	1.123	2.022	3.200	4.666

17. Find the first and second derivatives of at x = 15 from the table.

X	15	17	19	21	23	25
$\sqrt{x}$	3.873	4.123	4.359	4.583	4.793	5.000

18. From the table below, for what value of x, y is minimum? Also find this value of y.

X	3	4	5	6	7	8
Y	0.205	0.240	0.259	0.262	0.250	0.224

19. Find the first and second derivatives of the function y = f(x), tabulated below at the point x = 1.1.

X	1	1.2	1.4	1.6	1.8	2.0
Y	0.00	0.1280	0.5440	1.2960	2.4320	4.0000

20. Find the Force of Mortality $u_x = -\dfrac{1}{l_x}\dfrac{dl_x}{dx}$ at x = 50 yrs, using the table below:

X	50	51	52	53
l_x	73499	72724	71753	70599

CHAPTER 8

Numerical Integration

8.1 INTRODUCTION

In this chapter we introduce numerical integration. Which is a process or technique for determining the approximate value of the integral $\int_a^b f(x)dx$ where [a, b] is finite and f(x) is not known explicitly. Geometrically this is equivalent to replacing the curve y = f(x) which encloses the area formed between the ordinates x = a and x = b and the x – axis by y = f(x).

The definite integral $\int_a^b f(x)dx$ is obtained analytically when the integrand function f(x) is either explicitly known or is in a simple form or in the form given by some rules. Many practical problems involve integrals which cannot be solved by mathematical analysis, or the function may be too complicated for efficient computation.

As in the case or numerical differentiation; integration problems are worked out by first approximating the integral by a suitable interpolation formula. The approximation is usually known as numerical integration or quadrature. A quadrature formula is said to be of closed type if lower limit and upper limit as of the integration are taken interpolating as points; otherwise it is referred to be of open type.

A mechanical quadrature formula is said to have a degree of precision k (k being a positive integer), if it is exact i.e., the error is zero for an arbitrary polynomial of degree k + 1 for which it is not exact, i.e. the error is not zero. Although numerical integration can often yield exact solutions especially if the function under consideration is a simple polynomial, more often, our solution will only be approximate. This is especially true when integrating combinations of transcendental functions for which no solution can be found.

To find the integral $\int_a^b f(x)dx$ the interval [a, b] of integration is divided into number of sub–interval and then add the results. The formula thus obtained is called composite rule corresponding to that quadrature formula.

8.2 GAUSS - LEGENDRE QUADRATURE FORMULA

Consider the definite Integral $I = \int_a^b f(x)dx$

Let $y = f(x)$ take the values $y_0, y_1, y_2, \ldots, y_n$ corresponding to the values $x = x_0, x_1, x_2, \ldots, x_n$ and divide the interval $[a, b]$ into n equal sub–intervals of width h so that $x_0 = a$, $x_1 = x_0 + h$ $x_2 = x_0 + 2h, \ldots, x_n = x_0 + nh = b$.

Now $\qquad I = \int_a^b f(x)dx = \int_{x_0}^{x_0+nh} f(x)dx$

Let $x = x_0 + rh$ then $dx = h\, dr$. Also when $x = x_0$, we have $r = 0$, when $x = x_0 + nh$ we have $r = n$

Therefore Now $I = \int_0^n f(x_0 + rh)h\,dr$

From Newton's forward interpolation formula, we have

$$I = h\int_0^n \left[y_0 + r\,\Delta y_0 + \frac{r(r-1)}{2!}\Delta^2 y_0 + \frac{r(r-1)\,(r-2)}{3!}\Delta^3 y_0 + \ldots\right]dr$$

Integrating term by term, we get

$$I = h\left[y_0 r + \frac{r^2}{2}\Delta y_0 + \frac{1}{2!}\left(\frac{r^3}{3} - \frac{r^2}{2}\right)\Delta^2 y_0 + \frac{1}{3!}\left(\frac{r^4}{4} - r^3 + r^2\right)\Delta^3 y_0 + \ldots\right] \text{ where}$$

r takes values 0 and n

Hence we get

$$I = nh\left[y_0 + \frac{n}{2}\Delta y_0 + \frac{1}{2!}\left(\frac{n^2}{3} - \frac{n}{2}\right)\Delta^2 y_0 + \frac{1}{3!}\left(\frac{n^3}{4} - n^2 + n\right)\Delta^3 y_0 + \frac{1}{4!}\right.$$

$$\left(\frac{n^4}{5} - \frac{3n^3}{2} + \frac{11n}{3} - 3\right)\Delta^3 y_0 + \ldots$$

The above formula is known as general quadrature formula or Newton-Cotes quadrature formula

It can also be written as follows:

$$I = \int_a^b f(x)dx =$$

$$h\left[ny_0 + \frac{n^2}{2}\Delta^2 y_0 + \frac{2n^3 - 3n^2}{12}\Delta^2 y_0 + \frac{n^4 - 4n^3 + 4n^2}{24}\Delta^3 y_0\right.$$

$$\left. + \frac{6n^5 - 45n^4 + 110n^3 - 90n^2}{720}\Delta^4 y_0 + \ldots\right] \qquad\qquad \ldots\ldots(8.1)$$

We now deduce three quadrature formulae

Trapezoidal Rule: By substituting n = 1 in the quadrature formula and neglecting second and higher order differences we get

$$\int_{X_0}^{x_0+h} f(x)dx = h\left[\,y_0+\tfrac{1}{2}\Delta y_0\right]=\tfrac{h}{2}\left[\,y_0+y_1\right]$$

Similarly we get

$$\int_{X_0+h}^{x_0+2h} f(x)dx = h\left[\,y_1+\tfrac{1}{2}\Delta y_1\right]=\tfrac{h}{2}\left[\,y_1+y_2\right]$$

. . .   .

$$\int_{X_0+(n-1)h}^{x_0+nh} f(x)dx = h\left[\,y_{n-1}+\tfrac{1}{2}\Delta y_n\right]=\tfrac{h}{2}\left[\,y_{n-1}+y_n\right]$$

Adding these integrals we get

$$\int_{X_0}^{x_0+h} f(x)dx + \int_{X_0+h}^{x_0+2h} f(x)dx +\ldots+ \int_{X_0+(n-1)h}^{x_0+nh} f(x)dx$$

$$= \int_{X_0}^{x_0+nh} f(x)dx = \tfrac{h}{2}\left[\,y_0+ 2(y_1+y_2+\ldots+ y_n)\right]$$

This formula is known as Trapezoidal rule

The error term is $\dfrac{(b-a)^3}{12} f''(\xi)$

Where $x_0 < \xi < x_1$.

We assume here $f''(\xi)$ is the largest value of the second order derivatives.

This rule can be applied for any number of equal sub-intervals.

The degree of precision of this formula is one.

Example 1: Estimate the value of $\int_{\frac{\pi}{4}}^{\frac{\pi}{2}} \sin x \, dx$ by using trapezoidal rule

Solution: let $h = \dfrac{\pi}{4}$

Then $\quad \int_{\frac{\pi}{4}}^{\frac{\pi}{2}} \sin dx = \dfrac{1}{2}\dfrac{\pi}{4}\left[\sin\dfrac{\pi}{4} + \sin\dfrac{\pi}{2}\right] = \dfrac{1}{2}\dfrac{\pi}{4}\left[\dfrac{1}{\sqrt{2}}+1\right] = 0.6704$

Example 2: Evaluate $\int_{0}^{\pi} \sin dx$ taking 4 intervals

Solution: we have n = 4 and f(x) = sin x $h = \dfrac{\pi-0}{4} = \dfrac{\pi}{4}$

x	0	$\dfrac{\pi}{4}$	$\dfrac{\pi}{2}$	$\dfrac{3\pi}{4}$	π
Sin x	0	$\dfrac{1}{\sqrt{2}}$	1	$\dfrac{1}{\sqrt{2}}$	0

Applying Trapezoidal rule we get

$$\int_0^\pi \sin \, dx = \frac{h}{2}\,[\,y_0 + y_4 +(\,y_1+y_2+y_3)] \; = \frac{1}{2}\frac{\pi}{4}\,[0 + 0 + 2\,(\tfrac{1}{\sqrt{2}} +1+\tfrac{1}{\sqrt{2}}) = 1.89$$

Geometrical Significance:

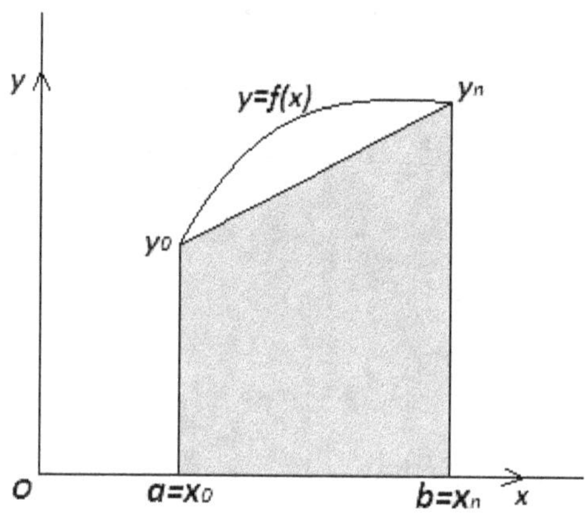

Fig. 8.1 Trapezoidal method

The geometrical significance of trapezoidal rule lies on the fact that the curve $y = f(x)$ is replaced by n straight lines joining the points (x_0,y_0) and $(x_1,y_1),(x_1,y_1)$ and (x_2,y_2).... as so on. The area is bounded by the curve $y=f(x)$, the ordinates $x = x_0,....x = x_n$ and the x-axis and the value is equal to total area of the n trapeziums obtained. Trapezoidal rule is also known as trapezium rule.

The error involved in trapezoidal rule is of order $O(h^2)$

Simpson's one – third rule:

By taking n = 2, in the quadrature formula and neglecting terms containing $\Delta^3 y_0, \Delta^4 y_0 \ldots we\;\; get$

$$\int_{x_0}^{x_0 +2h} f(x)\,dx = 2h\,[\,y_0 + \Delta y_0 + \tfrac{1}{6}\Delta^2 y_0\,]$$

But $\qquad\qquad \Delta y_0 = y_1 - y_0 \,,\; \Delta^2 y_0 = y_2 - 2y_1 + y_0\,,$ therefore

$$\int_{x_0}^{x_0 +2h} f(x)\,dx = \frac{h}{3}\,[\,y_0 + 4\,y_1 + y_2]$$

Similarly we get $\qquad \int_{x_0+2h}^{x_0 +4h} f(x)\,dx = \frac{h}{3}\,[\,y_2 + 4\,y_3 + y_4]$

$$\ldots \; \ldots \qquad\qquad \ldots \qquad\qquad\qquad \ldots$$

$$\int_{x_0+(n-2)h}^{x_0+nh} f(x)dx = \frac{h}{3}[\ y_{n-2}+4\ y_{n-1}+y_n]\ \text{adding all these integrals}$$

we get

$$\int_{x_0}^{x_0+nh} f(x)dx = \frac{h}{3}[(y_0+y_n)+4\ (y_1+y_3+\ldots+y_{n-1})+2(y_2+y_4+\ldots$$

$$+\ y_{n-2})]$$

$$\int_{x=0}^{n} f(x)dx = \frac{h}{3}[(y_0+y_n)+4\ (y_1+y_3+\ldots+y_{n-1})+2(y_2+y_4+\ldots+y_{n-2})]$$

This formula is known as Simpson's $\frac{1}{3}$ rule

The error term is $E_s = -\dfrac{h^5}{90} f^{iv}(\xi)$

Where $x_0 < \xi < x_2$

We can use Simpson's one third rule only when the total number of equal sub-intervals of (a,b) is even i.e., the total number of points is odd.

The degree of precision is three for Simpson's one third rule.

Geometrical Interpretation of Simpson's one third

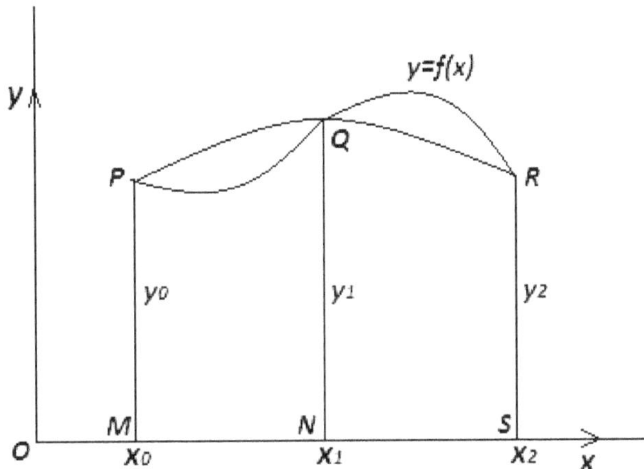

Fig. 8.2 Simpson's one-third rule

The formula is also known as parabolic rule. In Simpson's one-third rule, the whole area bounded by the curve $y = f(x)$, $x-$ axis and lines $x = a$

Example 3: Evaluate $\int_0^\pi \sin x\ dx$ dividing $[0,\ ]$ into 4 equal intervals

Solution: We have n = 4, therefore h $= \dfrac{\pi-0}{4} = \dfrac{\pi}{4}$

We get the following values

x	0	$\dfrac{\pi}{4}$	$\dfrac{\pi}{2}$	$\dfrac{3\pi}{4}$	π
Sin x	0	0.71	1	0.71	0

Hence we get $\int_0^\pi \sin x \, dx = \dfrac{h}{3}[(y_0+y_n) + 4(y_1+y_3+\ldots+y_{n-1})$
$$+ 2(y_2+y_4+\ldots+y_{n-2})]$$

$$\int_{x=0}^n f(x)dx = \frac{1}{3}\frac{\pi}{4}[0+4(0.71+0.71)+2(1)+0] = 0.21$$

Evaluate $\quad \int_0^6 \dfrac{dx}{1+x^2}$ by using Simpson's $\dfrac{1}{3}$ rule

Solution: By dividing the interval [0, 6] six equal parts we get

x	0	1	2	3	4	5	6
f(x)	1	0.5	0.2	0.1	0,0588	0.0385	0.027

By Simpson's rule

$$\int_0^6 \frac{dx}{1+x^2} = \frac{h}{3}[(y_0+y_6) + 4(y_1+y_3+\ldots+y_5)+ 2(y_2+y_4)]$$

$$= \frac{1}{3}[(1+0.027) + 4(0.5+0.1+0.9385) + 2(0.2+0.0588)]$$

$$= \frac{1}{3}[1.027 + 2.554 + 0.5176]$$

$$= 1.3662$$

Use Simpson's rule to evaluate $\int_0^3 \dfrac{1}{1+x} dx$ by taking $n = 6$

Solution: We have $h = \dfrac{3-0}{6} = \dfrac{1}{2}$

x	0	0.5	1.0	1.5	2.0	2.5	3
y	1	0.667	0.50	0.400	0.333	0.286	0.250

by Simpson's rule

$$\int_0^3 \frac{1}{1+x} dx = \frac{h}{3}[(y_0+y_6) + 4(y_1+y_3+\ldots+y_5) + 2(y_2+y_4)]$$

$$= \frac{0.5}{3}[(1+0.250) + 4(0.667+0.400+0.286) + 2(0.500+0.333]$$

$$= \frac{0.5}{3}[(1.250) + 5.412 + 1.666] = 1.388$$

Simpson's three-eight rules

We assume t

When the interval [a, b] is divided into of six equidistant subintervals then we can use this rule i.e. n = 6, 12, 18, 24 and so on. The degree of precision of Weddle, s rule is five.

Composite eddies' Rule: In case the interval [a, b] is divided into 6m (=n) equal subintervals by 6m + 1 points $a = x_0, x_1 \ldots x_{6m} = b,$. Then we apply the Weddle's rule for numerical integration to each of the subintervals $[x_0, x_6], [x_6, x_{12}], \ldots [x_{6m-6}, x_{6m}]$ and get the composite Weddle's rule for numerical integration as

$$\int_{x_0}^{x_0} y\,dx = \frac{3h}{10}[y_0 + 5y_1 + y_2 + 6y_3 + y_4 + 5y_5 + y_6]$$

$$\int_{x_0}^{x_{12}} y\,dx = \frac{3h}{10}[y_6 + 5y_7 + y_8 + 6y_9 + y_{10} + 5y_{11} + y_{12}]$$

$$\ldots = \quad \ldots\ldots\ldots$$

$$\int_{x_{6m-6}}^{6m} y\,dx = \frac{3h}{10}[y_{6m-6} + 5y_{6m-5} + y_{6m-4} + 6y_{6m-3} + y_{6m-2} + 5y_{6m-1} + y_{6m}]$$

$$I_W^C = \frac{3h}{10}[y_0 + y_n + (y_2 + y_4 + y_6 + y_8 + \ldots y_{n-2})$$

$$+ 5(y_1 + y_7 + y_{13} + \ldots + y_{n-1}) + 6(y_3 + y_9 + y_{15} + \ldots + y_{n-3})$$

$$+ 2(y_6 + y_{12} + \ldots + y_{n-6})$$

And $E_W^C = -\dfrac{nh^7}{840} f^{ui}(\xi)$ $\qquad\qquad\qquad$ .....(8.2)

$$x_0 < \xi < x_n$$

8.3 NEWTON'S COTES FORMULA [CLOSED TYPE]

The interpolating points $x_i (i = 0, 1, \ldots n)$ are equi-spaced of step length $h = \dfrac{b-a}{n}$ so that

$x_i = x_0 + ih \, (h > 0, i = 0, 1, \ldots n)$ and $x_0 = a, x_n = b.$ Then if we integrate by replacing it with Lagrange's interpolating polynomial then it is called Newton's – Cote's formula for numerical integration.

$$I = \int_a^b f(x)dx = \int_a^b \sum_{r=0}^{n} \frac{w(x)}{w'(x_r)(x - x_r)} y_r dx \sqrt{a^2 + b^2}$$

$$= \sum_{r=0}^{n} \int_{x_0}^{x_n} \frac{(x - x_0)\ldots(x - x_n)}{(x_r - x_0)\ldots(x_r - x_{r-1})(x_r - x_{r+1})\ldots(x_r - x_n)(x - x_r)} y_r \, dx \qquad .....(8.3)$$

Let us substitute $x = x_0 + sh$ and $x - x_i = (s-i)h$. Hence (8.3) becomes,

$$I = \sum_{r=0}^{n} \int_{x_0}^{x_0+nh} \frac{sh(s-1)h...(s-n)h}{rh(r-1)h...1.h(-1)h...(-n+1)h(s-r)h} dx$$

$$= \sum_{r=0}^{n} (-1)^{n-r} \frac{h}{r!(n-r)!} \int_0^n \frac{s(s-1)...(s-n)}{s-r} ds\, y_r$$

$$= \sum_{r=0}^{n} H_r^{(n)} y_r \qquad\qquad .....(8.4)$$

Where

$$H_r^{(n)} = \frac{(-1)^{n-r}(b-a)}{nr!(n-r)!} \int_0^n \frac{s(s-1)....(s-n)}{s-r} ds \qquad\qquad .....(8.5)$$

$$H_r^{(n)} = (b-a)C_r^{(n)} \qquad\qquad .....(8.6)$$

$C_r^{(n)}$ are called Cote's co-efficient.

$$I = (b-a)\sum_{r=0}^{n} C_r^{(n)} y_r \qquad\qquad .....(8.7)$$

The above formula is known as Newton's Cote's formula. Since a and b are included it is called closed type.

The error committed in Newton's Cote's formula is

$$E = \int_{x_0}^{x_n} (x-x_0)(x-x_1)...(x-x_n) \frac{f^{n+1}(\xi)}{(n+1)!} dx \quad (x_0 < \xi < x_n)$$

$$= \frac{h^{n+2}}{(n+1)!} \int_0^n s(s-1)...(s-n)f^{n+1}(\xi)dx \qquad\qquad .....(8.8)$$

We now deduce the three quadrature formulae. We here do not deduce the formulae explicitly.

(a)   Trapezoidal Rule:

$n = 1$

$$H_0^{(1)} = \frac{(-1)^{1-0}h}{0!1!} \int_0^1 \frac{s(s-1)}{s} ds = \frac{h}{2}$$

Therefore

$$I_T = H_0^{(1)} y_0 + H_1^{(1)} y_1 = \frac{h}{2}(y_0 + y_1) \qquad\qquad .....(8.9)$$

The composite trapezoidal rule:

$$I_T^C = \frac{h}{2}[y_0 + 2(y_1 + ... + y_{n-1}) + y_n] \qquad\qquad .....(8.10)$$

(b) Simpson's one third rule:

$$n=2, h=\frac{b-a}{2}$$

Therefore

$$I_T = \frac{h}{2}[y_0 + 2(y_1 + \ldots + y_{n-1}) + y_n] \qquad \ldots(8.11)$$

The composite trapezoidal rule:

$$I_T^C = \frac{h}{2}[y_0 + 2(y_1 + \ldots + y_{n-1}) + y_n] \qquad \ldots(8.12)$$

$$n=2, h=\frac{b-a}{2}$$

$$H_0^{(2)} = \frac{(-1)^{2-0}h}{2!0!} \int_0^2 \frac{s(s-1)(s-2)}{s} ds = \frac{h}{3}$$

$$H_1^{(2)} = \frac{(-1)^{2-1}h}{1!1!} \int_0^2 \frac{s(s-1)(s-2)}{s-1} ds = \frac{4h}{3}$$

$$H_2^{(2)} = \frac{(-1)^{2-2}h}{2!0!} \int_0^2 \frac{s(s-1)(s-2)}{s-2} ds = \frac{h}{3}$$

Hence

$$I_S = H_0^{(2)}y_0 + H_1^{(2)}y_1 + H_2^{(2)}y_2$$

$$= \frac{h}{3}y_0 + \frac{4h}{3}y_1 + \frac{h}{3}y_2$$

$$= \frac{h}{3}(y_0 + 4y_1 + y_2)$$

$$I_S^C = \frac{h}{3}[y_0 + 4(y_1 + y_3 + y_5\ldots) + 2(y_2 + y_4\ldots) + y_n]$$

(c) Weddles's rule

$$n = 6, h = \frac{b-a}{6}$$

In the interval [a, b] we get

$$H_0^6 = 3h/10, \; H_1^6 = 3h/2, \; H_2^6 = 3h/10, \; H_3^6$$

$$= 9h/5, H_4^6 = 3h/10, H_5^6 = 3h/2, H_6^6 = 3h/10 \quad \text{we get}$$

$$IW = \frac{3h}{10}y_0 + \frac{3h}{2}y_1 + \frac{3h}{10}y_2 + \frac{9h}{5}y_3 + \frac{3h}{10}y_4 + \frac{3h}{2}y_5 + \frac{3h}{10}y_6$$

$$= \frac{3h}{10}[y_0 + 5y_1 + y_2 + 6y_3 + y_4 + 5y_5 + y_0]$$

and composite Weddle's rule is

$$I_w^c = \frac{3h}{10}[(y_0 + y_n) + (y_2 + y_4 + \ldots + y_{n-2}) + 5(y_1 + y_5 + \ldots + y_{n-5} + y_{n-1}) + 6(y_3 + y_9 + y_{15} + \ldots + y_{n-3})$$

$$+ 2(y_6 + y_{12} + y_{18} + \ldots + y_{n-6})]$$

8.4 NEWTON COTES CO-EFFICIENT (CLOSE TYPE)

n	C_0^n	C_1^n	C_2^n	C_3^n	C_4^n	C_5^n	C_6^n
1	$\frac{1}{2}$	$\frac{1}{2}$					
2	$\frac{1}{6}$	$\frac{4}{6}$	$\frac{1}{6}$				
3	$\frac{1}{8}$	$\frac{3}{8}$	$\frac{3}{8}$	$\frac{1}{8}$			
4	$\frac{7}{90}$	$\frac{32}{90}$	$\frac{12}{90}$	$\frac{32}{90}$	$\frac{7}{90}$		
5	$\frac{19}{288}$	$\frac{75}{288}$	$\frac{50}{288}$	$\frac{50}{288}$	$\frac{75}{288}$	$\frac{19}{288}$	
6	$\frac{41}{840}$	$\frac{216}{840}$	$\frac{27}{840}$	$\frac{272}{840}$	$\frac{27}{840}$	$\frac{216}{840}$	$\frac{41}{840}$

8.5 ROMBERG INTEGRATION

This method is highly suitable for computer implementation. It eliminates the error in the trapezoidal of integration

An estimate of the maximum error in an integral obtained by the trapezoidal is $\frac{-1}{12}h\Delta^2 f_0$, but the actual error is of the form $a_1h^2 + a_2h^4 + a_3h^6 + \ldots$

Let

$$I = \int_a^b f(x)dx, h = b - a$$

$$I = I_{est} + a_1h^2 + a_2h^4 \qquad \ldots(8.13)$$

Where Iest is the estimated value of the integral.

Using the trapezoidal rule, we obtain the estimates of I for h, h/2, h/4,….

Let I_1, I_2... denote these estimates. Let us neglect h^4, h^6.... (if h is not large) we get

$$I = I_{j1} + a_1 \frac{h^2}{2^n} \qquad \qquad .....(8.14)$$

Where j = 1,2,... ; n = 0, 1, 2....

Eliminating a_i, we get the improved values of I which take general for

$$I_{j2} = \frac{4I_{j1} - I_{(j-1)1}}{3} j = 2,3,4... \qquad \qquad .....(8.15)$$

Now taking $I = I_{est} + a_2 h^4$ and proceedings as above we get the new set of values I given by

$$I_{j3} = \frac{16I_{j2} - I_{(j-1)2}}{15} j = 3,4,... \qquad \qquad .....(8.16)$$

Continuing this process, it can be shown that in case of trapezoidal rule.

$$I_{jn} = \frac{4^{n-1} I_{j(n-1)} - I_{(j-1)(n-1)}}{4^{n-1} - 1} j = n, n+1,... \qquad \qquad .....(8.17)$$

By the continuous application of the process we can form a sequence of columns with error terms of increasing order:

h^2	h^4	h^6	h^8	...
I_{11}				
I_{21}	I_{21}			
I_{31}	I_{32}	I_{33}		
I_{41}	I_{42}	I_{43}	I_{44}	
⋮	⋮	⋮	⋮	⋮

Romberg's method is suitable for computer implementation

Romberg's method is suitable for computer implementation since successive values can be compared to check when the process converges.

Example 4: Find the value of $\log 2^{1/3}$ from $\int_0^1 \frac{x^2}{1+x^3} dx$ using Simpson's 1/3rd rule where

h = 0.25

Solution: We divide the interval [0,1] in 4 sub-i

$$x_0 = 0, x_1 = 0.25, x_2 = 0.50, x_3 = 0.75, x_4 = 1.00.$$

$$f(x) = \frac{x^2}{1+x^3}$$

$$f(x_0) = 0$$

$$f(x_1) = 0.06154$$

$$f(x_2) \quad = 0.22222$$

$$f(x_3) \quad = 0.395604$$

$$f(x_4) \quad = 0.500000$$

Now,

$$I_S \quad = \frac{0.25}{3}[0 + 4(0.06154 + 0.395604) + 2 \times (0.22222) + 0.50000]$$

$$= 0.231085$$

Now,

$$\int_0^1 \frac{x^2}{1+x^3}dx \quad = \frac{1}{3}\log(1+x^3)]_0^1$$

$$= \frac{1}{3}\log 2$$

$$\log 2^{1/3} \quad = 0.231085$$

Example 5: Evaluate $\int_0^1 \frac{dx}{1+x}$ by Simpson's composite rule taking eleven ordinates and hence find the value of $\log_e 2$ correct up to five significant figures.

Solution: Here $f(x) = \frac{1}{1+x}$ $a = 0, b = 1, n = 10 h = 0.1$

Computational table is:

x	$y_i(i=0,10)$	$y_i \mid i = 1,3,5,7,9$	$y_i \mid i = 2,4,6,8$
0			
0.1			
0.2			
0.3		0.9090909	
0.4			0.8333333
0.5		0.76923076	0.7142857
0.6		0.66666666	0.625
0.7		0.55882353	0.555555
0.8		0.52631578	
0.9			
1.0			
Toal	1.5000000	3.45953934	2.7281745

$$I_s = \frac{0.1}{3}[1.500000 + 4 \times (3.45953934) + 2 \times (2.7281745)]$$

$$= \frac{0.1}{3} \times 20.79450636$$

$$= .693150212$$

$$= .69315$$

(Correct up to 5 significant figures)

Example 6: Compute the value of π from the formula $\dfrac{\pi}{4} = \int_0^1 \dfrac{dx}{1+x^2}$ using composite Trapezoidal rule with 10 subinterval and compare it with the exact

Solution: n=10, a = 0, b = 1, $h = \dfrac{1-0}{10} = 0.1 \, f(x) = \dfrac{1}{1+x^2}$

Example 7: Calculate the value $\int_0^1 \dfrac{x}{1+x} \, dx$ correct up to three significant figures taking six intervals by Trapezoidal rule.

Solution: Here we have $f(x) = \dfrac{x}{1+x}$,

$$a = 0, \ b = 1 \text{ and } n = 6$$

$$\therefore \qquad h = \frac{b-a}{n} = \frac{1-0}{6} = \frac{1}{6}$$

x	0	1/6	2/6	3/6	4/6	5/6	6/6 = 1
y-f(x)	0.00000	0.14286	0.25000	0.33333	0.40000	0.45454	0.50000
	y_0	y_1	y_2	y_3	y_4	y_5	y_6

The Trapezoidal rule can be written as

$$I = \frac{h}{2} \, [(y_0 + y_6) + 2 \, (y_1 + y_2 + y_3 + y_4 + y_5)]$$

$$= [\frac{1}{12} \, (0.00000 + 0.50000) \, 2 + (0.14286 + 0.2500 + 0.33333 +$$

$$0.40000 + 0.45454)]$$

$$= 0.30512$$

$$\therefore \qquad I = 0.30512, \text{ correct to three significant figures.}$$

Example 8: Find the value of $\int_0^1 \dfrac{dx}{1+x^2}$ taking 5 subintervals by Trapezoidal rule, correct to five significant figures.

Also compare it with its exact value.

Solution: Here $f(x) = \dfrac{x}{1+x^2}$

$a = 0$, $b = 1$ and $n = 5$.

$\therefore$
$$h = \frac{1-0}{5} = \frac{1}{5} + 0.2$$

x	0.0	0.2	0.4	0.6	0.8	1
y-f(x)	1.000000	0.961538	0.862069	0.735294	0.609756	0.500000
	y_0	y_1	y_2	y_3	y_4	y_5

Using trapezoidal rule we get

$$I = \int_0^1 \frac{dx}{1+x^2} = \frac{h}{2} \left[(y_0 + y_5) + 2\,(y_1 + y_2 + y_3 + y_4) \right]$$

$$= \frac{0.2}{2} \left[(1.000000 + 0.500000)\,2 + (0.961538 + 0.862069 + \right.$$

$$\left. 0.735294 + 0.609756) \right] = 0.783714,$$

$\therefore$ $I = 0.78373$, correct to five significant figures.

The exact value

$$= \int_0^1 \frac{1}{1+x^2} dx = \left[\tan^{-1} x \right]_0^1$$

$$= \tan^{-1} 1 - \tan^{-1} 0 = \frac{\pi}{4} = 0.7853981$$

$$\int_0^1 \frac{1}{1+x^2} dx = 0.78540$$

Correct to five significant figures.

Hence we have the error is = 0.78540 - 0.78373 = 0.00167

and Absolute error = 0.00167.

Example 9: Find the value of $\int_{1}^{5} \log_{10} x\, dx$, taking 8 subintervals correct to four decimal places by Trapezoidal rule.

Solution: Here we have

$$f(x) = \log 10\ x$$

$$a = 1, b = 5 \text{ and } n = 8,$$

$$\therefore \quad h = \frac{b-a}{n} = \frac{5-1}{8} = 0.5$$

x	1.0	1.5	2.0	2.5	3.0	3.5	4.0	4.5	5.0
−f(x)	0.00000	0.17609	0.30103	0.39794	0.47712	0.54407	0.60206	0.65321	0.69897
	y_0	y_1	y_2	y_3	y_4	y_5	y_6	y_7	y_8

Using Trapezoidal rule we can write

$$I = \frac{h}{2}\left[(y_0 + y_8) + 2\,(y_1 + y_2 + y_3 + y_4 + y_5 + y_6 + y_7)\right]$$

$$= \frac{0.5}{2} - [(0.00000 + 0.69897) + 2\,(0.17609 + 0.30103 + 0.39794$$

$$+\ 0.47712 + 0.54407 + 0.60206 + 0.65321)]$$

$$= 1.7505025$$

or $\qquad I = \int_{1}^{5} \log_{10} x\, dx = 1.75050.$

Example 10: Find the value $\int_{0}^{0.6} e^{x}\, dx$, taking $n = 6$, correct to five significant figures by Simpson's one-third rule.

Solution: We have $f(x) = e^{x}$

$$a = 0, b = 0.6, n = 6.$$

$$h = \frac{b-a}{n} = \frac{0.6 - 0}{6} = 0.1$$

x	0.0	0.1	0.2	0.3	0.4	0.5	0.6
y = f(x)	1.0000	1.10517	1.22140	1.34986	1.49182	1.64872	1.82212
	y_0	y_1	y_2	y_3	y_4	y_5	y_6

The Simpson's rule is .

$$I = \frac{h}{3}\ [(y_0 + y_6) + 4\ (y_1 + y_3 + y_5) + 2(y_2 + y_4)]$$

$$= \frac{0.1}{3}\ [(1.00000 + 1.82212) + 4(1.10517 + 1.34986 + 1.64872) +$$

$$2(1.22140 + 1.49182)]$$

$$= \frac{0.1}{3}\ [(2.82212) + 4(4.10375) + 2(2.71322)]$$

$$= 0.8221186 = 0.82212$$

or $\qquad I = 0.82212.$

Example 11: The velocity of a train which starts from rest is given by the following table, the time being reckoned 'n minutes from the start and the speed in km/hour.

t (minutes)	2	4	6	8	10	12	14	16	18	20
v (km/hr)	16	28.8	40	46.4	51.2	32.0	17.6	8	3.2	0

Estimate approximately the total distance run in 20 minutes.

Solution: $\qquad\qquad v = \dfrac{ds}{dt} \Rightarrow ds = v.\ dt$

$\Rightarrow \qquad\qquad\qquad \int ds = \int v.dt$

$$s = \int_{0}^{20} v.dt$$

The train starts from rest,

$\therefore$ The velocity $v = 0$ when $t = 0.$

The given table of velocities can be written as:

t	0	2	4	6	8	10	12	14	16	18	20
v	0	16	28.8	40	46.4	51.2	32.0	17.6	8	3.2	0
	y_0	y_1	y_2	y_3	y_4	y_5	y_6	y_7	y_8	y_9	y_{10}

$$h = \frac{2}{60}\,hrs = \frac{1}{30}\,hrs$$

The Simpson's rule is

$$s = \int_0^{20} v.dt = \frac{h}{3}\left[(y_0 + y_{10}) + 4(y_1 + y_3 + y_5 + y_7 + y_9) + 2(y_2 + y_4 + y_6 + y_8)\right]$$

$$= \frac{1}{30 \times 3}\left[(0 + 0) + 4(16 + 40 + 51.2 + 17.6 + 3.2) + 2(28.8 + 46.4 + 320 + 8)\right]$$

$$= \frac{1}{90}\left[0 + 4 \times 128 + 2 \times 115.2\right] = 8.25 \text{km}.$$

Hence the distance runs by the train in 20 minutes = 8.25 km.

Example 12: A tank in discharging water through an orifice at a depth of x meter below the surface of the water whose area is Am^2. The following are the values of x for the corresponding values of A.

A	1.257	1.39	1.52	1.65	1.809	1.962	2.123	2.295	2.462	2.650	2.827
x	1.50	1.65	1.80	1.95	2.10	2.25	2.40	2.55	2.70	2.85	3.00

Using the formula $(0.018)\ T = \int_{1.5}^{3.0} \frac{A}{\sqrt{x}}dx$, calculate T the time in seconds for the level of the water to drop from 3.0 m to 15 m above the orifice.

Solution: We have h = 0.15,

The table of values of x and the corresponding values of $\frac{A}{\sqrt{x}}$ is

x	1.50	1.65	1.80	1.95	2.10	2.25	2.40	2.55	2.70	2.85	3.00
$y = \dfrac{A}{\sqrt{x}}$	1.025	1.081	1.132	1.182	1.249	1.308	1.375	1.438	1.498	1.571	1.632

Using Simpson's rule, we get

$$\int_{1.5}^{3.0} \frac{A}{\sqrt{x}}dx = \frac{h}{3}\left[(y_0 + y_{10}) + 4(y_1 + y_3 + y_5 + y_7 + y_9) + 2(y_2 + y_4 + y_6 + y_8)\right]$$

$$= \frac{0.15}{3}\left[(1.025 + 1.632) + 4(1.081 + 1.182 + 1.308 + 1.438 + 1.571)\right.$$

$$\left. + 2(1.132 + 1.249 + 1.375 + 1.498)\right] = 1.9743$$

or $\qquad \int_{1.5}^{3} \frac{A}{\sqrt{x}}\ dx = 1.9743.$

Using aw formula (0.018} $T = \int_{1.5}^{3} \dfrac{A}{\sqrt{x}}\, dx$, we get

$$(0.018)T = 1.9743$$

$$T = \dfrac{1.9743}{0.018} = 110 \text{ sec (approximately)}$$

$$T = 110 \text{ sec. Hence}$$

Example 13: Evaluate $\int_{0}^{1} \dfrac{1}{1+x^2}\, dx$, by taking seven ordinates.

Solution: We have $n + 1 = 7 \Rightarrow n = 6$

The points of division are $0, \dfrac{1}{6}, \dfrac{2}{6}, \dfrac{3}{6}, \dfrac{4}{6}, \dfrac{5}{6}, 1.$

x	0	1/6	2/6	3/6	4/6	5/6	1
$y = \dfrac{1}{1+x^2}$	1.0000000	0.9729730	0.9000000	0.8000000	0.6923077	0.5901639	0.5000000

Here $h = \dfrac{1}{6}$, the Simpson's three-eighths rule is

$$I = \dfrac{3h}{8}\, [(y_0 + y_6) + 3(y_1 + y_2 + y_4 + y_5) + 2(y_3)]$$

$$= \dfrac{3}{6 \times 8}\, [(1 + 0.5000000) + 3(0.9729730 + 0.9000000 + 0.6923077$$

$$+ 0.5901639 + 2\,(0.8000000)]$$

$$= \dfrac{1}{16}\, [15000000 + 9.4663338 + 1.6000000]$$

$$= 0.7853959.$$

Example 14:   Calculate $\int_{0}^{\pi/2} e^{\sin x}\, dx$, correct to four decimal places.

Solution: We divide the range in three equal points with the division points

$$x_0 = 0,\ x_1 = \dfrac{\pi}{6},\ x_2 = \dfrac{\pi}{3},\ x_3 = \dfrac{\pi}{2}$$

where $\qquad h = \dfrac{\pi}{6}$

The table of values of the function is

x	0	$\dfrac{\pi}{6}$	$\dfrac{\pi}{3}$	$\dfrac{\pi}{2}$
$Y = e^{\sin x}$	1	1.64872	2.36320	2.71828
	y_0	y_1	y_2	y_3

By Simpson's three-eighths rule, we get

$$I = \int_0^{\pi/2} e^{\sin x}\, dx = \frac{3h}{8}\left[(y_0 + y_1) + 3(y_1 + y_2)\right]$$

$$= \frac{3}{8}\frac{\pi}{6}\left[(1 + 2.71828) + 3(1.64872 + 236320)\right]$$

$$= \frac{\pi}{16}\left[(3.71828 + 12.03576)\right] = 0.091111$$

$$I = \int_0^{\pi/2} e^{\sin x}\, dx = 0.091111.$$

Example 15: Compute the integral $\displaystyle\int_0^{\pi/2} \sqrt{1 - 0.162\sin^2\phi}\,d\phi$ by Weddle's rule.

Solution: Here, we have

$$y = f(\phi) = \sqrt{1 - 0.162\sin^2\phi}\ ,$$

$$a = 0,\ b = \frac{\pi}{2}$$

Taking n = 12, we get

$$h = \frac{b - a}{n} = \frac{\dfrac{\pi}{2} - 0}{12} = \frac{\pi}{24}$$

f	y = f(f)		f	y = f(f)	
0	1.000000	y_0	$\dfrac{6\pi}{24}$	0.958645	y_6
$\dfrac{\pi}{24}$	0.998619	y_1	$\dfrac{7\pi}{24}$	0.947647	y_7
$\dfrac{2\pi}{24}$	0.994559	y_2	$\dfrac{8\pi}{24}$	0.937283	y_8
$\dfrac{3\pi}{24}$	0.988067	y_3	$\dfrac{9\pi}{24}$	0.928291	y_9

f	y = f(f)		f	y = f(f)	
$\dfrac{4\pi}{24}$	0.979541	y_4	$\dfrac{10\pi}{24}$	0.9213322	y_{10}
$\dfrac{5\pi}{24}$	0.969518	y_5	$\dfrac{11\pi}{24}$	0.916930	y_{11}
			$\dfrac{12\pi}{24} = \dfrac{\pi}{2}$	0.915423	y_{12}

By Weddle's rule, we have

$$I = \int_0^{\pi/2} \sqrt{1 - 0.162\sin^2 f}\, df$$

$$= \frac{3h}{10}\left[(y_0 + y_{12}) + 5(y_1 + y_5 + y_7 + y_{11})\right] + \frac{3h}{10}\left[(y_2 + y_4 + y_8 + y_{10})\right.$$

$$+ 6(y_3 + y_9) + 2y_6]$$

$$= \frac{3\pi}{240}\left[(1.000000 + 0.915423) + 5(0.998619 + 0.969518 + 0.947647 + \right.$$

$$0.916930)]$$

$$+ \frac{3\pi}{240}\left[(994559 + 0.979541 + 0.937283 + 0.9213322)\right] + \frac{3\pi}{240}$$

$$[6(0.988067 + 0.92829]) + 2(0.958645)]$$

Hence $I = 1.505103504.$

Example 16: Find the value of $\displaystyle\int_4^{5.2} \log_e x\, dx$ by Weddte's rule.

Solution: Here $f(x) = \log_e n$, $a = x_0 = 4$, $b = x_n = 5.2$ taking $n = 6$ (a multiple of six) we have

$$h = \frac{5.2 - 4}{6} = 0.2$$

x	4.0	4.2	4.4	4.6	4.8	5.0	5.2
Y = f(x)	1.3863	1.4351	1.4816	1.5261	1.5686	1.6094	1.6457

Weddie's rule is

$$I = \int_4^{5.2} \log_e x\, dx = \frac{3h}{10}\left[y_0 + 5y_1 + y_2 + 6y_3 + y_4 + 5y_5 + y_6\right]$$

$$= \frac{3 \times (0.2)}{10}\left[1.3863 + 7.1755 + 1.4816 + 9.1566 + 1.5686 + 8.0470\right.$$

$$+ 1.6487]$$

$$= 0.06(30.4643]$$

$$= 1.827858$$

$$\int_{4}^{5.2} \log_e x = 1.827858.$$

EXERCISE 1

1. Evaluate $\int_{0}^{1} x^3 \, dx$ by Trapezoidal rule.

2. Evaluate $\int_{0}^{1} (4x - 3x^2) \, dx1$ taking 10 intervals by Trapezoidal rule.

3. Given that $e^0 = 1$, $e^1 = 2.72$, $e^2 = 7.39$, $e^3 = 20.09$, $e^4 = 54.60$, find an approximation value of $\int_{0}^{4} e^x$ by Trapezoidal rule.

4. Evaluate $\int_{0}^{1} \sqrt{1 - x^3} \, dx$ by (i) Simpson's rule and (ii) Trapezoidal rule, taking six interval correct to two decimal places.

5. Evaluate $\int_{0}^{\frac{\pi}{2}} \sqrt{\sin x} \, dx$ taking $x = 6$, correct to four significant figures by (i) Simpson's one-third rule and (ii) Trapezoidal rule.

6. Evaluate $\int_{1}^{2} \frac{dx}{x}$ taking 4 subintervals, correct to five decimal places (i) Simpson's one-third rule (ii) Trapezoidal rule.

7. Compute by Simpson's one-third rule, the integral $\int_{0}^{1} x^2 \, (1 - x) \, dx$ correct to three places of decimal, taking step length equal to $0 - 1$.

8. Evaluate $\int_{0}^{1} \sin x^2 \, dx$ by (i) Trapezoidal rule, and (ii) Simpson's one-third rule, correct to four decimals taking $x = 10$.

9. Calculate approximate value of $\int_{-3}^{3} \sin x^4 \, dx$ by using (i) Trapezoidal rule, and (ii) Simpson's rule, taking $n = 6$.

10. Find the value of $\displaystyle\int_0^{\pi/2} \sqrt{\cos x}\ dx$ by (i) Trapezoidal rule, and (if) Simpson's one-third rule taking $x = 6$.

11. Compote $\displaystyle\int_1^{15} e^x\ dx$ by (i) Trapezoidal rule, and (ii) Simpson's one-third rule taking $x = 10$.

12. Evaluate $\displaystyle\int_0^{0.5} \frac{x}{\cos x}\ dx$ taking $n = 10$, by (i) Trapezoidal rule, and (ii) Simpson's one-third rule.

13. Evaluate $\displaystyle\int_0^{0.4} \cos x\ dx$ taking four equal intervals by (i) Trapezoidal rule, and (ii) Simpson's one-third rule.

14. Evaluate $\displaystyle\int_0^{\pi/2} \sqrt{\cos x}\ dx$ by Weddle's rule taking $n = 6$.

15. Evaluate $\displaystyle\int_0^{1} \frac{x^2+2}{x^2+1}\ dx$ by Weddle's rule, correct to four decimals taking $n=12$.

16. Evaluate $\displaystyle\int_0^{2} \frac{1}{1+x^2}\ dx$ by using Weddle's rule taking twelve intervals.

17. Evaluate $\displaystyle\int_{0.4}^{16} \frac{x}{\sinh x}\ dx$ taking thirteen ordinates by Weddle's rule correct to five decimals.

18. Using Simpson's rule evaluate $\displaystyle\int_0^{\pi/2} \sqrt{2+\sin x}\ dx$ with seven ordinates.

19. Using Simpson's rule evaluate $\displaystyle\int_1^{2} \sqrt{x-1/x}\ dx$ with five ordinates.

20. Using Simpson's rule evaluate $\displaystyle\int_2^{6} \frac{1}{\log_e x}\ dx$ taking $n = 4$.

21. A river is 80 unit wide. The depth at a distance x unit from one bank d is given by the following table:

X	0	10	20	30	40	50	60	70	80
D	0	4	7	9	12	15	14	8	3

Find the area of cross-section of the river.

22. Find the approximate value of $\displaystyle\int_{0}^{\pi/2} \sqrt{\cos\theta}\ d\theta$ using Simpson's rule with six intervals.

23. Evaluate $\displaystyle\int_{0.5}^{0.7} x^{\frac{1}{2}}e^{-x}\ dx$ approximately by using a suitable formula for at least 5 points.

24. Evaluate $\displaystyle\int_{0}^{1} \sqrt{\sin x + \cos x}\ dx$, correct to two decimal places using seven ordinates.

25. Use Simpson's three-eighths rule to obtain an approximate value of
$$\int_{0}^{0.3} (1-8x^3)^{1/2}\ dx.$$

26. Find the value of $\displaystyle\int_{0}^{1/2} \frac{dx}{\sqrt{1-x^2}}$, using Weddle's rule.

27. Prove that $\displaystyle\int_{-1}^{1} f(x)dx = \frac{1}{12}\,|\,13f(1) - f93) - f(-3)\,|$

28. If $u_x = a + bx + cx^2$, prove that $\displaystyle\int_{1}^{3} u_x dx = 2u_2 + \frac{1}{12}(u_0 - 2u_2 - u_4)$ and hence

    approximate the value for $\displaystyle\int_{-1/2}^{1/2} e^{\frac{-x^2}{10}}\,dx$

29. If $f(x)$ is a polynomial in x of degree 2 and $u_{-1} = \displaystyle\int_{-3}^{-1} f(x)dx$, $u_0 = \displaystyle\int_{-1}^{1} f(x)dx$, $u_1 = \displaystyle\int_{-1}^{3} f(x)dx$

    then show that $f(0) = \dfrac{1}{2}\left[u_0 - \dfrac{\Delta^2 u_{-1}}{2u}\right]^{-1}$.

ANSWERS

1. 0.260	2. 0.995	3. 58.00
4. 0.83	5. 1.187, 1.170	6. 0.69326, 0.69702
7. 0.083	8. 0.3112, 0.3103	9. 115,98
10. 1.170,1.187	11. 1.764,1.763	12. 0.133494, 0.133400
13. 0.3891,0.3894	14. 1.18916	15. 1.7854

16. 1.1071	17. 1.1020	18. 2.545
19. 1.007	20. 3.1832	21. 710 Sq units
22. 1.1872	23. 0.08409	24. 1.14
25. 0.2899	26. 0.52359895	

Example 17: If y_x is a polynomial in x of the third degree, find an expression for $\int_0^2 y_x dx$ in terms of y_0, y_1, y_2 and y_3. Use this results to show that:

$$\int_1^2 y_x dx = \frac{1}{24}\left[-y_0 + 13y_1 + 13y_2 - y_3\right]$$

Solution:

We have

$$y_x = \frac{(x-1)(x-2(x-3)}{(-1)(-2)(-3)}y_0 + \frac{x(x-2)(x-3)}{(1)(-2)(-3)}y_1 + \frac{x(x-1)(x-3)}{(2)(1)(-1)}y_2 + \frac{x(x-1)(x-2)}{(3)(1)(1)}y_3$$

$$= \frac{x^3 - 6x^2 + 11x - 6}{-6}y_0 + \frac{x^3 - 5x^2 + 6x}{6}y_1 + \frac{x^3 - 4x^2 + 3x}{-2}y_2 + \frac{x^3 - 3x^2 + 2x}{6}y_3$$

$\therefore$ We get

$$\int_0^1 y_x dx = \left[-\frac{1}{6}\left(\frac{x^4}{4} - 2x^3 + \frac{11x^2}{2} - 6x\right)y_0 + \frac{1}{2}\left(\frac{x^4}{4} - \frac{5}{3}x^3 + 3x^2\right)y_1\right]_0^1$$

$$+ \left[-\frac{1}{2}\left(\frac{x^4}{4} - \frac{4}{3}x^3 + \frac{3x^2}{2}\right)y_2 + \frac{1}{6}\left(\frac{x^4}{4} - x^3 + x^2\right)y_0\right]_0^1$$

$$= \left(-\frac{1}{6}\right)\left(-\frac{9}{4}\right)y_0 + \left(\frac{1}{2}\right)\left(\frac{19}{12}\right)y_1 - \left(\frac{1}{2}\right)\left(\frac{5}{12}\right)y_2 + \frac{1}{6}\left(\frac{1}{4}\right)y_3$$

Similarly $$\int_0^2 y_x dx = \left(-\frac{1}{6}\right)(-2)y_0 + \left(\frac{1}{2}\right)\left(\frac{8}{3}\right)y_1 \cdot \left(\frac{1}{2}\right)\left(\frac{8}{3}\right)y_{-2} + \left(\frac{1}{6}\right)(0)y_3$$

Subtracting, we get $\int_1^2 y_x dx = \dfrac{1}{24}\ [-y_0 + 13y_1 + 13y_2 - y_3].$

Example 18: Show that

$$\int_0^1 y_x \, dx = \frac{1}{12}(5u_1 + 8u_0 - u_{-1})$$

Solution: We have

x	−1	0	1
u_x	u_{-1}	$'u_0$	u_1

Using Lagrange's formula, we get

$$u_x = \frac{(x-0)(x-1)}{(-1-0)(-1-1)}u_{-1} + \frac{(x+1)(x-1)}{(0+1)(0-1)}u_0 + \frac{(x+0)(x-0)}{(1+1)(1-0)}u_1$$

$$= \frac{x^2-x}{2}u_{-1} - (x^2-1)u_0 + (x^2+x)u_1$$

$$\int_0^1 u_x \, dx = \frac{1}{2}u_{-1}\int_0^1 (x^2-x)dx - u_0\int_0^1 (x^2-1)dx + \frac{1}{2}u_1\int_0^1 (x^2+x)dx$$

$$= -\frac{1}{12}u_{-1} + \frac{2}{3}u_0 + \frac{5}{12}u_1$$

$$= \frac{1}{12}(5u_{-1} + 8u_0 - u_{-1})$$

Example 19: Evaluate the integral of $f(x) = 1 + e^{-x} \sin 4x$ over the interval $[0, 1]$ using exactly five functional evaluations.

Solution: Taking $h = \frac{1}{4}$ and applying Boole's rule, we get

$$\int_0^1 f(x)dx = \frac{1}{4} \times \frac{2}{45}\left[7f(0) + 32f\left(\frac{1}{4}\right) + 12f\left(\frac{1}{2}\right) + 32f\left(\frac{3}{4}\right) + 7f(1)\right]$$

$$= \frac{1}{90}[7 \times 1.0000 + 32 \times 1.65534 + 12 \times 1.55152] + \frac{1}{90}[32 \times 1.06666 + 7 \times 0.72159]$$

$$= 1.30859$$

Example 20: Using Romberg's method compute $I = \int_0^{1.2} \frac{1}{1+x}dx$ correct to 4 decimal places.

Solution: Here $f(x) = \dfrac{1}{1+x}$

We can take h = 0.6, 0.3, 0.15

i.e., $h = 0.6, \dfrac{h}{2} = 0.3, \dfrac{h}{4} = 0.15$

x	0	0.15	0.30	0.40	0.60	0.75	0.90	1.05	1.20
f(x)	1	0.8695	0.7692	0.6896	0.6250	0.5714	0.5263	0.48780	0.4545

Using Trapezoidal rule with h = 0.6, we get

$$I(h) = I(0.6) = I_1 = \frac{0.6}{2}\,[(1 + 0.4545) + 2 \times 0.6256] = 0.8113$$

With $h = \dfrac{0.6}{2} = 0.3$, we get

$$I(h/2) = I(0.3) = I_2$$

$$= \frac{0.3}{2}\,[(1 + 0.4545) + 2 \times (0.7692 + 0.625 + 05263)]$$

$$= 0.7943$$

with $h = \dfrac{0.6}{4} = 0.15$, we get

$$I(h/4) = I(0.15) = I_3$$

$$= \frac{0.15}{2}\,[(1 + 0.4545) + 2 \times (0.8695 + 0.7692 + 0.6896)]$$

$$+ \frac{0.15}{2}\,[2 \times (0.6250 + 05714 + 05263 + 0.4878)]$$

$$= 0.7899.$$

Now $I(h, h/2) = I(0.6, 0.3)$

Therefore $I(0.6, 0.3) = \dfrac{1}{3}\,[4 \times I(0.3) - I(0.6)]$

$$= \frac{1}{3}\,[4 \times 0.7943 - 0 - 81131 = 0.7886.$$

Similarly $I(h/2, h/4) = I(0.3, 0.15)$

Therefore $I(0.3, 0, 15) = \dfrac{1}{3}\,[4 \times I(0.15) - I(0.3)]$

$$= \frac{1}{3}\,[4 \times 0.7899 - 0.7943] - 0.7884.$$

We get $I(h, h/2, h/4) = I(06, 0.3, 0.15)$

Hence
$$I(0.6, 0.3, 0.15) = \frac{1}{3}[4 \times I(0.15, 03) - I(03, 0.6)]$$

$$= \frac{1}{3}[4 \times 0.7884 - 0.7886] = 0.7883$$

The table of these values is

0.8113		
0.7948	0.7886	
0.7899	0.7884	0.7883

Hence we get
$$I = \int_{0}^{1.2} \frac{1}{1+x} dx = 0.7883$$

EXERCISE 2

1. If $H_0, H_1, H_2, ..., H_n$ are Cotes coefficients, show that

   a. $H_0 + h_1 + H_2 + ... + H_n = nh$

   b. $H_r = H_{n-r}$

2. Using Cotes formula, show that $\int_{x_0}^{x_2} f(x)dx = (x_2 - x_0)\left(\frac{1}{6}y_0 + \frac{4}{6}y_1 + \frac{1}{6}y_2\right)$ and also

   show that $c_0^2 = \frac{1}{6}, c_1^2 = \frac{1}{2}, c_2^2 = \frac{1}{6}$, where c_0^2, c_1^2, c_2^2 are Cotes numbers.

3. Using Romberg's method prove that $\int_{0}^{1} \frac{1}{1+x} dx = 0.6931$.

4. Apply Romberg's method to show that $\int_{0}^{1} \sin x \, dx = 1$.

5. Apply Romberg's method to evaluate $\int_{4}^{5.2} \log xk \, dk$ given that

x	4.0	4.2	4.4	4.6	4.8	5.0	5.2
$\log_e x$	1.3863	1.4351	1.4816	1.526	1.5686	1.6094	1.6486

6. Use Romberg's method and show that $\int_{0}^{1} \frac{dx}{1+x^2} = 0.7855$.

Example 21: Evaluate the integral $I = \int_1^2 \int_1^2 \dfrac{dxdy}{x+y}$ using Trapezoidal rule with h = k=0.5.

Solution: Using Trapezoidal rule, we get

$$I = \int_1^2 \int_1^2 \frac{dxdy}{x+y}$$

$$= \frac{1}{16}\, [f(1,1) + f(2,\,1) + f(1,\,2) + f(2,\,2)]$$

$$+ \frac{1}{16}\left[2\left[f\left(\frac{3}{2},1\right) + f\left(1,\frac{3}{2}\right) + f\left(\frac{3}{2},2\right)\right] + 4f\left(\frac{3}{2},\frac{3}{2}\right)\right]$$

$$= \frac{1}{16}\left[0.5 + \frac{1}{3} + \frac{1}{3} + 0.25 + 2\left[0.4 + 0.4 + \frac{2}{7} + \frac{2}{7}\right] + \frac{4}{3}\right]$$

$$= 0.343304.$$

Example 22: Using the table of values given below evaluate the integral of $f(x, y) = e^y \sin x$ over the interval $0 \le x \le 0.2,\ 0 \le y < 0.2$

(a)   by the Trapezoidal rule with h = k - 0.2, and

(b)   by Simpson's one-third rule with h = k = 0.1

y \ x	0.0	0.1	0.2
0.0	0.0	0.998	0.1987
0.1	0.0	0.1103	0.2196
0.2	0.0	0.1219	0.2427

Solution:

(a)   Applying Trapezoidal rule, we get

$$-I = \frac{(0.2)^2}{4}\, [0 + 0.1987 = 0.2427] = 0.004414.$$

(b)   By Simpson's rule, we get

$$I = \frac{(0.2)^2}{4}\, [1.0 + 4(0.998) + 1(0.1987) + 4.0 + 16(0.1103)]$$

$$+ \frac{(0.2)^2}{4}\, [4(02196) + 1.0 + 4(0.1219\} + 1(0.2427\}]$$

$$= 0.004413.$$

Example 23 Evaluate $\int_0^{0.5} \int_0^{0.5} \dfrac{xy}{1+xy} \, dxdy$ using Simpson's rule for double integrals with both equal to 0.25.

Solution: Taking $n = k = 0.25$,

We have $\quad x_0 = 0, \ x_1 = 0.25, \ x_2 = 0.5$

$$y_0 = 0, \ y_1 = 0.25, \ y_2 = 0.5$$

$$f(0, 0) = 0, \ f(0, 0.25) = 0, \ f(0, 0.5) = 0$$

$$f(0.25, 0) = 0, \ f(0.25, 0.25) = 0.05878525, \ f(0.25, 0.5) = 0.110822$$

$$f(0.5, 0) = 0. \ f(0.5, 0.25) = 0.110822, \ f(0.5, 0.5) = 0.197923$$

Applying, Trapezoidal rule, we get

$$I = \frac{1}{9.44}\left[f(0,0) + 4f\left(\frac{1}{4},0\right) + f\left(\frac{1}{2},0\right)\right] + 4\left\{ f\left(0,\frac{1}{4}\right) + 4f\left(\frac{1}{4},\frac{1}{4}\right) + f\left(\frac{1}{2},\frac{1}{4}\right)\right\}$$

$$+ f\left(0,\frac{1}{2}\right) + 4f\left(\frac{1}{4},\frac{1}{2}\right) + f\left(\frac{1}{2},\frac{1}{2}\right)$$

$$= \frac{1}{144}\,[0+0+0+4(0+0.235141+0.110822)+0+0.443288+0.197923]$$

$$= 0.014063$$

Example 24: Find the value of $\log_e 2$ from $\int_0^1 \dfrac{1}{1+x}dx$ using Euler-Maclaurin formula.

Solution: Taking $\quad y = \dfrac{1}{1+x}$, and $n = 10$

We have $\quad x_0 = 0, \ x_n = 1, \ h = 0.1$

$$y' = \frac{-1}{(1+x)^2}, \ y'' = \frac{2}{(1+x)^3}, \ y''' \ \frac{-6}{(1+x)^4}$$

$$y_0 = \frac{1}{1+x_0} = \frac{1}{1+0} = 1, \ y_1 = \frac{1}{1+x_1} = \frac{1}{1+0.1} = \frac{1}{1.1} \ldots\ldots, x_n = \frac{1}{1+1} = \frac{1}{2}$$

Example 25: Use the Euler – Maclaurin expansion to prove

$$\sum_{x=1}^{n} x^2 = \frac{n(n+1)(2n+1)}{6}$$

Solution: We have

$$y = f(x) = x^2$$

$$y' = f'(x) = 2x$$

$$y'' = 2, \ y''' = 0,....$$

Taking $h = 1$, we get $x_0 = 1$, $x_n = n$, $y_0 = 1$, $y_n = n^2$.

From Euler – Maclaurin formula, we have

$$y_0 + y_1 + ...y_n = \sum_{x=1}^{n} x^2$$

$$= \frac{1}{h}\int_{x_0}^{x_n} f(x)dx + \frac{1}{2}(y_n + y_0) + \frac{1}{2}(y'_n - y'_n) + ...$$

$$= \int_{1}^{n} x^2 dx + \frac{1}{2}(n^2 + 1) + \frac{1}{2}(2n - 2)$$

$$= \frac{1}{3}(n^2 - 1) + \frac{1}{2}(n^2 + 1) + \frac{1}{6}(n - 1)$$

$$= \frac{1}{6}(2n^3 - 2 + 3n^2 + 3 + n - 1)$$

$$= \frac{2n^3 + 3n^2 + n}{6} \ n(n + 1)\ (2n + 1).$$

Hence Proved.

EXERCISE 3

1. Evaluate $\displaystyle\int_{1}^{2}\int_{1}^{2} \frac{dxdy}{x+y}$ using the Trapezoidal rule with $h = k = 0.25$.

2. Evaluate the double integral $\displaystyle\int_{0}^{1}\left(\int_{0}^{2} \frac{2xy}{(1+x^2)(1+y^2)} dy\right) dx$ using (i) the Trapezoidal rule with $h = k = 0$.(ii) the Simpon's rule $h = k = 0.25$.

3. Evaluate the double integral $\displaystyle\int_{1}^{5}\left(\int_{1}^{5} \frac{dx}{(x^2 + y^2)^{1/2}}\right) dy$ using the Trapezoidal rule with two and four subintervals.

4. Using the table of values given below evaluate the integral of $f(x, y) = e^y \sin x$ over the interval $0 \le x \le 0.2, 0 \le y \le 0.2$

   (a)   by die Trapezoidal rule with $h = k = 0.1$.

   (b)   by Simpson's one-third rule with $h = k = 0.1$.

y \ x	0.0	0.1	0.2
0	0.0	0.0998	0.1987
0.1	0.0	0.1103	0.2196
0.2	0.0	0.1219	0.2427

5. Integrate the following functions over the given domains by the Trapezoidal formula, using the indicated spacing.

   (a) $f(x, y) = \sqrt{1-xy}$; $0 \le x \le 1.0 \le y \le 1$, with $h = k = 0.5$.

   (b) $f(x, y) = \sin x \cos y$; $0 \le x \le \dfrac{\pi}{2}$ with $h = k = \dfrac{\pi}{4}$.

6. Find the value of the double integral $I = \displaystyle\int_{2}^{3.2}\int_{1}^{3.6} \dfrac{1}{x+y}\,dy\,dx.$

7. Use Euler-Maclaurin formula to prov that $\displaystyle\sum_{x=1}^{n} x^3 = \dfrac{n^2(n+1)^2}{4}$

8. Use Euler-Maclaurin formula to show that $\dfrac{1}{51^2} + \dfrac{1}{53^2} + ... + \dfrac{1}{99^2} = 0.00499$

ANSWERS

1. 00340668   2. (i) 0.31233 (ii) 0.31772   3. n = 2, I = 4.I34; n = 4, I= 3.997

4. 0.004413   5. (a) 0.8308 (b) 0.8988   6. 0.48997

Example 26: Solve $\dfrac{dy}{dx} = x + y$, $y(I) = 0$, numerically up to $x = 1.2$, with $h = 0.1$.

Solution: We have $x_0 = 1$, $y_0 = 0$ and

$$\frac{dy}{dx} = y' = x + y \Rightarrow y'_0 = 1 + 0 + 1$$

$$\frac{d^2y}{dx^2} = y'' = 1 + y' \Rightarrow y''_0 = 1 + 1 + = 2$$

$$\frac{d^3y}{dx^3} = y'''' = y'' \Rightarrow y'''_0 = 2$$

$$\frac{d^4y}{dx^4} = y^{IV} = y'' \implies y^{IV}_0 = 2$$

$$\frac{d^5y}{dx^5} = y^{V} = y^{IV} \implies y^{V}_0 = 2$$

$$\vdots$$
$$\vdots$$

Substituting the above values in

$$y_2 = y_1 + (h/1!)\, y_1^{1} + (h^2/2!)\, y_1^{11} + (h^3/3!)\, y_1^{111} + \ldots, \text{ we get}$$

we get
$$y_1 = y_1 + (0.1) + \frac{(0.1)^2}{2}\,2 + \frac{(0.1)^3}{6}\,2 + \frac{(0.1)^4}{24}\,2 + \frac{(0.1)^5}{120}\,2 + \ldots$$

or
$$y_1 = 0.11033847$$

i.e.,
$$y_1 = y(0.1) \approx 0.110$$

Now
$$x_1 = x_0 + h = 1 + 0.1 = 1.1$$

We have
$$y'_1 = x_1 + y_1 = 1.1 + 0.110 = 1.21$$

$$y''_1 = 1 + y'_1 = 1 + 1.21 = 2.21$$

$$y'''_1 = y''_1 = 2.21$$

$$y_1^{IV} = 2.21$$

$$y_1^{V} = 2.21$$

$$\vdots$$
$$\vdots$$

Substituting the above values in (1), we get

$$y_2 = 0.110 + (0.1)\,(1.21) + \frac{(0.1)^2}{2}\,(2.2i) + \frac{(0.1)^3}{6}\,(2.21)$$

$$+\frac{(0.1)^4}{24}\,(2.21) + \frac{(0.1)^5}{120}\,(2.21),$$

Hence
$$y_2 = 0.24205$$

i.e.,
$$y(1.2) = 0.242.$$

Example 27: Given $\dfrac{dy}{dx}$ 1 + xy with the initial condition that y = 1, x = 0. Compute y(0.1) correct to four places of decimal by using Taylor's series method.

Solution: Given $\dfrac{dy}{dx} = 1 + xy$ and $y(0) = 1$

$\therefore \qquad y_1(0) = 1 + 0 \times 1 = 1.$

Differentiating the given equation w.r.t. x, we get

$$\frac{d^2y}{dx^2} = y + x\frac{dy}{dx}$$

$$y_0'' = 1 + 0 \times 1 = 1 + 0 = 1$$

Similarly

$$\frac{d^3y}{dx^3} = x\frac{d^2y}{dx^2} + 2\frac{dy}{dx}$$

or $\qquad y_0''' = 2,$

and

$$\frac{d^4y}{dx^4} = x\frac{d^3y}{dx^3} + 3\frac{d^2y}{dx^2}$$

or $\qquad y_0^{iv} = 3$

From Taylor's series method, we have

$$y_1 = 1 + hy_0 + \frac{h^2}{2}y_0'' + \frac{h^3}{3}y_0''' + \frac{h^4}{24}y_0^{iv} + \ldots$$

i.e.,

$$y(0.1) = 1 + (0.1)\,(1) + \frac{(0.1)^2}{2}1 + \frac{(0.1)^3}{6}2 + \frac{(0.1)^4}{24}3 + \ldots$$

$$= 1.1053425$$

Hence $\qquad y(0.1) = 1.1053$

correct to four decimal places.

CHAPTER 9

SOLUTION OF ORDINARY DIFFERENTIAL EQUATIONS

9.1 INTRODUCTION

In this chapter we introduce the numerical methods for solving ordinary differential equations .The classical initial value problem is to find a function y (x) which satisfies the first order differential equation $\dfrac{dy}{dx} = f(x, y)$ and with initial value $y(x_0) = y_0$. But in many cases the analytical methods cannot be applied to solve in one of the two forms

(i) A power series in x, from which the value of the function y can be obtained by direct substitution.

(ii) A set of tabulated values of x and y.

The methods due to Picard and Taylor gives the solution in a power series. The methods due to Euler, Runge – Kutta, Adams Moulton and Milne are step by step methods. The method in which the value y_{i+1} depends on the knowledge of the proceeding value y_i only is called Single step method. The method in which the value of y_{i+1} depends on the knowledge of the preceding values y_i, y_{i-1}, y_{i-1},... is called multi-step method.

9.2 PICARD'S METHOD

Consider the first order differential equation $\dfrac{dy}{dx} = f(x, y)$ with $y = y_0$ at $x = x_0$.

$$\text{We get} \qquad \int_{x_0}^{x} dy = \int_{x_0}^{x} f(x, y)dx$$

$$\Rightarrow \qquad y\Big|_{x_0}^{x} = \int_{x_0}^{x} f(x, y)dx$$

$$\Rightarrow \qquad y(x) - y(x_0) = \int_{x_0}^{x} f(x, y)dx$$

$$\Rightarrow \qquad y(x) = y_0 + \int_{x_0}^{x} f(x, y)dx \qquad\qquad \left(\because y(x_0) = y_0\right)$$

Integrating by taking initial limits y_0 to y and x_0 to x this approximation as $y_1 = y_0 + \int_{x_0}^{x} f(x, y)dx$ and 2nd approximation $y_2 = y_0 + \int_{x_0}^{x} f(x, y)dx$. Repeat the procedure until we get the desired accuracy.

Example 1: Using Picard's method of successive approximation obtain a solution up to the fourth approximation of the differential equation. $\dfrac{dy}{dx} = x + y$ where $y = 1$, when $x = 0$.

Solution: Here integrating the given equation between $x = 0$ to x we get

$$y_1 = 1 + \int_0^x (x + 1)dx$$

$$= 1 + \frac{x^2}{2} + x$$

$$y_2 = y_0 + \int_0^x (1 + x + \frac{x^2}{2} + x)dx$$

$$= 1 + x + x^2 + \frac{x^3}{6}$$

$$y_3 = y_0 + \int_0^x x + (1 + x + x^2 + \frac{x^3}{6})dx$$

$$= 1 + \int_0^x (1 + 2x + x^2 + \frac{x^3}{6})dx$$

$$= 1 + x + x^2 + \frac{x^3}{3} + \frac{x^4}{24}$$

$$y_4 = 1 + \int_0^x \left[(1 + x + x^2 + \frac{x^3}{3} + \frac{x^3}{24} + x \right] dx$$

$$= 1 + x + x^2 + \frac{x^3}{3} + \frac{x^4}{12} + \frac{x^5}{12}$$

9.3   TAYLOR'S SERIES METHOD

The solution of the given differential equation, is represented by a series.

$$y(x) = y_0 + (x - x_0)y_0' + \frac{(x - x_0)^2}{2!} y_0'' + \dots..$$

Now,
$$\frac{dy}{dx} = f(x, y)$$

$$\frac{d^2y}{dx^2} = f_x + f_y y'(x)$$

$$\frac{d^3y}{dx^3} = f_{xx} + 2f_{xy} y'(x) + f_{yy}\{y'(x)\}^2 + f_x y'(x)$$

And so on.

$$y(x_0 + h) = y(x_0) + h y'(x_0) + \frac{h^2}{2!} y''(x_0) + \dots$$

$$y(x_0 + 2h) = y(x_0 + h) + h y'(x_0 + h) + \frac{h^2}{2!} y''(x_0) + \dots$$

Continuing in this manner we obtain a discrete set of values y_n which are approximation to the actual values of y at the points $x_n = x_0 + nh$.

Example 2: Using Taylor's method solve $\dfrac{dy}{dx} = 1 + xy$ with y(0) = 2, find y (0.2)

Solution:

$$\frac{dy}{dx} = 1 + xy \qquad y(0) = 2$$

$$y' = 1 + xy \qquad y'\big|_0 = 1$$

$$y'' = y + xy' \qquad y''\big|_0 = 4$$

$$y''' = 2y' + xy'' \qquad y'''\big|_0 = 4$$

$$y^{IV} = 3y'' + xy''' \qquad y^{IV}\big|_0 = 6$$

If h = 0.1 then y(0.1)

$$= y_0 + \frac{y_0^{I}}{1!}h + \frac{y_0^{II}}{2!}h^2 + \frac{y_0^{III}}{3!}h^3 + \frac{y_0^{IV}}{4!}h^4 + \ldots$$

$$= 2 + 1(0.1) + \frac{2}{2!} \times (0.1)^2 + \frac{4}{6} \times (0.1)^3 + \frac{6}{24}(0.1)^4 + \ldots$$

$$= 2 + 0.1 + 0.1 + 000667 + 0.000025$$

$$= 2.11060$$

Again for x1 = 0 + 0.1

$$y^{I}\Big|_{0.1} = 1 + (0.1) \times 2.11069$$

$$= 1.21107$$

$$y^{II}\Big|_{0.1} = 2.1107 + 0.1(1.21107)$$

$$= 2.1107 + .121107$$

$$= 2.231807$$

$$y^{III}\Big|_{0.1} = 2 \times (1.21107) + (0.1) \times (2.231807)$$

$$= 2.6453207$$

$$y^{IV}\Big|_{0.1} = 3 \times 2.231807 + (0.1) \times (2.231807)$$

$$= 6.95995307$$

$$y(0.2) = y(x_1 + h)$$

$$= y(0.1 + 0.1)$$

$$= y_1 + \frac{y_1^{I}}{1!}h + \frac{y_1^{II}}{2!}h^2 + \frac{y_1^{III}}{3!}h^3 + \frac{y_1^{IV}}{4!}h^4$$

$$= 2.11069 + 1.21107 \times (0.1) + \frac{2.231807}{2!} \times (0.1)^2$$

$$+ \frac{2.6453207}{6} \times (0.1)^3 + \frac{6.95995307}{24} \times (0.1)^4$$

$$= 2.11069 + 0.121107 + \frac{0.02231807}{2}$$

$$+\frac{0.0026453207}{6}+\frac{0.000695995307}{24}$$

$$= 2.11069 + 0.121107 + 0.001116903 + 0.0004409 + 0.000028$$

$$= 2.243435$$

$$= 2.243\,(\text{up to three decimal places})$$

Note: if we take $h = 0.2$ then

$$y(0.2) = y_0 + \frac{y_0^I}{1!}(0.2)^2 + \frac{y_0^{II}}{2!}(0.2)^2 + \frac{y_0^{III}}{3!}(0.2)^3 + \frac{y_0^{IV}}{4!}(0.2)^2 + \dots$$

$$= 2 + 1\times(0.2) + \frac{2}{2}(0.04) + \frac{4}{6}(0.008) + \frac{6}{24}(0.0016)$$

$$= 2.243067$$

$$= 2.243\left(\text{upto three decimal places}\right)$$

Example 3: Solve $\dfrac{dy}{dx} = x + y, y(1) = 0$ numerically up to $x = 1.2$ with $h = 0.1$ by Taylor's series method. Compare the final result with value of the explicit solution.

Solution: The given equation is of the following form

$$\frac{dy}{dx} = x + y \quad y(1) = 0$$

$$y^I\big|_0 = 1 \qquad y^{II}\big|_0 = 2$$

$$y^{III}\big|_0 = \quad 2 \quad y^{IV}\big|_0 = 2$$

Hence

$$y(1.1) = y_0 + \frac{y_0^I}{1!}(0.1) + \frac{y_0^{II}}{2!}(0.1)^2 + \frac{y_0^{III}}{3!}(0.1)^3 + \frac{y_0^{IV}}{4!}(0.1)^4$$

$$= 0 + 1(0.1) + \frac{2}{2}\times(0.01) + \frac{2}{6}(0.001) + \frac{2}{24}(0.001)$$

$$= 0.1 + 0.01 + 0.000333 + 0.000008$$

$$= 0.11033.$$

Again,

$$y^I\big|_{1.1} = 1.1 + 0.11033$$

$$= 1.21033$$

$$y^{II}\big|_{1.1} = \quad 1 + y^I\big|_{1.1}$$

$$= 1 + 1.21033$$

$$y''' \big|_{1.1} = y'' \big|_{1.1}$$

$$= 2.21033$$

$$y^{IV} \big|_{1.1} = y''' \big|_{1.1}$$

$$= 2.21033$$

Hence

$$y(1.2) = 0.11033 + 1.21033 \times (0.1) + \frac{2.21033}{2} \times (0.01)$$

$$+ \frac{2.21033}{6} \times (0.001) + \frac{2.21033}{24} \times (0.0001)$$

$$= 0.11033 + 0.121033 + 0.0110516 + 0.00036838 + 0.00000921$$

$$= 0.24279219$$

$$= 0.2428 \left(\text{Correct upto 4 decimals places} \right)$$

Now

$$\frac{dy}{dx} = x + y \text{ and } y = 0 \text{ at } x = 1$$

$$\frac{dy}{dx} - y = x$$

IF is

$$e^{-\int dx} = e^{-x}$$

$$ye^{-x} = \int e^{-x} x \, dx$$

$$= -e^{-x} x + \int e^{-x} dx$$

$$= -e^{-x} x - e^{-x} + k$$

$$= -(x+1) e^{-x} + k$$

Now at x = 1, y = 0, Therefore $k\, e - 2 = 0$, $k = \dfrac{2}{e}$

$$y = -(x+1) + \frac{2}{3} e^x \text{ at } x = (1.2)$$

$$y = -(1.2+1) + \frac{2e^{1.2}}{e}$$

$$= -2.2 + 2.442805516$$

$$= 0.242805516$$

$$= 0.2428 \,(\text{Correct upto 7 decimal place})$$

Example 4: Compute y(1.1) and y (1.2) correct up to 5 decimal places using Taylor series method, when y (x) satisfies the equation $\dfrac{dy}{dx} = xy$ with y (1.0) = 2.

Solution:

$$\frac{dy}{dx} = xy \quad y = 2, x = 1$$

$$y'\Big|_1 = 1 \times 2 = 2$$

$$y''\Big| = y + xy'$$

$$y''\Big|_1 = 2 + 1.2 = 4$$

$$y''' = 2y' + xy''$$

$$y'''\Big|_1 = 2 \times 2 + 1 \times 4 = 8$$

$$y^{IV} = 3xy'' + xy'''$$

$$y^{IV}\Big|_1 = 3 \times 4 + 1 \times 8 = 20$$

$$y(1.1) = y_1 + \frac{y_1^{I}}{1!}(0.1) + \frac{y_1^{II}}{2!}(0.1)^2 + \frac{y_1^{III}}{3!}(0.1)^3 + \frac{y_1^{IV}}{4!}(0.1)^4$$

$$= 2.2 + 2 \times (.1) + \frac{4}{2}(.1)^2 + 1.3333 \times 0.001 + \frac{5}{6} \times 0.0001$$

$$= 2 + .2 + 0.02 + 0.0013333 + 0.00008333$$

$$= 2.22141663$$

$$= 2.22142 \ (\text{Correct upto 5 decimal places})$$

$$y'\Big|_{1.1} = (1.1) \times 2.22142$$

$$= 2.443562$$

$$y''\Big|_{1.1} = 2.22142 + (1.1) \times 2.443562$$

$$= 4.9093382$$

$$y'''\Big|_{1.1} = 2 \times (2.443562) + (1.1) \times 4.9093382$$

$$= 10.894622$$

$$y^{IV}\Big|_{1.1} = 3 \times 4.9093382 + (1.1) \times 10.8964622$$

$$= 26.71412302$$

$$y(1.2) = y\big|_{1.1} + y'_{1.1} \times (0.1) + \frac{y''_{1.1}}{2!} \times (0.1)^2 + \frac{y'''_{1.1}}{3!} \times (0.1)^3 + \frac{y^{IV}_{1.1}}{4!} \times (0.1)^4 + ...$$

$$= 2.22142 + 2.443562 \times 0.1 + \frac{4.9093382}{2} \times (0.01) + \frac{10.8964622}{6} \times 0.001$$

$$\frac{26.71412302}{24} \times (.0001)$$

$$= 2.22142 + 0.2443562 + 0.02454669 + 0.001816077 + 0.000111309$$

$$= 2.492250276$$

$$= 2.49225 \, (\text{Correct up to 5 decimal places})$$

Example 5: Solve by Taylor series method $\dfrac{dy}{dx} = 2x + 3y^2$ when $y = 0$, $x = 0$ at $x = 0.2$.

Solution:

$$\frac{dy}{dx} = 2x + 3y^2 \qquad y'\big|_0 = 0$$

$$y^{11} \qquad = \qquad 2 + 6y'$$

$$y''\big|_0 \qquad = \qquad 2$$

$$y''' \qquad = \qquad 6y''$$

$$y'''\big|_0 \qquad = \qquad 6 \times 2 = 12$$

$$y^{IV} \qquad = \qquad 6y'''$$

$$y^{IV}\big|_0 \qquad = \qquad 6 \times 12 = 72$$

Let h = 0.2

$$y(0.2) = y_0 + \frac{y'_0}{1!}(h) + \frac{y''_0}{2!}(h)^2 + \frac{y'''_0}{3!}(h)^3 + \frac{y^{IV}_0}{4!}(h)^4$$

$$= 0 + 0 + \frac{2}{2}(0.2)^2 + \frac{12}{6}(0.2)^3 + \frac{72}{24}(0.2)^4$$

$$= 0.04 + 2(0.008) + 3(0.0016)$$

$$= 0.04 + 0.016 + 0.0048$$

$$= 0.04208$$

9.4 EULER'S METHOD

Let us consider the first order first degree differential equations as $\dfrac{dy}{dx} = f(x,y)$ with

$y(x_0) = y_0$ We divide the range [x0, xn] into n − equal partial intervals by the points $x_0, x_1, ..., x_{r-1}, x_{r+1}, ... x_n$ where $x_r = x_0 + rh (r = 1, 2, ...n)$ and $h = x_r - x_{r-1}$ = step length Now, $f(x,y) = f(x_{r-1}, y_{r-1})$ in $(x_{r-1} \leq x \leq x_r)$ and integrating in the range $[x_{r-1}, x_r]$ we get, the Euler's iteration formula as;

$$\int_{x_{r-1}}^{x_r} dy = \int_{x_{r-1}}^{x_r} f(x,y)dx$$

$$y_r = y_{r-1} + \int_{x_{r-1}}^{x_r} f(x,y)dx$$

$$= y_{r-1} + f(x_{r-1}, y_{r-1}) \int_{x_{r-1}}^{x_r} dx$$

$$= y_{r-1} + hf(x_{r-1}, y_{r-1})$$

Hence it can be written in the form

$$y_1 = y_0 + hf(x_0, y_0)$$

$$y_2 = y_1 + hf(x_1, y_1)$$

$$... = ...$$

$$y_n = y_{n-1} + hf(x_{n-1}, y_{n-1})$$

Example 6: Given $\dfrac{dy}{dx} + \dfrac{y}{x} = \dfrac{1}{x^2}, y(1) = 1$. Evaluate y(1.2) Euler's method.

Solution: h = 0.1

$$\frac{dy}{dx} = \frac{1}{x^2} - \frac{y}{x} = \frac{1 - xy}{x^2}$$

Then

$$y_1 = y_0 + hf(x_0, y_0)$$

$$= 1 + (0.1)\frac{1 - 1 \times 1}{1^2}$$

$$= 1$$

$$y_2 = y_1 + (0.1)\frac{1 - 1.1 \times 1}{(1.1)^2}$$

$$= 1 + (0.1)\frac{(-0.1)}{1.21}$$

$$= 1 - \frac{0.01}{1.21}$$

$$= \frac{1.21 - 0.01}{1.21}$$

$$= \frac{1.2}{1.21}$$

$$= 0.991736$$

Example 7: Solve by using Euler's method the following differential equation for $x = 1$, taking $h = 0.2$; $\frac{dy}{dx} = xy, y = 1$ when $x = 0$.

Solution:
$$\frac{dy}{dx} = xy = f(x, y)$$

$$y_1 = y_0 + hf(x_0, y_0)$$

$$= 1 + (0.2)(0 \times 1)$$

$$= 1$$

$$y_2 = y_1 + hf(x_1, y_1)$$

$$= 1 + (0.2)[(0.2)x]$$

$$= 1 + 0.04$$

$$= 1.04$$

$$y_3 = y_2 + hf(x_2, y_2)$$

$$= 1.04 + (0.2)[0.4 \times 1.04]$$

$$= 1.04 + 0.0832$$

$$= 1.1232$$

Example 8: Given $\frac{dy}{dx} = \frac{y - x}{y + x}$ with initial condition $y = 1$ at $x = 0$ find y for $x = 0.1$ by Euler's method, correct up to 4 decimal places, taking step length $h = 0.02$.

Solution:
$$\frac{dy}{dx} = \frac{y - x}{y + x}; \ y(0) = 1$$

$$h = 0.02 \qquad y(0.02) = y(0) + (0.02)\frac{1-0}{1+0}$$

$$= 1.02 + (0.02 \times 0.961538$$

$$= 1.039231$$

$$y(0.06) = y(0.04) + (0.02)\frac{1.039231 - 0.04}{1.039231 + 0.04}$$

$$= 1.039231 + 0.018517462$$

$$= 1.057748$$

$$y(0.08) = 1.057748 + (0.02)\frac{1.057748 - 0.06}{1.057748 + 0.06}$$

$$= 1.057748 + 0.017852825$$

$$= 1.0756008$$

$$y(0.1) = 1.075601 + (0.02)\frac{1.075601 - 0.08}{1.07601 + 0.08}$$

$$= 1.075601 + 0.01723088$$

$$= 1.0928318$$

$$= 1.0928 \,(\text{Correct upto 4 decimal places})$$

Example 9: Using Euler's method obtain the solution of $\dfrac{dy}{dx} = x - y$ with $y\,(0) = 1$ and $h = 0.2$ at $x = 0.4$

Solution: $\qquad \dfrac{dy}{dx} = x - y, y_0 = +1, x_0 = 0, h = 0.2$

$$y_1 = y_0 + hf(x_0, y_0)$$

$$= 1 + (0.2)[0 - 1]$$

$$= 1 - 0.2$$

$$= 0.8$$

$$y_2 = y_1 + hf(x_1, y_1)$$

$$= 0.8 + (0.2)[0.2 - 0.8]$$

$$= 0.8 - 0.12$$

$$= 0.68$$

Example 10: Use Euler's method, evaluate y (2) correct up to three decimal places from

$\dfrac{dy}{dx} = \dfrac{1}{2}(x+y)$ with y (0) = 2 taking h = 0.5

Solution: $\qquad \dfrac{dy}{dx} = \dfrac{1}{2}(x+y), y_0 = 2, x_0 = 0$

$$y_1 = y_0 + hf(x_0, y_0)$$

$$= 2 + (0.5)\dfrac{1}{2}(2+0)$$

$$= 2 + 0.5 = 2.5$$

$$y_2 = y_1 + hf(x_1, y_1)$$

$$= 2.5 + (0.5)\dfrac{1}{2}[0.5 + 2.5]$$

$$= 2.5 + 0.75$$

$$= 3.25$$

$$y_3 = y_2 + hf(x_2, y_2)$$

$$= 73.25 + (0.5)\dfrac{1}{2}(3.25 + 1)$$

$$= 3.25 + 1.0625$$

$$= 4.3125$$

$$y_4 = y_3 + hf(x_3, y_3)$$

$$= 4.3125 + (0.5)\dfrac{1}{2}(4.3125 + 1.5)$$

$$= 4.3125 + 1.453125$$

$$= 5.765625$$

$$= 5.766 \ (\text{Correct up to three decimal place})$$

9.5 MODIFIED EULER'S METHOD

We start with the initial value y0 at x0 and

$$y_1^{(1)} = y_0 + hf(x_0, y_0)$$

The second approximation is

$$y_1^{(2)} = y_0 + h[f(x_0, y_0) + f(x_1, y_1^{(1)})]/2$$

In this way we get

$$y_1^{(n+1)} = y_0 + h[f(x_0, y_0) + f(x_1, y_1^{(n)})]/2$$

Hence

$$y_{j+1}^{(n+1)} = y_j + h[f(x_j, y_j) + f(x_{j+1}, y_{j+1}^{(n)})]/2 \qquad \ldots\ldots(9.5.1)$$

Example 11: Given $\dfrac{dy}{dx} + \dfrac{y}{x} = \dfrac{1}{x^2}, y(1) = 1,$ Evaluate $y(1.2)$ by modified Euler's method.

Solution: Let $x_0 = 1, x_1 = 1.2, h = 0.2$ given $y_0 = 1$ to find y, $\dfrac{dy}{dx} + \dfrac{y}{x} = \dfrac{1}{x^2} = f(x, y)$

$$f(x_0, y_0) = \frac{1-1}{1^2} = 0$$

$$y_1^1 = y_0 + hf(x_0, y_0) = 1 + (0.2).0 = 1$$

$$y_1^2 = y_0 + h[f(x_0, y_0) + f(x_1, y_1^1)]/2$$

$$= 1 + (0.2)[0 + \frac{1 - 1.2 \times 1}{(1.2)^2}]/2$$

$$= 1 - 0.0138888$$

$$= 0.986111$$

$$y_1^3 = y_0 + (0.2)[f(x_0, y_0) + f(x_1, y_1^2)]/2$$

$$= 1 + (0.2)[0 + \frac{1 - 1.2 \times 0.986111}{(1.2)^2}]/2$$

$$= 1 - 0.012731472$$

$$= 0.97727$$

$$y_1^4 = y_0 + (0.2)[f(x_0, y_0) + f(x_1, y_1^3)]/2$$

$$= y_0 + (0.2)[0 + \frac{1 - 1.2 \times 0.97727}{(1.2)^2}]/2$$

$$= 0.98800$$

$$y_1^5 = y_0 + (0.2)[f(x_0, y_0) + f(x_1, y_1^4)]/2$$

$$= 1 + (0.2)[0 + \frac{1 - (1.2) \times 0.988}{(1.2)^2}]/2$$

$$= 0.987111$$

Hence $y(1.2) = 0.987$ correct up to three significant figures.

Example 12: Given $\dfrac{dy}{dx} + \dfrac{y}{x} = \dfrac{1}{x^2}$, Evaluate $y(1.2)$ by modified Euler method correct up to 4 decimal places with $h = 0$.

Solution:

$$x_0 = 1, y(1) = 1, x_1 = 1.1; f(x,y) = \frac{1-xy}{x^2}, y_1 = y(1.1)$$

$$y_1^{(1)} = y_0 + hf(x_0,y_0) = 1 + (0.1) \times 0 = 1$$

$$y_1^{(2)} = y_0 + h/2[f(x_0,y_0) + f(x_1,y_1^1)]/2$$

$$= 1 + \frac{0.1}{2}\left[0 + \frac{1-1.1\times1}{(1.1)^2}\right] = 0.99587$$

$$y_1^{(3)} = y_0 + h/2[f(x_0,y_0) + f(x_1,y_1^2)]/2$$

$$= 1 + \frac{0.1}{2}\left[0 + \frac{1-1.1\times0.99587}{(1.1)^2}\right] = 0.99606$$

$$y_1^{(4)} = y_0 + h/2\left[f(x_0,y_0) + f(x_1,y_1^3)\right]$$

$$= 1 + \frac{0.1}{2}\left[0 + \frac{1-1.1\times0.99606}{(1.1)^2}\right]$$

$$= 0.99605$$

Hence $\qquad y_1^2 = y_1^3 = 0.9961$

Now $\qquad y(1.2) = y_2$

$$y_2^{(1)} = y_1 + hf(x_1,y_1)$$

$$= 0.9961 + 0.1\left[\frac{1-(1.1)\times0.9961}{(1.1)^2}\right]$$

$$= 0.98819$$

$$y_2^{(2)} = y_1 + \frac{0.1}{2}\left[f(x_1,y_1) + f(x_2,y_2^{(1)})\right]$$

$$= 0.9961 + 0.05\left[\frac{1-(1.1)\times0.9961}{(1.1)^2} + \frac{1-(1.2)\times0.98819}{(1.2)^2}\right]$$

$$= 0.9961 + 0.05\left[(-0.0791) + (-0.129047)\right]$$

$$= 0.98569$$

$$y_2^{(3)} = y_1 + \frac{0.1}{2}\left[f(x_1, y_1) + f(x_2, y_2^{(2)})\right]$$

$$= 0.985797$$

Hence $\qquad y_2 = y(1.2) = 0.9858$

Example 13: Solve by Euler's modified method, the following differential equation for x = 0.02 by taking step length h = 0.01. $\dfrac{dy}{dx} = x^2 + y$ y = 1, when x = 0

Solution:

$$\frac{dy}{dx} = x^2 + y = f(x, y); x_0 = 0, y_0 = 1$$

$$[y_1 = y(0.01), \ h = 0.01, x_1 = 0.001]$$

$$y_1^{(1)} = y_0 + hf(x_0, y_0)$$

$$= 1 + 0.01[0^2 + 1] = 1.01$$

$$y_1^{(2)} = y_0 + \frac{h}{2}[f(x_0, y_0) + f(x_1, y_1^{(1)})]$$

$$= 1 + \frac{0.01}{2}[1 + [(0.01)^2 + 1.01]]$$

$$= 1.00505$$

$$y_1^{(3)} = 1 + \frac{0.01}{2}[f(x_0, y_0) + f(x_1, y_1^{(2)})]$$

$$= 1 + \frac{0.01}{2}[1 + [(0.01)^2 + 1.00505]]$$

$$= 1 + 0.010026$$

$$= 1.010026$$

$$y_1^{(4)} = 1 + \frac{0.01}{2}[f(x_0, y_0) + f(x_1, y_1^{(3)})]$$

$$= 1 + \frac{0.01}{2}[1 + [(0.01)^2 + 1.010026]]$$

$$= 1 + 0.01005063$$

$$= 1.01005063$$

$$= 1.0101 \ (\text{Correct up to 4 decimal places})$$

Therefore y (0.01) = y_1 = 1.0101

Now $y_2 = y(0.02)$, $h = 0.01$, $x_2 = 0.02$

$$y_2^{(1)} = y_1 + hf(x_1, y_1)$$

$$= 1.0101 + 0.01[(0.01)^2 + 1.0101]$$

$$= 1.020202$$

$$y_2^{(2)} = y_1 + h/2[f(x_1, y_1) + f(x_2, y_2^{(1)})]$$

$$= 1.0101 + \frac{0.01}{2}[(0.1)^2 + 1.0101 + (0.012)^2 + 1.0202]$$

$$= 1.0101 + 0.005[1.0102 + 1.0206]$$

$$= 1.0101 + 0.010154$$

$$= 1.020254$$

$$y_2^{(3)} = y_1 + h/2[f(x_1, y_1) + f(x_2, y_2^{(2)})]$$

$$= 1.0101 + \frac{0.01}{2}[1.0102 + (0.02)^2 + (1.02025)]$$

$$= 1.0101 + 0.01015425$$

$$= 1.02025425$$

Therefore $\quad y_2^{(3)} = y_2^{(2)} = y_2 = 1.02025$ (Correct up to 5 decimal places.)

9.6 RUNGE - KUTTA METHOD

This method is widely used, because it does not require computations of higher derivatives. A large number of modification of Runge – Kutta method have been made to get rid of some of the difficulties particularly regarding the estimation of error and regulating the round off errors. We first consider Runge Kutta method of order 2.

$$\left. \begin{array}{l} y_{n+1} = y_n + \dfrac{(k_1 + k_2)}{2} \\[2mm] \text{where } k_1 = hf(x_n, y_n) \\[2mm] k_2 = hf(x_n + h, y_n + k_1) \end{array} \right\} \qquad \qquad .....(9.2)$$

n = 0, 1, 2, ...

We next consider Runge Kutta Method of order 4.

$$y_{n+1} = y_n + k$$

Where

$$k - (k_1 + 2k_2 + 2k_3 + k_4)/6$$

Where

$$
\begin{aligned}
k_1 &= hf(x_n, y_n) \\
k_2 &= hf(x_n + h/2, y_n + k_1/2) \\
k_3 &= hf(x_n + h/2, y_n + k_2/2) \\
k_4 &= hf(x_n + h, y_n + k_3)
\end{aligned}
\qquad .....(9.3)
$$

n = 0,1,2....

Example 14: Use Runge Kutta method, to calculate y (0.2) and y(0.4) for the equation $\dfrac{dy}{dx} = -xy, y(0) = 1.$

Solution:

$$\frac{dy}{dx} = -xy = f(x, y), x_0 = 0, y(0) = 1, h = 0.2$$

$$k_1 = (0.2)f(x_0, y_0)$$

$$= 0$$

$$k_2 = (0.2)f(0 + \frac{0.2}{2}, 1 + \frac{0}{2})$$

$$= (0.2)[-(0.1) \times 1]$$

$$= -\,0.02$$

$$k_3 = (0.2)f(0 + \frac{0.2}{2}, 1 - \frac{0.02}{2})$$

$$= -0.0198$$

$$k_4 = (0.2)f(0 + 0.2, 1 - 0.0198)$$

$$= -\,0.039208$$

$$k = \frac{1}{6}(k_1 + 2k_2 + 2k_3 + k_4)$$

$$= -\,0.019801$$

$$y(0.2) = y_0 + k$$

$$= 0.98019867$$

Again, $x_0 = 0.2$, $y_0 = 0.98019867$, $h = 0.2$

$$k_1 = (0.2)f(x_0, y_0)$$

$$= -0.03921$$

$$k_2 = (0.2)f(x_0 + h/2, y_0 + k_1/2)$$

$$= -0.05764$$

$$k_3 = (0.2)f(x_0 + h/2, y_0 + k_2/2)$$

$$= -0.05708272$$

$$k_4 = (0.2)f(x_0 + h, y_0 + k_3)$$

$$= -0.073849276$$

$$\therefore \quad k = \frac{1}{6} = \left(k_1 + 2k_2 + 2k_3 + k_4\right) = -0.05708412$$

$$y(0.4) = y_0 + k = 0.92311455; 0.923114.$$

Example 15: Solve initial value problem $10\,\dfrac{dy}{dx} = x^2 + y^2; y(0) = 1$ for $h = 0.1$, Calculate $y(0.2)$ by using Runge Kutta fourth order method and find the solution correct upto 4 places of decimal.

Solution:

$$f(x,y) = \frac{x^2 + y^2}{10}; y_0 = 1, x_0 = 0, h = 0.1$$

$$f(x,y) = \frac{x^2 + y^2}{10}; y_0 = 1, x_0 = 0, h = 0.1$$

$$k_3 = (0.1)f(x_0 + h/2, y_0 + k_2/2)$$

$$= (0.1)\left[\frac{(0.05)^2 + (1.0050625)^2}{10}\right]$$

$$= 0.01012650$$

$$k_4 = (0.1)f(x_0 + h, y_0 + k_3)$$

$$= 0.01030356$$

$$k = \frac{1}{6}(k_1 + 2k_2 + 2k_3 + k_4)$$

$$= \frac{1}{6}(0.06080656) = 0.01013443$$

$$\therefore \quad y(0.1) = y_0 + k = 1 + 0.01013443 = 1.0101344$$

$$k_1 = (0.1)f(x_0, y_0)$$

$$= 0.1 \times \frac{1}{10}$$

$$= 0.1 \times 0.1$$

$$= 0.01$$

$$k_2 = (0.1)f(x_0 + h/2, y_0 + k_1/2)$$

$$= (0.1)\left[\frac{(0.05)^2 + (1.005)^2}{10}\right]$$

$$= 0.010125$$

Now $x_0 = 0.1$ $y_0 = 1.0101344$, $h = 0.1$

$$k_1 = hf(x_0, y_0)$$

$$= (0.1)\left[\frac{(0.1)^2 + (1.0101344)^2}{10}\right]$$

$$= 0.01030372$$

$$k_2 = hf(x_0 + h/2, y_0 + k_1/2)$$

$$= (0.1)\left[\frac{(0.15)^2 + (1.015286)^2}{10}\right]$$

$$= 0.010533$$

$$k_3 = hf(x_0 + h/2, y_0 + k_2/2)$$

$$= (0.1)\left[\frac{(0.15)^2 + (1.0154009)^2}{10}\right]$$

$$= 0.010535$$

$$k_4 = hf(x_0 + h, y_0 + k_3)$$

$$= (0.1)\left[\frac{(0.2)^2 + (1.020669)^2}{10}\right]$$

$$= 0.010818$$

$$k = \frac{1}{6}(k_1 + 2k_2 + 2k_3 + k_4) = 0.0104665$$

$$y(0.2) = y_0 + k = 1.0101344 + 0.0104665$$

$$= 1.0206009$$

Example 16: Find $y(1.1)$ using Runge Kutta method of the fourth order, given that $\frac{dy}{dx} = y^2 + xy, y(1) = 1$ and $h = 0.1$.

Solution: $f(x,y) = y^2 + xy, x_0 = 1, y_0 = 1, h = 0.1$

$$k_1 = (0.1)f(x_0, y_0)$$

$$= (0.1)[1^2 + 1.1]$$

$$= (0.1) \times 2$$

$$= 0.2$$

$$k_2 = (0.1)f(x_0 + h/2, y_0 + k_1/2)$$

$$= (0.1)[(1.1)^2 + (1.1) \times (1.05)] = 0.2365$$

$$k_3 = (0.1)f\left(x_0 + \frac{h}{2}, y_0 + \frac{k_2}{2}\right) + [1.11825)^2 + (1.11825)(1.05)]$$

$$= (0.1)[(1.11825)^2 + (1.11825)(1.05)]$$

$$= 0.24246$$

$$k_4 = (0.1)f(x_0 + h, y_0 + k_3)$$

$$= (0.1)[(1.24246)^2 + (1.24246) \times (1.1)] = 0.29104$$

$$k = \frac{1}{6}(k_1 + 2k_2 + 2k_3 + k_4)$$

$$= 0.241493$$

$$\therefore \quad y(1.1) = y_0 + k$$

$$= 1 + 0.241493$$

$$= 1.241493$$

Example 17: Find the values of $y(0.1)$, $y(0.2)$ and $y(0.3)$ using Runge Kutta method of the fourth order, given that $\frac{dy}{dx} = xy + y^2, y(0) = 1$.

Solution: Here $x_0 = 0$, $y_0 = 1$, $h = 0.1$ and $f(x,y) = xy + y^2$

$$k_1 = 0.1(0 + 1) = 0.1$$

$$k_2 = (0.1)[(0.05)(1.05) + (1.05)^2] = 0.1155$$

$$k_3 = (0.1)[(0.05)(1.05775) + (1.05775)^2] = 0.1172$$

$$k_4 = (0.1)[(0.1)(1.1172) + (1.1172)^2] = 0.13598$$

$$k = \frac{1}{6}[k_1 + 2k_2 + 2k_3 + k_4] = \frac{0.7014}{6} = 0.1169$$

$$y(0.1) = y_0 + k = 1 + 0.1169 = 1.1169$$

$$y(0.1) = 1.1169, \quad x_1 = 0.1, \quad h = 0.1$$

$$k_1 = (0.1)[(0.1)(1.1169) + (1.1169)^2] = 0.1359$$

$$k_2 = (0.1)[(0.15)(1.1849) + (1.1849)^2] = 0.1582$$

$$k_3 = (0.1)[(0.15)(1.196) + (1.196)^2] = 0.1610$$

$$k_4 = (0.1)[(0.2)(1.2779) + (1.2779)^2] = 0.1889$$

$$k = \frac{1}{6}[k_1 + 2k_2 + 2k_3 + k_4] = 0.1605$$

$$y(0.2) = 1.2774$$

$$y(0.2) = 1.2774, \quad x_2 = 0.2 \quad h = 0.1$$

$$k_1 = (0.1) \times [(0.2)(1.2774) + (1.2774)^2] = 0.1887$$

$$k_2 = (0.1)[(0.25)(1.3718) + (1.3718)^2] = 0.2225$$

$$k_3 = (0.1)[(0.25)(1.3886) + (1.3886)^2] = 0.2275$$

$$k_4 = (0.1)[(0.3)(1.5049) + (1.5049)^2] = 0.2716$$

$$k = \frac{1}{6}[k_1 + 2k_2 + 2k_3 + k_4] = 0.2267$$

$$y(0.3) = 1.5041$$

Example 18: Using 4th order Runge Kutta method evaluate the value of y when x = 1.2 given $\dfrac{dy}{dx} + \dfrac{y}{x} = \dfrac{1}{x^2}, y(1) = 1.$

Solution: $\qquad x_0 = 1, y_0 = 1, f(x,y) = \dfrac{1-xy}{x^2}, h = 0.2$

$$k_1 = hf(x_0, y_0)$$

$$= (0.2)\left[\frac{1 - 1 \times 1}{1^2}\right]$$

$$= 0$$

$$k_2 = hf(x_0 + h/2, y_0 + k_1/2)$$

$$= (0.2)\left[\frac{1 - (1.1) \times 1}{(1.1)^2}\right]$$

$$= -0.016529$$

$$k_3 = hf(x_0 + h/2, y_0 + k_2/2)$$

$$= (0.2)\left[\frac{1 - (1.1) \times 0.9917355}{(1.1)^2}\right]$$

$$= -0.015026$$

$$k_4 = hf(x_0 + h, y_0 + k_3)$$

$$= (0.2)\left[\frac{1 - (1.2) \times (0.98497)}{(1.2)^2}\right] = -0.025273$$

$$k = \frac{1}{6}(k_1 + 2k_2 + 2k_3 + k_4)$$

$$= -0.0147305$$

$$y(1.2) = 1 - 0.0147305$$

$$= 0.9852695$$

$$= 0.98527 \ (\text{Correct upto 5 decimal places})$$

Then $x_0 = 0.05, y_0 = y(0.05) = 1.0525, h = 0.05$

$$k_1 = hf(x_0, y_0)$$

$$= (0.05)(0.05 + 1.0525)$$

$$= 0.055125$$

$$k_2 = hf(x_0 + h/2, y_0 + k_1/2)$$

$$= (0.05)[0.075 + 1.08006]$$

$$= 0.057753$$

$$k_3 = hf(x_0 + h/2, y_0 + k_2/2)$$

$$= (0.05)[.075 + 1.08138]$$

$$= 0.057819$$

$$k_4 = hf(x_0 + h, y_0 + k_3)$$

$$= (0.05)[0.10 + 1.1103]$$

$$= 0.0605$$

$$\therefore \quad k = \frac{1}{6}(k_1 + 2k_2 + 2k_3 + k_4) = 0.05779$$

$$\therefore \quad y(0.1) = y_0 + k = 1.1103$$

$$k_4 = hf(x_0 + h, y_0 + k_3)$$

$$= (0.05)[0.1 + 1.11032]$$

$$= 0.06052$$

$$k = \frac{1}{6}(k_1 + 2k_2 + 2k_3 + k_4)$$

$$= 0.057798$$

$$y(0.1) = 1.0525 + 0.057798$$

$$= 1.110298$$

$$= 1.1103 \ (\text{Correct up to 4 decimal figures})$$

Example 19: Use Range – Kulta Method to compute y (0.4) from $\dfrac{dy}{dx} = xy$ with y (0) = 2 taking h = 0.1

Solution: First we take $x_0 = 0$, $y_0 = 2$ and f (x, y) = xy with h = 0.1

$$k_1 = hf(x_0, y_0) = 0.00$$

$$k_2 = hf\left(x_0 + \frac{h}{2}, y_0 + \frac{k_1}{2}\right)$$

$$= (0.2)f(0.05, 2)$$

$$= (0.1) \times 0.05 \times 2$$

$$= 0.01$$

$$k_3 = hf\left(x_0 + \frac{h}{2}, y_0 + \frac{k_2}{2}\right)$$

$$= (0.1)f(0.05, 2.005)$$

$$= 0.1 \times 0.05 \times 2.005$$

$$= 0.010025$$

$$k_4 = hf(x_0 + h, y_0 + k_3)$$

$$= (0.1)\,f(0.1, 2.010025)\ 0.1 \times 0.1 \times 2.010025$$

$$= 0.02010025$$

$$\therefore \quad k = \frac{1}{6}(k_1 + 2k_2 + 2k_3 + k_4)$$

$$= \frac{1}{6}\,(0.06015025) = 0.010025041$$

$$\therefore \quad y(0.1) = y_0 + k$$

$$= 2 + 0.010025401$$

$$= 2.01003$$

Again, $x_1 = 0.1$, $y_1 = 2.01004$, $h = 0.1$

$$k_1 = hf(x_1, y_1) = 0.1 \times 0.1 \times 2.01003 = 0.0201003$$

$$k_2 = hf\left(x_1 + \frac{h}{2}, y_1 + \frac{k_1}{2}\right)$$

$$= (0.1)\,f(0.15, 2.01003 + 0.01005) = 0.1 \times 0.15 \times 2.02008 = 0.0303012$$

$$k_3 = h_f\left(x_1 + \frac{h}{2}, y_1 + \frac{k_2}{2}\right)$$

$$= (0.1)f(0.15, 2.01003 + 0.015151) = 0.1 \times 0.15 \times 2.025181 = 0.03038$$

$$k_4 = hf(x_1 + h, y_1 + k_3)$$

$$= (0.1)f(0.2, 2.01003 + 0.03038) = 0.1 \times 0.2 \times 2.04041 = 0.04081$$

$$\therefore \quad k = \frac{1}{6}(k_1 + 2k_2 + 2k_3 + k_4) = 0.030378$$

$$\therefore \quad y_2 = y(0.2) = y_1 + k = 2.01003 + 0.030378 = 2.040408$$

Here $x_2 = 0.2$, $y_2 = 2.040408$, $h = 0.1$

$$k_1 = hf(x_2, y_2) = 0.1 \times 0.2 \times 2.040408 = 0.0408082$$

$$k_2 = hf\left(x_2 + \frac{h}{2}, y_2 + \frac{k_1}{2}\right)$$

$$= (0.1)f(0.25, 2.040408 + 0.0204041)$$

$$= 0.1 \times 0.25 \times 2.0608121$$

$$= 0.0515203$$

$$k_3 = hf\left(x_2 + \frac{h}{2}, y_2 + \frac{k_2}{2}\right)$$

$$= (0.1)f(0.25, 2.040408 + 0.015760)$$

$$= 0.1 \times 0.25 \times 2.066168 = 0.0516542$$

$$k_4 = hf\left(x_2 + h, y_2 + k_3\right)$$

$$= (0.1)f(0.3, 2.040408 + 0.0516542)$$

$$= 0.1 \times 0.3 \times 2.0920622 = 0.062762$$

$$\therefore \ k = \frac{1}{6}(k_1 + 2k_2 + 2k_3 + k_4) = \frac{0.3099192}{6} = 0.0516532$$

$$\therefore \ y_3 = y(0.3) = y_2 + k = 2.040408 + 0.0516532 = 2.0920612$$

Again $\qquad\qquad x_3 = 0.3, \ y_3 = 2.0920612, \ h = 0.1$

$$\therefore \ k_1 = hf(x_3, y_3) = 0.1 \times 0.3 \times 2.0920612 = 0.0627618$$

$$k_2 = hf\left(x_3 + \frac{h}{2}, y_3 + \frac{k_1}{2}\right)$$

$$= (0.1)f(0.35, 2.0920612 + 0.031381)$$

$$= 0.1 \times 0.35 \times 2.1234421$$

$$= 0.0743205$$

$$k_3 = hf\left(x_3 + \frac{h}{2}, y_3 + \frac{k_2}{2}\right)$$

$$= (0.1)f(0.35, 2.0920612 + 0.0371602)$$

$$= 0.1 \times 0.35 \times 2.1239921 = 0.074523$$

$$k_4 = hf(x_3 + h, y_3 + k_3)$$

$$= (0.1)f(0.4, 2.0920612 + 0.074523)$$

$$= 0.1 \times 0.4 \times 2.1665842$$

$$= 0.086663$$

$$\therefore k = \frac{1}{6}(k_1 + 2k_2 + 2k_3 + k_4) = 0.074519$$

$$\therefore y_4 = y(0.4) = y_3 + k = 2.1665802 \; ; 2.16658 \text{ (Correct up to 5 decimal places)}$$

Example 20: Use Range – Kutta Method to compute y (0.2) from $\dfrac{dy}{dx} = \dfrac{1}{x+y}$ with y (0) = 1 taking h = 0.1

Solution: First we take $x_0 = 0$, $y_0 = 1$ and $f(x,y) = \dfrac{1}{x+y}$ with h = 0.1

$$k_1 = hf(x_0, y_0) = (0.1) \times \frac{1}{0+1} = 0.1$$

$$k_2 = hf\left(x_0 + \frac{h}{2}, y_0 + \frac{k_1}{2}\right)$$

$$= (0.1)f(0.05, 1.05)$$

$$= (0.1) \times \frac{1}{(0.05 + 1.05)}$$

$$= 0.09091$$

$$k_3 = hf\left(x_0 + \frac{h}{2}, y_0 + \frac{k_2}{2}\right)$$

$$= (0.1) \times f(0.05, 1.04545)$$

$$= (0.1) \times \frac{1}{(0.05 + 1.4545)}$$

$$= 0.09129$$

$$k_4 = hf(x_0 + h, y_0 + k_3)$$

$$= (0.1) \times f(0.1, 1.09129)$$

$$= (0.1) \times \frac{1}{(0.1 + 1.09129)} = 0.083943$$

$$\therefore \quad k = \frac{1}{6}(k_1 + 2k_2 + 2k_3 + k_4)$$

$$= 0.091390$$

$$\therefore \quad y(0.1) = y_0 + k = 1 + 0.091390 = 1.0914$$

Now, $\quad x_1 = 0.1,\ y_1 = 1.0914\ \ h = 0.1$

$$k_1 = hf(x_1, y_1) = (0.1) \times \frac{1}{(0.1 + 1.0914)} = \frac{0.1}{1.1914} = 0.083043$$

$$k_2 = hf\left(x_1 + \frac{h}{2}, y_1 + \frac{k_1}{2}\right)$$

$$= (0.1) \times f(0.15, 1.0914 + 0.04197)$$

$$= (0.1) \times \frac{1}{(0.15 + 1.13337)} = 0.07792$$

$$k_3 = hf\left(x_1 + \frac{h}{2}, y + \frac{k_2}{2}\right)$$

$$= (0.1) \times \frac{1}{(0.15 + 1.13036)}$$

$$= 0.07810$$

$$k_4 = hf\left(x_1 + h, y_1 + k_3\right)$$

$$= (0.1) \times \frac{1}{(0.2 + 1.1695)}$$

$$= 0.07302$$

$$\therefore \quad k = \frac{1}{6}(k_1 + 2k_2 + 2k_3 + k_4) = 0.078167$$

$$\therefore \quad y(0.2) = y_1 + k = 1.0914 + 0.078167 = 1.169567$$

$$= 1.1696 \text{ (Correct up to 4 decimal gleam)}$$

9.7 ADAM MOULTON METHOD

We have

$$y_{n+1} = y_n + \int_{x_n}^{x_{n-1}} f(x, y(x))dx$$

Replacing f[x,y(x)] by a 4 point Newton's backward difference interpolation formula at $x_{n+1}, x_n, x_{n-1}, x_{n-2}$ we have $y_{n+1} = y_n + \dfrac{h}{24}(9f_{n+1} + 12f_n - 5f_{n-1} + f_{n-2})$ .....(9.4)

The above rule is known as Adam – Moulton predictor formula. The corrector formula is

$$y_{n+1}^k = y_n + \frac{h}{24}[9f(x_{n+1}, y_{n+1}^k) + 19f_n - 5f_{n-1} + f_{n-2}]$$

9.8 MILNE'S METHOD

Let us consider the differential equation of the form $\dfrac{dy}{dx} = f(x,y)$ with $y(x_0) = y_0$. Integrating both sides of the equation taking limits from x_0 to x_4 and using Newton's forward difference formula we get. $y_{k+1} = y_{k-3} + \dfrac{4h}{3}[2f_{k-2} - f_{k-1} + 2f_k]$ Which is Milne's Predictor formula.

Milne's Corrector formula is

$$y_{k+1}^c = y_{k-1} + \frac{h}{3}[f_{k-1} + 4f_k + f_{k+1}^p]$$

Where $y_{k+1}^p = y_{k+1}$ in the predictor formula.

Example 21: Solve $y^1 = 1 + xy^2, y(0) = 1$ for x = 0.4 using Milne's predictor corrector method when it is given that y(0.1) = 1.105, y(0.2) = 1.223, y(0.3) = 1.354.

Solution: The differential equation yields
$y^1(0) = y_0^1 = 1, y^1(0.1) = y_1^1 = 1.1221, y^1(0.2) = y_2^1 = 1.2992, y^1(0.3) = y_3^1 = 1.55$

The Milne's predictor formula gives $y_4 = y_0 + \dfrac{4 \times 0.1}{3}[2 \times (1.1221) - 1.2992 + 2 \times (1.55)]$

$= 1.324$

Putting this value in the differential equations we get $y_4^1 = 1 + .4 \times (1.324)^2 = 1.701$

The Milne's Corrector formula gives a new approximation.

$$y_4 = y_2 + \frac{0.1}{3}[1.2992 + 4 \times 1.55 + 1.701] = 1.53$$

Again

$$y_4^1 = 1 + (0.4) \times (1.53)^2 = 1.936$$

Reapplying the corrector formula we get

$$y_4 = y_2 + \frac{0.1}{3}[1.2992 + 4 \times 1.55 + 1.936]$$

$$= 1.538$$

$$y_4^1 = 1 + (.4) \times (1.538)^2$$

$$= 1.946$$

$$y_4 = y_2 + \frac{0.1}{3}[1.2992 + 4 \times 1.55 + 1.946]$$

$$= 1.538$$

Example 22: Use Adam-Moultons Method to find y(1.4) from the differential equation

$$\frac{dy}{dx} = x^2 + y^2, y(1) = 0$$

Solution:

$$x = 1.0; y = 0; y^1 = x^2 + y^2 = 1.0$$

$$x = 1.1; y = 0.11072; y^1 = 1.22226$$

$$x = 1.2; y = 0.24631; y^1 = 1.50067$$

$$x = 1.3; y = 0.41357; y^1 = 1.86104$$

$$y_n = 0.41357, y_{n-1} = 0.24631, y_{n-2} = 0.11072, \ y_{n-3} = 0.0$$

Then

$$y_{(1.4)}^p = 0.41357 + \frac{0.1}{24}[0.55 \times 1.86104 - 59 \times 1.50067 + 37 \times 1.22226 - 9]$$

$$= 0.62208$$

$$y_{(1.4)}^c = 0.41357 + \frac{0.1}{4}[9 \times (1.96 + 0.38698) + 19 \times 1.86104 - 5 \times 1.50067$$

$$+ 1.22226]$$

$$= 0.62274$$

9.9 ERROR IN DIFFERENT DIFFERENTIAL EQUATION METHODS

The error terms of different differential equation

(i) Taylor series method $= \dfrac{h^{k+1}}{(k+1)!} f_{(\xi)}^{(k+1)}, x_k < \xi < x_k + 1$

(ii) Euler's method $\dfrac{h^2}{2} y^{11}(c) x_0 < c < x_0 + h$.

(iii) Modified Euler's method: the error is of $0(h3)$

(iv) Runge Kutta method of order 2. $\dfrac{h^3}{12}[f_{xx} + 2f.f_{xy} + f_{yy}f^2 - 2f_x f_y - 2f_y^2 f]$ which is of order (h^3)

(v) Runge kutta method of order is $0(h5)$

(vi) Adam Moulton method $-19h^5 f^4(s)1720,\ x_{n-2} < s < x_{n+1}$

(vii) Milnes method $14h^5 f^4(\eta)/45,\ x_{n-3} < \eta < x_{n+1}$

EXERCISE

1. What are corrector and predictor method?

Solution:

These methods are two formula of multistep numerical integration. The predictor formula is the usual open type explicit formula derived by using polynomials which interpolate at the points: $x_n, x_{n-1}, ...x_{n-m}$ where as the corrector is a closed type or implicit formula derived by using interpolating polynomial using the points $x_{n+1}, x_n ......, x_{n-p}$.

2. Solve by Euler's method and modified Euler method:

(i) $y^1 = 2x + y, y(0) = 0, y(0.5), h = 0.1$

(ii) $y^1 = \dfrac{1+y}{x}, y(2) = 3, y(2.8), h = 0.2$

(iii) $y^1 = x^2 - y, y(1) = 0, y(1.6), h = 0.1$

[*Ans:* 1.95, 4.60, 0.81]

3. Solve by Runge Kutta method.

(i) $\dfrac{dy}{dx} = x - y^2, y(0)=1;\ y(0.2), y(0.4)$ (ii) $\dfrac{dy}{dx} = \dfrac{1}{x+y}, y(0) =1,\ x = 0.5, 1.0, 1.5, 2.0$

Solution:

(i) 0.851, 0.780

(ii) $y(0.5) = 1.3571$

$y(1.0) = 1.5837$

$y(1.5) = 1.7565$

$y(2.0) = 1.8950$

CHAPTER 10

Inverse Interpolation

10.1 INTRODUCTION

Inverse Interpolation is defined as the process of finding the value of the argument corresponding to a given value of the function lying between two tabulated functional values.

In this chapter, we introduce the methods to find the value of the argument x for a given value of y by the following methods

1. Lagrange's inverse interpolation formula
2. Successive approximation method or iteration method
3. Reversion of series method

10.2 LAGRANGE'S INVERSE INTERPOLATION FORMULA

In this section we derive a formula for finding the value of x for a given value of y from Lagrange's interpolation formula

Let $y = f(x)$ be a function which takes the values $y_0, y_1, \ldots .y_n$ corresponding to the values $x_0, x_1, \ldots x_n$ of x. The Lagrange's formula for finding the value of y for a given

value of x is $y = \dfrac{(x-x_1)(x-x_2)(x-x_3)\ldots ..(x-x_n)}{..(x_0-x_1).(x_0-x_2).(x_0-x_3) \quad \ldots \quad (.x_0-x_n)} y_0$

$$+ \dfrac{(x-x_0)(x-x_2)(x-x_3)\ldots ..(x-x_n)}{(x_1-x_0).(x_1-x_2).(x_1-x_3) \quad \ldots \quad (x_1-x_n)} y_1 + \ldots +$$

$$\dfrac{(x-x_0)(x-x_2)(x-x_3)\ldots (x-x_{n-1})}{(x_n-x_0).(x_n-x_2).(x_n-x_3) \ldots (x_n-x_{n-1})} y_n$$

Replacing y by   x in the above formula we get

$$X = \dfrac{(y-y_1)(y-y_2)(y-y_3)\ldots ..(y-y_n)}{..(y_0-y_1).(y_0-y_2).(y_0-y_3) \quad \ldots \quad (.y_0-y_n)} x_0$$

$$+ \dfrac{(y-y_0)(y-y_2)(y-y_3)\ldots ..(y-y_n)}{(y_1-y_0).(y_1-y_2) \ldots (y_1-y_n)} x_1 + \ldots + \dfrac{(y-y_0)(y-y_2)(y-y_3)\ldots (y-y_{n-1})}{(y_n-y_0)(y_n-y_2).(y_n-y_3) \ldots (y_n-y_{n-1})} x_n$$

Which is the Lagrange's inverse formula for finding the value of x for a given value of y

Example 1: Find the value of x for y = 5 from the table

x	1	3	4
y	3	12	19

Solution: We have $x_0 = 1, x_1 = 3, x_2 = 4$

$y_0 = 3, y_1 = 12, x_2 = 19$ $X = \dfrac{(y-y_1)(y-y_2)}{.(y_0-y_1).(y_0-y_2).} x_0 + \dfrac{(y-y_0)(y-y_2)}{(y_1-x_0).(y_1-x_2)} x_1 + \dfrac{(y-y_0)(y-y_1)}{(y_2-y_0)(y_2-y_1).} x_2$

Substituting the values of x and y in the above formula we get $X = \dfrac{(5-12)(5-19))}{.(3-12).(3-19)} \times 1 +$

$+ \dfrac{(5-3)(5-19)}{(12-3).(12-19)} \times 3 + \dfrac{(5-3)(5-12)}{(19-3)(19-12).} \times 4 = 0.6805 + 1.3333 - 0.5000 = 1.5138$

Example 2: Find the value of x for which f(x) = 59 given the data f(1) = -1 , f(3) = 23 , f(5) = 119

Solution:

x	0	1	3
y	1	3	13

We have $x_0 = 0, x_1 = 1, x_2 = 3$ and $f_0 = -1, f_1 = 23, f_2 = 19$ Substituting in

$X = \dfrac{(f-f_1)(f-f_2)}{.(f_0-f_1).(f_0-f_2).} x_0 + \dfrac{(f-f_0)(f-f_2)}{(f_1-f_0).(f_1-f_2)} x_1 + \dfrac{(f-f_0)(f-f_1)}{(f_2-f_0)(f_2-f_1).} x_2$

We get $\qquad x = \dfrac{(59-23)(59-119)}{.(-1-23).(-1-119).} (1)$

$\qquad + \dfrac{(59+1)(59-119)}{(23+1).(23-119)} (3) + \dfrac{(59+1)(59-119)}{(110-1)(119-23).} (5)$

$\qquad = -0.75 + 6.875 + 0.9575 = 4.875$

EXERCISE

1. Find the value of x for which x = 59 F(1) = -1 , F(3) = 23 , F (5) = 119

Ans: 4.875

2. Given are the values of x and y find x for which f(x) = 101

Y = f(x)	46	66	81	93
x	0	10	20	30

Ans: 37.82

3. By Lagrange's formula for inverse interpolation determine the value of t when A = 85 given that

t	2	5	8	4
A	94.8	87.9	81.3	68.7

Ans : 6.5928

4. Find the value of x when y = 19 given

x	0	1	2
y	0	1	20

Ans : 14.6666

5. Find the value of x when y = 0.3 given

x	0.4	0.6	0.8
y	0.3683	0.3332	0.2897

Ans: 0.0.07575

6. Find the value of x when y = 100 given that

x	3	5	7	9	11
y	6	24	58	108	108

Ans: 8.6556

CHAPTER 11

Curve Fitting

11.1 INTRODUCTION

In this chapter we introduce Curve fitting , It is the process of constructing a curve or mathematical function that has the best fit to a series of data points .The fitting of curves of a given type are generally not unique. The main objective is to find the parameters of a mathematical model that describes a set of data in a way that minimizes the difference between the model and the data.

The best fitting curve with minimum deviations from all data points can be obtained by the method of least squares.

11.2 THE METHOD OF LEAST SQUARES

The method of least squares assumes that the best fit curve of a given type is the curve that has the minimal sum of deviations i.e. least square error from a given set of data .

Suppose that the data points are $(x_1 ,y_1) , (x_2, y_2), \ldots , (x_n \cdot y_n)$ where x is the independent variable and y is the dependent variable. The fitting curve f(x) has the deviation (error) e_i from each data point, i.e.

$$e_1 = y_1 - f(x_1),$$

$$e_2 = y_2 - f(x_2),$$

.

.

.

$$e_n = y_n - f(x_n)$$

According to the method of least squares, the best fitting curve has the property that:

$$\sum_1^n e_i^2 = \sum_1^n [y_i - f(x_i)]2$$ Is minimum we now introduce the method of least squares using polynomials.

11.3 THE LEAST-SQUARES LINE

The least-squares line uses a straight line $Y = a + bx$ to approximate the given set of data, $(x_1, y_1), (x_2, y_2), \ldots, (x_n \cdot y_n)$ where $n \geq 2$. The best fitting curve $y = f(x)$ has the least square error. Let E denote the error where E is a function of two variables a and b.

Then $E = \sum_1^n e_i^2 = \sum_1^n [y_i - f(x_i)]\,2 = \sum_1^n [y_i - (a + b\,x_i)]\,2$ is minimum, where a and b are unknown coefficients. The necessary conditions for E to be minimum are

$$\frac{\partial E}{\partial a} = 0 \text{ and } \frac{\partial E}{\partial b} = 0 \text{ that is } \frac{\partial E}{\partial a} = 2\sum_1^n [y_i - (a + bx_i)]\,(-1) = 0 \text{ and } \frac{\partial E}{\partial b} = 2\sum_1^n (-x_i)[y_i - (a + bx_i)] = 0$$

Expanding the above equations we have

$$\sum_1^n y_i = a\sum_1^n 1 + b\sum_1^n x_i$$

$$\sum_1^n x_i y_i = a\sum_1^n x_i + b\sum_1^n x_i^2 \text{ Or } \sum_1^n y_i = n\,a + b\sum_1^n x_i \qquad \ldots (11.1)$$

$$\sum_1^n x_i y_i = a\sum_1^n x_i + b\sum_1^n x_i^2 \qquad \ldots (11.2)$$

The equations (1) and (2) are known as the normal equations. Solving the normal equations we get $\quad a = \dfrac{\sum y_i \sum x_i^2 - \sum x_i \sum x_i y_i}{n\sum x_i - (\sum x_i)2}$ and $b = \dfrac{n\sum x_i yi - \sum x_i \sum y_i}{n\sum x_i - (\sum x_i)2}$ Remark. The normal equations for fitting the least square line $Y = a\,x + b$ are $\sum_1^n y_i = nb + a\sum_1^n x_i$

$$\sum_1^n x_i y_i = b\sum_1^n x_i + a\sum_1^n x_i^2.$$

Example 1: Fit the straight line $y = \alpha x + b$ for the data given below

x	5	10	15	20	25
y	16	19	23	26	30

Solution: We have the following table

X_i	Y_i	x_i^2	$x_i y_i$
5	16	25	80
10	19	100	190
15	23	225	345
20	26	400	520
25	30	625	750
75	114	1375	1885

From the table we have $n = 5$, $\sum x_i = 75$, $\sum y_i = 114$, $\sum x_i^2 = 1375$, and $\sum x_i y_i = 1885$

The normal equations for fitting the least square line $y = a\,x + b$ are

$$\sum_1^n y_i = nb + a\sum_1^n x_i = \sum_1^n y_i$$

$$\sum_1^n x_i y_i = b\sum_1^n x_i + a\sum_1^n x_i^2 = \sum_1^n x_i y_i$$

Substituting the values in normal equations, we get 75a + 5b = 114 and 1375 a + 75b = 1885 solving these equations, we get a = 0.7, b = 12.3 therefore the required least square line is y = 0.7 x + 12.3.

Example 2: Find the least square line y = a + b x for the data points (-1, 10), (0,9), (1,7), (2,5), (3,4), (4,3), (5,0) and (6, -1).

Solution: Here

X_i	Y_i	x_i^2	$x_i y_i$
-1	10	1	-10
0	9	0	0
1	7	1	7
2	5	4	10
3	4	9	12
4	3	16	12
5	0	25	0
6	-1	36	-6
20	37	92	25

From the table we have n = 8, $\sum x_i$ = 20, $\sum y_i$ = 37, $\sum x_i^2$ = 92, and $\sum x_i y_i$ = 25

The normal equations are n a + b $\sum_1^n x_i = \sum_1^n y_i$ .....(11.3)

$$a\sum_1^n x_i + b \sum_1^n x_i^2 = \sum_1^n x_i y_i$$ .....(11.4)

Putting the values in normal equations

$$8a + 20b = 37$$

and $$20a + 97b = 25$$

and solving these equations, we get

$$a = 8.6428571$$

$$b = -1.6071429$$

hence the required least square line is

$$y = 8.6428571 + (-1)\ 1.6071429x$$

I.e., $$y = -1.6071429x + 8.6428571$$

11.4 FITTING A PARABOLA BY THE METHOD OF LEAST SQUARES

let$(x_1, y_1), (x_2, y_2) , \ldots, (x_n . y_n)$, (where n $\geq$ 2) denote a set of n observations of two variables of two variables x and y . The best fitting Parabola y = a + bx +c x^2 has the least error. Let E denote the error where E is a function of two variables a, b and c.

Then $E = \sum_1^n e_i^2 = \sum_1^n [y_i - f(x_i)]^2 = \sum_1^n [y_i - f(a + b\,x_i + c\,x_i^2)]^2$ is minimum, where a, b, c are unknown coefficients. The necessary conditions for E to be minimum are $\dfrac{\partial E}{\partial a} = 0$ and $\dfrac{\partial E}{\partial b} = 0$, $\dfrac{\partial E}{\partial c} = 0$ that is we have $\dfrac{\partial E}{\partial a} = \sum_1^n 2[y_i - (a + b\,x_i + c\,x_i^2)]$ $\qquad$ (-1) $= 0$

on simplifying we get

$$na + b\sum x_i + c\sum x_i^2 = \sum y_i \qquad \qquad \dots\dots(11.5)$$

$$\frac{\partial E}{\partial b} = \sum_1^n 2[y_i - (a + b\,x_i + c\,x_i^2)]\,(-x_i) = 0$$

on simplifying we get

$$a\sum x_i + b\sum x_i^2 + c\sum x_i^3 = \sum x_i y_i \qquad \qquad \dots\dots(11.6)$$

$$\frac{\partial E}{\partial c} = \sum_1^n 2[y_i - (a + b\,x_i + c\,x_i^2)]\,(-x_i^2) = 0$$

on simplifying we get

$$a\sum x_i^2 + b\sum x_i^3 + c\sum x_i^4 = \sum x_i^2 y_i \qquad \qquad \dots\dots(11.7)$$

Hence the normal equations for fitting the least square curve $y = a + bx + c\,x^2$ (Parabola) are

$$na + b\sum x_i + c\sum x_i^2 = \sum y_i$$

$$a\sum x_i + b\sum x_i^2 + c\sum x_i^3 = \sum x_i y_i$$

$$a\sum x_i^2 + b\sum x_i^3 + c\sum x_i^4 = \sum x_i^2 y_i$$

Dropping the suffix we can write

$$na + b\sum x + c\sum x^2 = \sum y$$

$$a\sum x + b\sum x^2 + c\sum x^3 = \sum xy$$

$$a\sum x^2 + b\sum x^3 + c\sum x^4 = \sum x^2 y$$

We can find the values of a, b and c by solving these equations. Substituting the values of a, b, c in $y = a + bx + c\,x^2$ we get the required equation of the curve.

Example 3: Fit a least square parabola $y = a + b\,x + c\,x^2$ by the method of least squares for the data given below

x	0	1	2	3	4
y	1	1.8	1.3	2.5	2.3

Solution: We have the following table

x	y	xy	x^2y	x^2	x^3	x^4
0	1	0	0	0	0	0
1	1.8	1.8	1.8	1	1	1
2	1.3	2.6	5.2	4	8	16
3	2.5	7.5	22.5	9	27	81
4	2.3	9.2	36.8	16	64	256
$\sum x = 10$	$\sum y = 8.9$	$\sum xy = 21.1$	$\sum x^2y = 66.3$	$\sum x^2 = 30$	$\sum x^3 = 100$	$\sum x^4 = 354$

The required normal equations are

$$5a + 10b + 30c = 8.9$$

$$10a + 30b + 100c = 21.1$$

$$30a + 100b + 354c = 66.3$$

Solving these equations we get $a = 1.078$, $b = 0.414$, $c = -0.021$

Therefore the equation of the curve is $y = 1.078 + 0414x - 0021x^2$

Example 4: Find a formula for the line of the form $y = a + bx + cx^2$ which will fit the following data:

x	0	0.1	0.2	0.3	0.4	0.5	0.6	0.7	0.8	0.9
y	3.1950	3.2299	3.2532	3.2611	3.2516	3.2282	3.1807	3.1266	3.0594	2.9759

Solution: The normal equations are

$$10a + 4.5b + 2.85c = 31.7616$$

$$4.5a + 2.85b + 2.025c = 14.0896$$

$$2.85a + 2.0256 + 1.5333c = 8.82881$$

and solving these equations we obtain

$$a = 3.1951, b = 0.44254, c = -0.76531.$$

$\therefore$ The required equation is

$$y = 3.1951 + 0.44254x - 0.76531x^2.$$

Example 5: Find a second degree $y = ax^2 + bx + c$ in the least square sense for the following data

x	1	2	3	4	5
Y	10	12	13	16	19

Solution: Let $U = x - 3$, $V = y - 14$

We have the following table

x	y	U	V	UV	U²V	U²	U³	U⁴
1	10	−2	−4	8	−16	4	−8	16
2	12	−1	−2	2	−2	1	−1	1
3	13	0	−1	0	0	0	0	0
4	16	1	2	2	2	1	1	1
5	19	2	5	10	20	4	8	16
Total	-	0	0	22	4	10	0	34

The normal equations are

$$na + b\sum U + c\sum U^2 = \sum V$$

$$a\sum U + b\sum U^2 + c\sum U^3 = \sum UV$$

$$a\sum U^2 + b\sum U^3 + c\sum U^4 = \sum U^2 V$$

Substituting the values of a, b and c, the normal equations can be written as

$$10a + 5c = 0$$

$$10b = 22$$

$$34a + 10c = 4$$

Solving the above equations we get c = 0.29, b = 2.2 and a = − 0.58

The equation the Parabola is $V = a + b U + c U^2$

i.e. $y - 14 = a + b (x\text{-}3) + c (x - 3)^2$

or $y - 14 = -0.58 + 2.2 (x\text{-}3) + (0.29)(x - 3)^2$

Hence the required equation of the Parabola is

$$y = 9.43 + 0.46 x + 0.29x^2$$

Fitting the exponential curve of the form $y = ae^{bx}$:

We explain the method of fitting the curve of the form $y = ae^{bx}$ with the help of the following example

Example 6: Fit a least square curve of the form $y = ae^{bx}$ (a > 0) to the data given below:

X	1	2	3	4
Y	1.65	2.70	4.50	7.35

Solution: Consider $y = ae^{bx}$

Applying logarithms (with base 10) on both sides, we get

$$\log_{10} y = \log_{10} a + bx \log_{10} e \qquad\qquad .....(11.8)$$

taking $\log_{10} y = Y$, the equation (11.8) can be written as

$$Y = A + Bx \qquad\qquad .....(11.9)$$

Where $\qquad$ $Y = log_{10}\, y$ A = log_{10}a, B = log_{10}e $x = x$

Eqn. (11.9) is a linear equation in x and Y, the normal equations are

$$nA + B\sum x - \sum Y$$

$$A\sum x + B \sum x^2 = \sum xY$$

x	y	$Y = log_{10}y$	xY	x^2
1	1.65	0.2175	0.2175	1
2	2.70	0.4314	0.8628	4
3	4.50	0.6532	1.9596	9
4	7.35	0.8663	3.4652	16
Total 10	-	2.1684	6.5051	30

We have $\qquad$ 4A + 10B = 2.1684 $\qquad$ .....(11.10)

$\qquad\qquad$ 10A + 30B = 6.5051 $\qquad$ .....(11.11)

Solving the equations (11.10) and (11.11), we get

Now $\qquad$ A = 0.0001, B = 0.2168

$\qquad\qquad$ A = 0.0001 $\Rightarrow$ log10a = 0.0001 $\Rightarrow$ A = 1.0002

$\qquad\qquad$ B = 0.218 $\Rightarrow$ b log10e = 0.2168

or $\qquad\qquad$ $b = \dfrac{0.2168}{log_{10}e} = \dfrac{0.2169}{0.4343}$

or $\qquad\qquad$ b = 0.4992

Therefore the required curve is Y = (1.0002)e 0.4992x.

Remark : we can also take A = log_ea , B = log_ee = 1 and apply the method of least squares. The method is explained below

Example 6: Fit the curvey = a e^{bx} for the following data

x	0	2	4
y	8.12	10	31.82

Solution: consider the equation y = ae^{bx}

Taking logarithms on both sides we get $log_e\, y = (log_e a)\, b\, log_e e$

We have $\qquad$ A = log_ea, log_ee = 1

The normal equations are

$$nA + b\sum x = \sum Y$$

$$A\sum x + b\sum x^2 = \sum xY$$

We have   the following table

x	y	$Y = \log_e y$	xY	x^2
0	8.12	2.09	0	0
2	10	2.30	4.60	4
4	31.82	3.46	13.84	16
$\sum x = 6$	–	$\sum y = 7.85$	$\sum xY = 18.44$	$\sum x^2 = 30$

The normal equations are 3A + 6b = 7.85,

$\qquad$ 6A + 20 b = 18.44

Solving we get   A = 7.85

$\qquad a = e^A = 6.903$, b = 0.3425

Hence the required equation of the curve is $Y = 6.903\, e^{0.3425\, x}$

Where $Y = \log_e y$

EXERCISE

1. Find the least square line y = a + bx for the data

x	–4	–2	0	2	4
y	1.2	2.8	6.2	7.8	13.2

$\qquad\qquad\qquad$ **Ans:** y = 6.24 + 1.45x

2. Find the least square line y = a0 + a1x for the data.

x	1	2	3	4
y	0	1	1	2

$\qquad\qquad\qquad$ **Ans:** $y = -\dfrac{1}{2} + \dfrac{3}{5}x$

3. Find the least squares parabola for the points (-3, 3), (0, 1), (2, 1), (4, 3).

$\qquad\qquad$ **Ans:** $y = 0.850519 - 0.192495x + 0.178462x^2$

4. Find the least squares parabolic fit $y = ax^2 + bx + c$, for the following data:

x	–3	–1	1	3
Y	15	5	1	5

$\qquad\qquad$ **Ans:** $y = \dfrac{7}{8}x^2 - \dfrac{17}{10}x + \dfrac{17}{8}$

5. The pressure and volume of a gas are related by the equation $PV^\lambda = k$ (λ an k are constants). Fit this equation for the data given below:

P	0.5	1.0	1.5	2.0	2.5	3.0
V	1.62	1.00	0.75	0.62	0.52	0.046

$\qquad\qquad\qquad$ **Ans:** $pv^{1.4225} = 0.997$

6. Using the method of least squares fit a curve of the form $y = ab^x$ to the following data:

x	1	2	3	4
Y	4	11	35	100

Ans: $y = (1.3268). (2,948)^x$

7. Using the method of least squares, fit a relation of the form $y = ab^x$ to the following data:

x	2	3	4	5	6
y	1.44	172.8	207.4	248.8	298.5

Ans: $y = 9.986\ x^{1.2}$

8. Fit a least square curve of the form $y = ae^{bx}$ $(a > 0)$ to the data given below:

X	1	2	3	4
Y	1.65	2.70	4.50	7.35

Ans: $y = (1.0002).e^{0.4992x.}$

9. Find the least squares parabolic fit $y = a + bx + cx^2$.

x	-3	-1	1	3
y	15	5	1	5

Ans: $y = 2.125 - 1.70x + 0.875x^2$

10. Fit a least square curve of the form $y = ax^b$ for the following data where a and b are constants:

x	61	26	7	2.6
Y	350	400	500	600

Ans: $y = (701.94)\ x^{-0.1708}$

11. The observations from an experiment are as given below:

y	2	10	26	61
x	600	500	400	350

It is known that a relation of type $y = ae\ bx$ exists.

Find the best possible values of a and b   **Ans:** $y = 43.12777\ e^{-\ 0.0057056x}$

12. If P is the Pull required to lift a load W by means of a pulley block, find a liner law of the form $P = mW + c$, connecting P and W, using the data.

P	12	15	21	25
W	50	70	100	120

Where P and W are taken in kg-wt. Compute P when W = 150 kg.

Ans: $2.2759 + 0.1879\ W$, 30.4635 kg

13. Use the method of least squares and fit the line $y = a + bx$ for the following data

x	60	61	62	63	64
y	40	40	48	52	55

Ans: $y = 4x - 200.6$

CHAPTER 12

Solving of a System of Linear Equations and Matrices

12.1 INTRODUCTION

In this chapter we introduce the methods for solving systems of linear equations in n unknowns. When the number of equations is not too great and the number of equations is not too unwieldy, the method of exact elimination possesses some advantages. In this chapter we shall discuss few numerical methods for the solution of a system of 3 – equations with 3-unknown which are of the form.

12.2 GAUSS - ELIMINATION METHOD

Here we consider a system of 3 – equations with 3 unknown as:

$$\left. \begin{array}{l} a_{11}^{(1)}x_1 + a_{12}^{(1)}x_2 + a_{13}^{(1)}x_3 = b_1^{(1)} \\ a_{21}^{(1)}x_1 + a_{22}^{(1)}x_2 + a_{23}^{(1)}x_3 = b_2^{(1)} \\ a_{31}^{(1)}x_1 + a_{32}^{(1)}x_2 + a_{33}^{(1)}x_3 = b_3^{(1)} \end{array} \right| \qquad .....(12.1)$$

Let $a_{11}^{(1)} {}^1 0$. Then the elements of the second and third line is to be added to the corresponding elements of first line with multiplication of $-\dfrac{a_{21}^{(1)}}{a_{11}^{(1)}}$ and $-\dfrac{a_{31}^{(1)}}{a_{11}^{(1)}}$ respectively.

This will give the following three equations.

$$\left. \begin{array}{l} a_{11}^{(1)}x_1 + a_{12}^{(1)}x_2 + a_{13}^{(1)}x_3 = b_1^{(1)} \\ a_{22}^{(2)}x_2 + a_{23}^{(2)}x_3 = b_2^{(2)} \\ a_{32}^{(2)}x_2 + a_{33}^{(2)}x_3 = b_3^{(2)} \end{array} \right| \qquad .....(12.2)$$

With

$$a_{22}^{(2)} = a_{22}^{11} - \frac{a_{12}^{(1)}a_{21}^{(1)}}{a_{11}^{(1)}}; \ a_{23}^{(2)} = a_{23}^{(1)} - \frac{a_{13}^{(1)}a_{21}^{(1)}}{a_{11}^{(1)}}$$

$$a_{32}^{(2)} = a_{32}^{(1)} - \frac{a_{12}^{(1)} a_{31}^{(1)}}{a_{11}^{(1)}} ; \quad a_{33}^{(2)} = a_{33}^{(1)} - \frac{a_{13}^{(1)} a_{31}^{(1)}}{a_{11}^{(1)}}$$

$$b_{2}^{(2)} = b_{2}^{(1)} - \frac{b_{1}^{(1)} a_{21}^{(1)}}{a_{11}^{(1)}} ; \quad a_{3}^{(2)} = b_{3}^{(1)} - \frac{b_{1}^{(1)} a_{31}^{(1)}}{a_{11}^{(1)}}$$

Now, if $a_{22}^{(2)} \neq 0$ then we add the elements of third row by multiplying $- \dfrac{a_{32}^{(2)}}{a_{22}^{(2)}}$ and we get the following three equations.

$$\left. \begin{array}{r} a_{11}^{(1)} x_1 + a_{12}^{(1)} x_2 + a_{13}^{(1)} x_3 = b_1^{(1)} \\ a_{22}^{(2)} x_2 + a_{23}^{(2)} x_3 = b_2^{(2)} \\ a_{33}^{(3)} x_3 = b_3^{(3)} \end{array} \right| \qquad \ldots (12.3)$$

Where

$$a_{33}^{(3)} = a_{33}^{(2)} - \frac{a_{23}^{(2)} \times a_{32}^{(2)}}{a_{22}^{(2)}}$$

Hence by solving $x_3, x_2 x_1$ from (12.3) we get

$$x_3 = \frac{b_3^{(3)}}{a_{33}^{(3)}}$$

$$x_2 = \frac{1}{a_{22}^{(2)}} \left[b_2^{(2)} - \frac{a_{23}^{(2)} b_3^{(3)}}{a_{33}^{(3)}} \right]$$

And

$$x_1 = \frac{1}{a_{11}^{(1)}} \left[b_1^{(1)} - a_{12}^{(1)} x_2 - a_{13}^{(1)} x_3 \right]$$

Where x_3 and x_2 are given above.

This method is a direct method, moreover $a_{11}^{(1)}, a_{22}^{(2)}, a_{33}^{(3)}$ are called the pivotal element and if any one of them is equal to zero then the method fails and we have to arrange the equations in another form. Actually at the it h stage of elimination, i.e. when the first element of the row is to be the greatest element compared to the elements of the corresponding columns. Otherwise we have to bring that row (containing the greatest) in the first place and start the method. Total no. of operations: $n^3/3$.

Example 1: Solve the equations using Gauss – elimination method.

$$6x_1 + 3x_2 + 2x_3 = 6$$

$$6x_1 + 4x_2 + 3x_3 = 0$$

$$20x_1 + 15x_2 + 12x_3 = 6$$

Solution: Computational table:

Multiplier	X_1	X_2	X_3	b	$X_1 + X_2 + X_3 + b$
	20	15	12	0	47
	6	4	3	0	13
	6	3	2	6	17
−0.3		−0.5	−0.6	0	−1.1
−0.3		−1.5	−1.6	6	2.9
		−1.5	−1.6	6	2.9
		−0.5	−0.6	0	−1.1
$-\dfrac{1}{3}$			$-\dfrac{0.2}{3}$	−2	−2.067

The above equations will give

$$-\frac{0.2}{3}x_3 = -2$$

$$x_3 = 30$$

$$-1.5x_2 - 1.6 \times 30 = 0 \qquad \qquad \text{.....(12.4)}$$

$$-1.5x_2 = 48$$

$$x_2 = -48 \times \frac{2}{3}$$

$$= -32$$

$$20x_1 + 15x_2 + 12x_3 = 0 \qquad \qquad \text{.....(12.5)}$$

$$20x_1 = 480 - 360 = 120$$

$$x_1 = 6 \qquad \qquad \text{.....(12.6)}$$

$$x_1 = 6, x_2 = -32, x_3 = 30$$

Example 2: Solve the equation using Gauss elimination method.

$$0.34x_1 - 0.58x_2 + 0.94x_3 = 2$$

$$0.27x_1 + 0.42x_2 + 0.13x_3 = 1.5$$

$$0.2x_1 - 0.51x_2 + 0.54x_3 = 0.8$$

***Solution*:** Computational table

Multiplier	X_1	X_2	X_3	b	Check
	0.34	−0.58	0.94	2	2.70
	0.27	0.42	0.13	1.5	2.32
	0.2	−0.51	0.54	0.8	1.03
-0.794		0.88	−0.62	−0.088	0.18
-0.588		−0.17	−0.01	−0.38	−0.56
0.193			−0.13	−0.040	−0.53

$$- 0.13x_3 = - 0.40, x_3 = 3.1$$

$$+ 0.88x_2 - (0.62) \times (3.1) = - 0.088x_2 = 2.1$$

$$0.34x_1 - 0.58 \times (2.1) + 0.94 \times (3.1) = 2x_1 = 0.89$$

Therefore $\quad x_1 = 0.89, x_2 = 2.1, x_3 = 3.1$

Example 3: Solve the equations using Gauss-elimination method.

$$x_1 + x_2 + 4x_3 = 12$$

$$4x_1 + 11x_2 - x_3 = 33$$

$$8x_1 - 3x_2 + 2x_3 = 20$$

***Solution*:** Computational table is

Multiplier	X_1	X_2	X_3	b	Check
	8	−3	2	20	27
	4	11	−1	30	47
	1	1	4	12	18
−0.5		12.5	−2	23	33.5
−0.125		1.375	3.75	9.5	14.625
−0.11			3.97	6.97	10.94

Hence $\quad 3.97 \times 3 = 6.97, x_3 = 1.756$

$$12.5x_2 = 23 + 2 \times 1.756$$

$$x_2 = 2.12$$

$$8x_1 = 20 + 3 \times (2.12) - 2 \times (1.756)$$

$$x_1 = 2.856$$

Therefore $\quad x_1 = 2.856, x_2 = 2.12, x_3 = 1.756$

Example 4: Solve the following system of equation by Gauss-elimination method:

$$5x - y + z = 10$$

$$2x + 4y = 12$$

$$x + y + 5z = -1$$

Solution: $5x - y + z = 10$

$$2x + 4y = 12$$

$$x + y + 5z = -1$$

Multiplier	x_1	x_2	x_3	b	Check
	5	−1	1	10	15
	2	4	0	12	18
	1	1	5	−1	6
−0.1333		4.1333	−0.1333	10.667	14.6675
−0.0667		1.0667	4.9333	−1.667	4.3325
−0.258			4.968	−4.4190	0.5484

Hence $4.968z = -4.419$

$$z = -0.89$$

$$1.0667y + 4.9333z = -1.667$$

$$y = \frac{-1.667 + (0.89) \times 4.9333}{1.0667}$$

$$= 2.55$$

$$5x - y + z = 10$$

$$5x = 10 + 2.55 + 0.89$$

$$= 13.44$$

Therefore $x = 2.69, y = 2.55, z = -0.89.$

Example 5: Solve the following system of equation using Gaussian elimination method

$$x + y + z = 9$$

$$2x - 3y + 4z = 13$$

$$3x + 4y + 5z = 40$$

Solution: We arrange the above system with the following manner

$$3x + 4y + 5z = 40$$

$$2x - 3y + 4z = 13$$

$$x + y + z = 9$$

Multiplier	x	y	z	b	Check
	3	4	5	40	52
	2	−3	4	13	16
	1	1	1	9	12
− 2/3		−5.667	0.6667	−13.6667	− 18.6667
− 1/3		−0.333	−0.6667	−4.3333	− 5.3333
− 0.0588			−0.7059	−3.5296	− 4.2353

Therefore

$$-0.7059z = -3.5296$$

$$z = 5$$

$$-0.333y + (-0.6667) \times 5 = -4.333$$

$$-0.333y = -0.9998$$

$$y = 3$$

$$3x + 4y + 5z = 40$$

$$3x = 3$$

$$x = 1$$

Therefore $x = 1$, $y = 3$, $z = 5$.

Example 5: Solve the system of linear equation by Gauss –elimination method:

$$5x_1 - x_2 = 9$$

$$-x_1 + 5x_2 - x_3 = 4$$

Solution:
$$-x_2 + 5x_3 = -6$$

Multiplier	X_1	X_2	X_3	b	Check
	5	−1	0	9	13
	−1	5	−1	4	7
	0	−1	5	-6	−2
0.2		4.8	−1	5.8	9.6
0		−1	5	−6	−2
0.21			4.79	−4.79	0

Then

$$4.79x_3 = -4.79$$

$$x_3 = -1$$

$$4.8x_2 - x_3 = 5.8$$

$$4.8x_2 = 4.8$$

$$x_2 = 1$$

$$5x_1 - x_2 = 9$$

$$5x_1 = 10$$

$$x_1 = 2$$

Therefore $\qquad x_1 = 2, x_2 = 1, x_3 = -1$

12.3 MATRIX INVERSION METHOD

To compute the inverse of a matrix $A = (a_{ij})_{3 \times 3}$ by Gauss-elimination matrix is as follows:

$$\begin{bmatrix} a_{11}^{(1)} & a_{12}^{(1)} & a_{13}^{(1)} & 1 & 0 & 0 \\ a_{21}^{(1)} & a_{22}^{(1)} & a_{23}^{(1)} & 0 & 1 & 0 \\ a_{31}^{(1)} & a_{32}^{(1)} & a_{33}^{(1)} & 0 & 0 & 1 \end{bmatrix}$$

We get

$$\begin{bmatrix} a_{11}^{(1)} & a_{12}^{(1)} & a_{13}^{(1)} & 1 & 0 & 0 \\ 0 & a_{22}^{(2)} & a_{23}^{(2)} & -\dfrac{a_{21}^{(1)}}{a_{11}^{(1)}} & 1 & 0 \\ 0 & a_{32}^{(2)} & a_{33}^{(2)} & -\dfrac{a_{31}^{(1)}}{a_{11}^{(1)}} & 0 & 1 \end{bmatrix} \qquad \qquad .....(12.7)$$

Where

$$a_{22}^{(2)} = a_{22}^{(1)} - \frac{a_{12}^{(1)} a_{21}^{(1)}}{a_{11}^{(1)}}$$

$$a_{23}^{(2)} = a_{23}^{(1)} - \frac{a_{13}^{(1)} a_{21}^{(1)}}{a_{11}^{(1)}}$$

$$a_{32}^{(2)} = a_{32}^{(1)} - \frac{a_{12}^{(1)} a_{31}^{(1)}}{a_{11}^{(1)}}$$

$$a_{33}^{(2)} = a_{33}^{(1)} - \frac{a_{13}^{(1)} a_{31}^{(1)}}{a_{11}^{(1)}}$$

Now if we write the above augmented matrix as

$$\begin{bmatrix} a_{11}^{(1)} & a_{12}^{(1)} & a_{13}^{(1)} & 1 & 0 & 0 \\ 0 & a_{22}^{(2)} & a_{23}^{(2)} & b_{21}^{(1)} & 1 & 0 \\ 0 & a_{32}^{(2)} & a_{33}^{(2)} & b_{31}^{(1)} & 0 & 1 \end{bmatrix}$$

.....(12.8)

Where $\qquad b_{21}^{(1)} = -\dfrac{a_{21}^{(1)}}{a_{11}^{(1)}},\ b_{31}^{(1)} = -\dfrac{a_{31}^{(1)}}{a_{11}^{(1)}}.$

Then, for the second stage we have $\begin{bmatrix} a_{11}^{(1)} & a_{12}^{(1)} & a_{13}^{(1)} & 1 & 0 & 0 \\ 0 & a_{22}^{(2)} & a_{23}^{(2)} & b_{21}^{(1)} & 1 & 0 \\ 0 & 0 & a_{33}^{(3)} & b_{31}^{(2)} & b_{32}^{(2)} & 1 \end{bmatrix}$

Where

$$a_{33}^{(3)} = a_{33}^{(2)} - \frac{a_{32}^{(2)} a_{23}^{(2)}}{a_{22}^{(2)}}$$

$$a_{31}^{(2)} = \frac{a_{21}^{(1)} a_{32}^{(2)}}{a_{11}^{(1)} a_{22}^{(2)}} - \frac{a_{31}^{(1)}}{a_{11}^{(1)}}$$

$$b_{32}^{(2)} = - \frac{a_{32}^{(2)}}{a_{22}^{(2)}}$$

Hence then we get

$$I = \begin{bmatrix} 1 & 0 & 0 \\ b_{21}^{(1)} & 0 & 0 \\ b_{31}^{(2)} & b_{32}^{(2)} & 1 \end{bmatrix}$$

And by using back substitution we get,

$$\begin{bmatrix} a_{11}^{(1)} & a_{12}^{(1)} & a_{13}^{(1)} \\ 0 & a_{22}^{(2)} & a_{23}^{(2)} \\ 0 & 0 & a_{33}^{(3)} \end{bmatrix} \begin{bmatrix} x_{11} \\ x_{21} \\ x_{31} \end{bmatrix} = \begin{bmatrix} 1 \\ b_{21}^{(1)} \\ b_{31}^{(2)} \end{bmatrix}$$

$$\begin{bmatrix} a_{11}^{(1)} & a_{12}^{(1)} & a_{13}^{(1)} \\ 0 & a_{22}^{(2)} & a_{23}^{(2)} \\ 0 & 0 & a_{33}^{(3)} \end{bmatrix} \begin{bmatrix} x_{12} \\ x_{22} \\ x_{32} \end{bmatrix} = \begin{bmatrix} 0 \\ 0 \\ b_{32}^{(2)} \end{bmatrix}$$

$$\begin{bmatrix} a_{11}^{(1)} & a_{12}^{(1)} & a_{13}^{(1)} \\ 0 & a_{22}^{(2)} & a_{23}^{(2)} \\ 0 & 0 & a_{33}^{(3)} \end{bmatrix} \begin{bmatrix} x_{13} \\ x_{23} \\ x_{33} \end{bmatrix} = \begin{bmatrix} 0 \\ 0 \\ 1 \end{bmatrix}$$

Solving these three set of equations we get the inverse of the matrix A as

$$A^{-1} = \begin{bmatrix} x_{11} & x_{12} & x_{13} \\ x_{21} & x_{22} & x_{23} \\ x_{31} & x_{32} & x_{33} \end{bmatrix}$$

Example 7: Find the inverse of the matrix

$$A = \begin{bmatrix} 2 & -2 & 4 \\ 2 & 3 & 2 \\ -1 & 4 & -1 \end{bmatrix}$$

By using Gauss elimination method

Solution: Consider the augmented matrix

$$\begin{bmatrix} 2 & -2 & 4 & 1 & 0 & 0 \\ 2 & 3 & 2 & 0 & 1 & 0 \\ -1 & 4 & -1 & 0 & 0 & 1 \end{bmatrix}$$

$$= \begin{bmatrix} 2 & -2 & 4 & 1 & 0 & 0 \\ 0 & 5 & -2 & -1 & 1 & 0 \\ 0 & 3 & 1 & 1/2 & 0 & 1 \end{bmatrix}$$

$$= \begin{bmatrix} 2 & -2 & 4 & 1 & 0 & 0 \\ 0 & 5 & -2 & -1 & 1 & 0 \\ 0 & 0 & 11/5 & 11/10 & -3/5 & 1 \end{bmatrix}$$

Thus we have

$$\begin{bmatrix} 2 & -2 & 4 \\ 0 & 5 & -2 \\ 0 & 0 & 11/5 \end{bmatrix} \begin{bmatrix} x_{11} \\ x_{21} \\ x_{31} \end{bmatrix} = \begin{bmatrix} 1 \\ -1 \\ 11/10 \end{bmatrix}$$

$$x_{11} = -1/2, x_2 = 0, x_{31} = 1/2$$

$$\begin{bmatrix} 2 & -2 & 4 \\ 0 & 5 & -2 \\ 0 & 0 & 11/5 \end{bmatrix}\begin{bmatrix} x_{12} \\ x_{22} \\ x_{32} \end{bmatrix} = \begin{bmatrix} 0 \\ 1 \\ -3/5 \end{bmatrix}$$

$$x_{12} = 7/11, x_{22} = 1, x_{32} = 3/11$$

$$\begin{bmatrix} 2 & -2 & 4 \\ 0 & 5 & -2 \\ 0 & 0 & 11/5 \end{bmatrix}\begin{bmatrix} x_{13} \\ x_{23} \\ x_{33} \end{bmatrix} = \begin{bmatrix} 0 \\ 0 \\ 1 \end{bmatrix}$$

$$x_{13} = -8/11, x_{23} = 2/11, x_{33} = 5/11$$

Hence the matrix inverse of A is

$$A^{-1} = \begin{bmatrix} -1/2 & 7/11 & -8/11 \\ 0 & 1/11 & 2/11 \\ 1/2 & -3/11 & 5/11 \end{bmatrix}$$

Verification

$$\begin{bmatrix} 2 & -2 & 4 \\ 2 & 3 & 2 \\ -1 & 4 & -1 \end{bmatrix}\begin{bmatrix} -1/2 & 7/11 & -8/11 \\ 0 & 1/11 & 2/11 \\ 1/2 & -3/4 & 5/11 \end{bmatrix} = \begin{bmatrix} 1 & 0 & 0 \\ 0 & 1 & 0 \\ 0 & 0 & 1 \end{bmatrix}$$

12.4 GAUSS-JORDAN ELIMINATION METHOD

This method is a slight modification of the Gauss elimination method. In this method the final form of the matrix after elimination is a diagonal form. This method can be used either with or without pivoting.

From equation (12.8) we get the last step of Gauss elimination method i.e.,

$$\begin{aligned} a_{11}^{(1)}x_1 + a_{12}^{(1)}x_2 + a_{13}^{(1)}x_3 &= b_1^{(1)} \\ a_{22}^{(2)}x_2 + a_{23}^{(2)}x_3 &= b_2^{(2)} \\ a_{33}^{(3)}x_3 &= b_3^{(3)} \end{aligned} \qquad \dots(12.9)$$

Now from the above three equation Gauss – Jordan can be written as

$$a_{11}^{(1)}x_1 = b_1^{(1)} - \frac{a_{12}^{(1)}}{a_{22}^{(2)}}\left[b_2^{(2)} - \frac{a_{22}^{(2)}b_3^{(3)}}{a_{33}^{(3)}} \right] - \frac{a_{13}^{(1)}b_3^{(3)}}{a_{33}^{(3)}}$$

$$a_{22}^{(2)}x_2 = b_2^{(2)} - \frac{a_{23}^{(2)}b_3^{(3)}}{a_{33}^{(3)}}$$

$$a_{33}^{(3)}x_3 = b_3^{(3)} \qquad \dots(12.10)$$

Example 8: Solve the following system of linear equations by Gauss Jordan elimination method:

$$5x_1 - x_2 = 9$$

$$-x_1 + 5x_2 - x_3 = 4$$

$$-x_2 + 5x_3 = -6$$

Solution:

Multiplier	x_1	x_2	x_3	b	Check
	5	−1	0	9	13
	−1	5	−1	4	7
	0	-1	5	−6	−2
1/5		24/5	−1	29/5	48/5
0		−1	5	−6	−2
5/24			115/24	−115/24	0

The above computational table is same as Gauss elimination. Now for Gauss-Jordan we have the three underlined equations and it will give us

$$x_1 = 2$$
$$x_2 = 1$$
$$x_3 = -1$$

i.e.
$$\begin{bmatrix} 1 & 0 & 0 \\ 0 & 1 & 0 \\ 0 & 0 & 1 \end{bmatrix} \begin{bmatrix} x_1 \\ x_2 \\ x_3 \end{bmatrix} = \begin{bmatrix} 2 \\ 1 \\ -1 \end{bmatrix}$$

$$x_1 = 2, \ x_2 = 1, \ x_3 = -1.$$

Example 9: Use the Gauss Jordan method to solve the following system up to 4 decimal places.

$$0.732x_1 - 5.421x_2 + 1.013x_3 = 4.256$$

$$3.491x_1 + 2.203x_2 + 0.782x_3 = -7.113$$

$$0.961x_1 - 1.523x_2 + 4.265x_3 = 3.727$$

Solution:

Multiplier	x_1	x_2	x_3	b
	0.732	− 5.421	1.013	4.256
	3.491	2.203	0.782	− 7.113
	0.961	− 1.523	4.265	3.727

$-\dfrac{3.491}{0.732}$	0.732	-5.4221	1.013	4.256
		28.0564	-4.0491	-27.4104
$-\dfrac{0.961}{0.732}$		5.5939	2.9351	-1.8606
$-\dfrac{5.421}{28.0564}$	0.732	0	0.2306	-1.0402
	0	28.0564	-4.0491	-27.4104
$-\dfrac{5.5939}{28.0564}$			3.7424	3.6046
$-\dfrac{0.2306}{3.7424}$	0.732	0	0	-1.2623
	0	28.0564	0	-23.5104
$-\dfrac{4.0491}{3.7424}$	0	0	3.7424	3.6046

The above matrix is a diagonal matrix. Hence the solution is

$$x_1 = -1.7244$$

$$x_2 = -0.8380$$

$$x_3 = 0.9632$$

12.5 MATRIX INVERSION BY GAUSS - JORDAN METHOD

Let

$$A = \begin{bmatrix} a_{11} & a_{12} & a_{13} \\ a_{21} & a_{22} & a_{23} \\ a_{31} & a_{32} & a_{33} \end{bmatrix}$$

Where $|A| \neq 0$

Then we write the augmented matrix as

$$\begin{bmatrix} a_{11} & a_{12} & a_{13} & 1 & 0 & 0 \\ a_{21} & a_{22} & a_{23} & 0 & 1 & 0 \\ a_{31} & a_{32} & a_{33} & 0 & 0 & 1 \end{bmatrix}$$

Then the first step will be

$$\begin{bmatrix} 1 & a_{11}^{(1)} & a_{12}^{(1)} & a_{13}^{(1)} & 0 & 0 \\ 0 & a_{21}^{(1)} & a_{22}^{(1)} & a_{23}^{(1)} & 1 & 0 \\ 0 & a_{31}^{(1)} & a_{32}^{(1)} & a_{33}^{(1)} & 0 & 1 \end{bmatrix}$$

Where
$$a_{11}^{(1)} = \frac{a_{12}}{a_{11}}, a_{12}^{(1)} = \frac{a_{13}}{a_{11}}, \ a_{13}^{(1)} = \frac{1}{a_{11}}$$

and
$$a_{21}^{(1)} = a_{22} - \frac{a_{21} \times a_{12}}{a_{11}}, \ a_{22}^{(1)} = a_{23} - \frac{a_{21}a_{13}}{a_{11}}, \ a_{23}^{(1)} = 0 \qquad \frac{a_{21}}{a_{11}}$$

$$a_{31}^{(1)} = a_{32} - \frac{a_{31}a_{12}}{a_{11}}, \ a_{32}^{(1)} = a_{33} - \frac{a_{31}a_{13}}{a_{11}}, \ a_{33}^{(1)} = 0 - \frac{a_{31}}{a_{11}}$$

The second step will be

$$\begin{bmatrix} 1 & 0 & a_{11}^{(2)} & a_{12}^{(2)} & a_{13}^{(2)} & 0 \\ 0 & 1 & a_{21}^{(2)} & a_{22}^{(2)} & a_{23}^{(2)} & 0 \\ 0 & 0 & a_{31}^{(2)} & a_{32}^{(2)} & a_{33}^{(2)} & 1 \end{bmatrix}$$

$$a_{11}^{(2)} = a_{12}^{(1)} - \frac{a_{11}^{(1)}a_{22}^{(1)}}{a_{21}^{(1)}}, \ a_{12}^{(2)} = a_{13}^{(1)} - \frac{a_{11}^{(1)}a_{23}^{(1)}}{a_{21}^{(1)}}, a_{13}^{(2)} = 0 - \frac{a_{11}^{(1)}}{a_{21}^{(1)}}$$

$$a_{21}^{(2)} = \frac{a_{22}^{(1)}}{a_{21}^{(1)}}, \ a_{22}^{(2)} = \frac{a_{23}^{(1)}}{a_{21}^{(1)}}, \ a_{23}^{(2)} = \frac{1}{a_{21}^{(1)}}$$

$$a_{31}^{(2)} = a_{32}^{(1)} - \frac{a_{31}^{(1)}a_{22}^{(1)}}{a_{21}^{(1)}}, \ a_{32}^{(2)} = a_{33}^{(2)} = a_{33}^{(1)} - \frac{a_{31}^{(1)}a_{23}^{(1)}}{a_{21}^{(1)}}, \ a_{33}^{(2)} = 0 - \frac{a_{31}^{(1)}}{a_{21}^{(1)}}$$

The third and ultimate step is:

$$\begin{bmatrix} 1 & 0 & 0 & a_{11}^{(3)} & a_{12}^{(3)} & a_{13}^{(3)} \\ 0 & 1 & 0 & a_{21}^{(3)} & a_{22}^{(3)} & a_{23}^{(3)} \\ 0 & 0 & 1 & a_{31}^{(3)} & a_{32}^{(3)} & a_{33}^{(3)} \end{bmatrix}$$

$$a_{11}^{(3)} = a_{12}^{(2)} - \frac{a_{11}^{(2)}a_{32}^{(3)}}{a_{31}^{(2)}}, \ a_{12}^{(3)} = a_{13}^{(2)} - \frac{a_{11}^{(2)}a_{33}^{(2)}}{a_{31}^{(2)}}, a_{13}^{(3)} = 0 - \frac{a_{11}^{(2)}}{a_{31}^{(2)}}$$

$$a_{21}^{(3)} = a_{22}^{(3)} - \frac{a_{21}^{(2)}a_{32}^{(2)}}{a_{31}^{(2)}}, \ a_{22}^{(3)} = a_{23}^{(2)} - \frac{a_{21}^{(2)}a_{33}^{(2)}}{a_{31}^{(2)}}; a_{23}^{(3)} = 0 - \frac{a_{21}^{(2)}}{a_{31}^{(2)}}$$

$$a_{31}^{(3)} = \frac{a_{32}^{(2)}}{a_{31}^{(?)}}, a_{32}^{(3)} = \frac{a_{33}^{(2)}}{a_{31}^{(2)}}, \ a_{33}^{(3)} = \frac{1}{a_{31}^{(3)}}$$

Then the matrix

$$\begin{bmatrix} a_{11}^{(3)} & a_{12}^{(3)} & a_{13}^{(3)} \\ a_{21}^{(3)} & a_{22}^{(3)} & a_{23}^{(3)} \\ a_{31}^{(3)} & a_{32}^{(3)} & a_{33}^{(3)} \end{bmatrix}$$

Is the inverse matrix of

$$\begin{bmatrix} a_{11} & a_{12} & a_{13} \\ a_{21} & a_{22} & a_{23} \\ a_{31} & a_{32} & a_{33} \end{bmatrix}$$

Example 10: Find the inverse of the matrix by Gauss Jordan method.

$$\begin{bmatrix} 1 & 2 & 6 \\ 2 & 5 & 15 \\ 6 & 15 & 46 \end{bmatrix}$$

Solution:

$$\begin{bmatrix} 1 & 2 & 6 & 1 & 0 & 0 \\ 2 & 5 & 15 & 0 & 1 & 0 \\ 6 & 15 & 46 & 0 & 0 & 1 \end{bmatrix}$$

$$= \begin{bmatrix} 1 & 2 & 6 & 1 & 0 & 0 \\ 0 & 1 & 3 & -2 & 1 & 0 \\ 0 & 0 & 1 & -6 & 0 & 1 \end{bmatrix}$$

$$= \begin{bmatrix} 1 & 0 & 0 & 5 & -2 & 0 \\ 0 & 1 & 3 & -2 & 1 & 0 \\ 0 & 0 & 1 & 0 & -3 & 1 \end{bmatrix}$$

$$= \begin{bmatrix} 1 & 0 & 0 & 5 & -2 & 0 \\ 0 & 1 & 0 & -2 & 10 & -3 \\ 0 & 0 & 1 & 0 & -3 & 1 \end{bmatrix}$$

Hence the inverse of the matrix is

$$\begin{bmatrix} 5 & -2 & 0 \\ -2 & 10 & -3 \\ 0 & -3 & 1 \end{bmatrix}$$

Example 11: Find the inverse of the matrix by Gauss – Jordan method.

$$A = \begin{bmatrix} 1 & 3 & 3 \\ 1 & 1 & 3 \\ 1 & 3 & 4 \end{bmatrix}$$

Solution:

$$\begin{bmatrix} 1 & 3 & 3 & 1 & 0 & 0 \\ 1 & 4 & 3 & 0 & 1 & 0 \\ 1 & 3 & 4 & 0 & 0 & 1 \end{bmatrix}$$

$$= \begin{bmatrix} 1 & 3 & 3 & 1 & 0 & 0 \\ 0 & 1 & 0 & -1 & 1 & 0 \\ 0 & 0 & 1 & -1 & 0 & 1 \end{bmatrix} \rightarrow \begin{cases} R_1^1 = R_1 / 1 \\ R_2^1 = R_2 - \dfrac{1}{1}R_1 \\ R_3^1 = R_3 - \dfrac{1}{1}R_1 \end{cases}$$

$$= \begin{bmatrix} 1 & 0 & 3 & 4 & -3 & 0 \\ 0 & 1 & 0 & -1 & 1 & 0 \\ 0 & 0 & 1 & -1 & 0 & 1 \end{bmatrix} \rightarrow \begin{cases} R_1^1 = R_1 - \dfrac{3}{1}R_2 \\ R_2^1 = R_2 / 1 \\ R_3^1 = R_3 - 0/1R_2 \end{cases}$$

$$= \begin{bmatrix} 1 & 0 & 0 & 7 & -3 & -3 \\ 0 & 1 & 0 & -1 & 1 & 0 \\ 0 & 0 & 1 & -1 & 0 & 1 \end{bmatrix} \rightarrow \begin{cases} R_1^1 = R_1 - 3/1R_3 \\ R_2^1 = R_2 - \dfrac{0}{1}R_3 \\ R_3^1 = \dfrac{R_3}{1} \end{cases}$$

Therefore the inverse of the matrix is

$$\begin{bmatrix} 7 & -3 & -3 \\ -1 & 1 & 0 \\ -1 & 0 & 1 \end{bmatrix}$$

Example12: Solve the following system by matrix inversion method.

$$2x + y + z = 10$$

$$3x + 2y + 3z = 18$$

$$x + 4y + 9z = 16$$

Solution: First we find the inverse of the matrix

$$A = \begin{bmatrix} 2 & 1 & 1 \\ 3 & 2 & 3 \\ 1 & 4 & 9 \end{bmatrix}$$

Let

$$\begin{bmatrix} 2 & 1 & 1 & 1 & 0 & 0 \\ 3 & 2 & 3 & 0 & 1 & 0 \\ 1 & 4 & 9 & 0 & 0 & 1 \end{bmatrix}$$

$$= \begin{bmatrix} 1 & 1/2 & 1/2 & 1/2 & 0 & 0 \\ 0 & 1/2 & 3/2 & -3/2 & 1 & 0 \\ 0 & 7/2 & 17/2 & -1/2 & 0 & 1 \end{bmatrix} \rightarrow \begin{cases} R_1^1 = R_1/2 \\ R_2^1 = R_2 - \dfrac{3}{2}R_1 \\ R_3^1 = R_3 - \dfrac{1}{2}R_1 \end{cases}$$

$$= \begin{bmatrix} 1 & 0 & -1 & 2 & -1 & 0 \\ 0 & 1 & 3 & -3 & 2 & 0 \\ 0 & 0 & -2 & 10 & -7 & 1 \end{bmatrix} \rightarrow \begin{cases} R_1^1 = R_1 - R_2 \\ R_2^1 = R_2 \times 2 \\ R_3^1 = R_3 - 7R_2 \end{cases}$$

$$= \begin{bmatrix} 1 & 0 & 0 & -3 & 5/2 & -1/2 \\ 0 & 1 & 0 & 12 & -17/2 & 3/2 \\ 0 & 0 & 1 & -5 & 7/2 & -1/2 \end{bmatrix} \rightarrow \begin{cases} R_1^1 = R_1 - \dfrac{R_3}{2} \\ R_2^1 = R_2 + \dfrac{3}{2}R_3 \\ R_3^1 = \dfrac{-R_3}{2} \end{cases}$$

Then

$$A^{-1} = \begin{bmatrix} 3 & 5/2 & -1/2 \\ 12 & -17/2 & 3/2 \\ -5 & 7/2 & -1/2 \end{bmatrix}$$

Therefore

$$\begin{bmatrix} x \\ y \\ z \end{bmatrix} = \begin{bmatrix} -3 & 5/2 & -1/2 \\ 12 & -17/2 & 3/2 \\ -5 & 7/2 & -1/2 \end{bmatrix} \begin{bmatrix} 10 \\ 18 \\ 16 \end{bmatrix}$$

i.e
$$\begin{bmatrix} x \\ y \\ z \end{bmatrix} = \begin{bmatrix} 7 \\ -9 \\ 5 \end{bmatrix}$$

$\therefore \qquad x = 7,\ y = -9,\ z = 5$

12.6 LU FACTORIZATION METHOD

This method is leased on the logic that every square matrix can be expressed as a product of lower triangular and upper triangular matrix. This theorem is applicable if f the principal minors are non – singular.

Then, AX = B can be written as

LUX = B where L is the lower triangular and U is the upper triangular matrix.

To solve for X, we first write UX = Y and then LY = B.

Now, LY = B will give solution Y and then immediately X can be found out.

$$\begin{bmatrix} l_{11} & 0 & 0 \\ l_{21} & l_{22} & 0 \\ l_{31} & l_{32} & l_{33} \end{bmatrix} \begin{bmatrix} 1 & u_{12} & u_{13} \\ 0 & 1 & u_{23} \\ 0 & 0 & 1 \end{bmatrix} = \begin{bmatrix} a_{11} & a_{12} & a_{13} \\ a_{21} & a_{22} & a_{23} \\ a_{31} & a_{32} & a_{33} \end{bmatrix} \qquad \dots\dots(12.11)$$

From (12.11) we get, $l_{11} = a_{11},\ l_{21} = a_{21},\ l_{31} = a_{31}, l_{11}u_{12} = a_{12}, l_{11}u_{13} = a_{13}$

i.e
$$u_{12} = \frac{a_{12}}{a_{11}}, u_{13} = \frac{a_{13}}{a_{11}}$$

$$l_{21}u_{12} + l_{22} + a_{22}, l_{31}u_{12} + l_{32} = a_{32}$$

$$l_{21}u_{13} + l_{22} + u_{23} = a_{23}$$

And
$$l_{31}a_{13} + l_{32}a_{23} + l_{33} = a_{3} \qquad \dots\dots(12.12)$$

From the above equation (12.12) we can solve $l_{11}, l_{21}, l_{22}, l_{31}, l_{32}, l_{33}$ and u_{12}, u_{13} and u_{23}. Let us illustrate the above method in the following examples.

Example 13: Solve the following system of equation by LU – factorization method:

$$3x + 4y + 2z = 15$$

$$5x + 2y + z = 18$$

$$2x + 3y + 2z = 10$$

Solution: Let

$$L = \begin{bmatrix} l_{11} & 0 & 0 \\ l_{21} & l_{22} & 0 \\ l_{31} & l_{32} & l_{33} \end{bmatrix}$$

And

$$U = \begin{bmatrix} 1 & u_{12} & u_{13} \\ 0 & 1 & u_{23} \\ 0 & 0 & 1 \end{bmatrix}$$

And AX = B can be written as

$$LUX = B$$

$$LU = A$$

$$\begin{bmatrix} l_{11} & 0 & 0 \\ l_{21} & l_{22} & 0 \\ l_{31} & l_{32} & l_{33} \end{bmatrix} \begin{bmatrix} 1 & u_{12} & u_{13} \\ 0 & 1 & u_{23} \\ 0 & 0 & 1 \end{bmatrix} = \begin{bmatrix} 3 & 4 & 2 \\ 5 & 2 & 1 \\ 2 & 3 & 2 \end{bmatrix}$$

$$l_{11} = 3, l_{21} = 5, l_{31} = 2$$

$$l_{11}u_{12} = 4, \ u_{12} = \frac{4}{3}$$

$$l_{11}u_{13} = 2, \ u_{13} = \frac{2}{3}$$

$$l_{21}u_{12} + l_{22} = 2, 5 \times \frac{4}{3} + l_{22} = 2, l_{22} = 2 - \frac{20}{3} = -\frac{14}{3}$$

$$l_{21}u_{13} + l_{22}u_{23} = 1, 5 \times \frac{2}{3} - \frac{14}{3}u_{23} = 1. u_{23} = \frac{1}{2}$$

$$l_{31}u_{12} + l_{32} = 3, 2 \times \frac{4}{3} + l_{32} = 3, l_{32} = \frac{1}{3}$$

$$l_{31}u_{13} + l_{32}u_{23} + l_{33} = 2$$

$$2 \times \frac{2}{3} + \frac{1}{3} \times \frac{1}{2} + l_{33} = 2$$

$$l_{33} = 2 - \frac{4}{3} = \frac{1}{6}$$

$$= \frac{12 - 8 - 1}{6}$$

$$= \frac{1}{2}$$

$$LY = \begin{bmatrix} 15 \\ 18 \\ 10 \end{bmatrix}$$

$$= \begin{bmatrix} 3 & 0 & 0 \\ 5 & -14/3 & 0 \\ 2 & 1/3 & 1/2 \end{bmatrix} \begin{bmatrix} y_1 \\ y_2 \\ y_3 \end{bmatrix}$$

Therefore

$$3y_1 = 15; \, y_1 = 5$$

$$5y_1 - \frac{14}{3} y_2 = 18; \, y_2 = \frac{3}{2}$$

$$2y_1 + \frac{1}{3} y_2 + \frac{1}{2} y_3 = 10$$

$$5 \times 2 + \frac{1}{3} \times \frac{3}{2} + \frac{1}{2} y_3 = 10$$

$$y_3 = -1$$

$$\begin{bmatrix} 1 & 4/3 & 2/3 \\ 0 & 1 & 1/2 \\ 0 & 0 & 1 \end{bmatrix} \begin{bmatrix} x \\ y \\ z \end{bmatrix} = \begin{bmatrix} 5 \\ 3/2 \\ -1 \end{bmatrix}$$

$$x + \frac{4}{3} y + \frac{2}{3} z = 5$$

$$y + \frac{1}{2} z = \frac{3}{2}$$

$$z = -1$$

Therefore $y = 2, \, x = 3$

$$x = 3, \, y = 2, \, z = -1$$

Example 14: Solve the following system of equations by LU – factorization method:

$$8x_1 - 3x_2 + 2x_3 = 20$$

$$4x_1 + 11x_2 - x_3 = 33$$

$$6x_1 + 3x_2 + 12x_3 = 36$$

Solution: Hence by LU Factorization we can write LU = A

Now

$$\begin{bmatrix} l_{11} & 0 & 0 \\ l_{21} & l_{22} & 0 \\ l_{31} & l_{32} & l_{33} \end{bmatrix} \begin{bmatrix} 1 & u_{12} & u_{13} \\ 0 & 1 & u_{23} \\ 0 & 0 & 1 \end{bmatrix} = \begin{bmatrix} 8 & -3 & 2 \\ 4 & 11 & -1 \\ 6 & 3 & 12 \end{bmatrix}$$

$$l_{11} = 8, l_{21} = 4, l_{31} = 6, l_{11}u_{12} = -3, u_{12} = -\frac{3}{8}$$

$$l_{21}u_{12} + l_{22} = 11$$

$$4 \times -\frac{3}{8} + l_{22} = 11$$

$$l_{22} = 11 + \frac{3}{2} = \frac{25}{2}$$

$$l_{11}u_{13} = 2$$

$$u_{13} = \frac{1}{4}$$

$$l_{31}u_{12} + l_{32} = 3$$

$$l_{32} = \frac{21}{4}$$

$$l_{21}u_{13} + l_{22}u_{23} = -1$$

$$4 \times \frac{1}{4} + \frac{25}{2}u_{23} = -1$$

$$u_{23} = -\frac{4}{25}$$

$$l_{31}u_{13} + l_{32}u_{23} + l_{33} = 12$$

$$6 \times \frac{1}{4} + \frac{21}{4} \times -\frac{4}{25} + l_{33} = 12$$

$$l_{33} = \frac{567}{50}$$

$$l_{21}u_{12} + l_{22} = 11$$

$$4 \times -\frac{3}{8} + l_{22} = 11$$

$$l_{22} = 11 + \frac{3}{2} = \frac{25}{2}$$

$$l_{11}u_{13} = 2$$

$$u_{13} = \frac{1}{4}$$

$$l_{31}u_{12} + l_{32} = 3$$

$$l_{32} = \frac{21}{4}$$

$$l_{21}u_{13} + l_{22}u_{23} = -1$$

$$4 \times \frac{1}{4} + \frac{25}{2}u_{23} = -1$$

$$u_{23} = -\frac{4}{25}$$

$$l_{31}u_{13} + l_{32}u_{23} + l_{33} = 12$$

$$6 \times \frac{1}{4} + \frac{21}{4} \times -\frac{4}{25} + l_{33} = 12$$

$$l_{33} = \frac{567}{50}$$

Now,

$$\begin{bmatrix} 8 & 0 & 0 \\ 4 & 25/2 & 0 \\ 6 & 21/4 & \dfrac{567}{50} \end{bmatrix} \begin{bmatrix} y_1 \\ y_2 \\ y_3 \end{bmatrix} = \begin{bmatrix} 20 \\ 33 \\ 36 \end{bmatrix}$$

$$8y_1 = 20, y_1 = 5/2$$

$$4y_1 + \frac{25}{2}y_2 = 33, \; y_2 = \frac{46}{25}$$

$$6y_1 + \frac{21}{4}y_2 + \frac{567}{50}y_3 = 36, \; y_3 = 1$$

$$\therefore \quad \begin{bmatrix} 1 & -3/8 & 1/4 \\ 0 & 1 & -4/25 \\ 0 & 0 & 1 \end{bmatrix} \begin{bmatrix} x_1 \\ x_2 \\ x_3 \end{bmatrix} = \begin{bmatrix} 5/2 \\ 46/25 \\ 1 \end{bmatrix}$$

$$x_1 - \frac{3}{8}x_2 + \frac{1}{4}x_3 = \frac{5}{2}$$

$$x_2 - \frac{4}{25}x_3 = \frac{46}{25}$$

$$x_3 = 1$$

$$x_2 = 2, x_1 = 3$$

Therefore the solution is $x_1 = 3, x_2 = 2, x_3 = 1$

Example 15: Solve the following system of equations by LU decomposition method.

$$x_1 + x_2 - x_3 = 2$$

$$2x_1 + 3x_2 + 5x_3 = -3$$

$$3x_1 + 2x_2 - 3x_3 = 6$$

Solution: As AX = B can be written in this method as LUX = B and

$$A = \begin{bmatrix} 1 & 1 & -1 \\ 2 & 3 & 5 \\ 3 & 2 & -3 \end{bmatrix} = \begin{bmatrix} l_{11} & 0 & 0 \\ l_{21} & l_{22} & 0 \\ l_{31} & l_{32} & l_{33} \end{bmatrix} \begin{bmatrix} 1 & u_{12} & u_{13} \\ 0 & 1 & u_{23} \\ 0 & 0 & 1 \end{bmatrix}$$

Then,

$$l_{11} = 1, \, l_{21} = 2, l_{31} = 3, l_{11}u_{12} = 1, u_{12} = 1, l_{11}u_{13} = -1, u_{13} = -1$$

$$l_{21}u_{12} + l_{22} = 3, 2 \times 1 + l_{22} = 3, l_{22} = 1$$

$$l_{21}u_{13} + l_{22}u_{23} = 5, 2 \times (-1) + 1 \times u_{23} = 5, u_{23} = 7$$

$$l_{31}u_{12} + l_{32} = 2, 3 \times 1 + l_{32} = 2, l_{32} = -1$$

$$l_{31}u_{13} + l_{32}u_{23} + l_{33} = -3$$

$$3 \times (-1) + (-1) \times 7 + l_{33} = -3$$

$$l_{33} = 7$$

We get LY = B

$$\begin{bmatrix} 1 & 0 & 0 \\ 2 & 1 & 0 \\ 3 & -1 & 7 \end{bmatrix} \begin{bmatrix} y_1 \\ y_2 \\ y_3 \end{bmatrix} = \begin{bmatrix} 2 \\ -3 \\ 6 \end{bmatrix}$$

Now we get from the above condition

$$y_1 = 2$$

$$2y_1 + y_2 - = -3$$

$$\therefore \quad y_2 = -7$$

$$3y_1 - y_2 + 7y_3 = 6$$

$$6 + 7 + 7y_3 = 6$$

$$y_3 = -1$$

Hence UX = Y will give

$$\begin{bmatrix} 1 & 1 & -1 \\ 0 & 1 & 7 \\ 0 & 0 & 1 \end{bmatrix} \begin{bmatrix} x_1 \\ x_2 \\ x_3 \end{bmatrix} = \begin{bmatrix} 2 \\ -7 \\ -1 \end{bmatrix}$$

$$x_1 + x_2 - x_3 = 2$$

$$x_2 + 7x_3 = -7$$

$$x_3 = -1$$

Therefore $x_2 = 0, x_1 = 1$

Hence the solution of the above three linear equation is $x_1 = 1, x_2 = 0, x_3 = -1$

Example 16: Solve the following system of equation by LU Factorization method.

$$2x - 6y + 8z = 24$$

$$5x + 4y - 3z = 2$$

$$3x + y + 2z = 16$$

Solution: We take LU such that

$$LUX = AX = B$$

Where

$$\begin{bmatrix} l_{11} & 0 & 0 \\ l_{21} & l_{22} & 0 \\ l_{31} & l_{32} & l_{33} \end{bmatrix} \begin{bmatrix} 1 & u_{12} & u_{13} \\ 0 & 1 & u_{23} \\ 0 & 0 & 1 \end{bmatrix} = \begin{bmatrix} 2 & -6 & 8 \\ 5 & 4 & -3 \\ 3 & 1 & 2 \end{bmatrix}$$

We get $l_{11} = 2, l_{21} = 5, l_{31} = 3, l_{11}u_{12} = -6, u_{12} = -3, l_{11}u_{13} = 8, u_{13} = 4$

$$l_{21}u_{12} + l_{22} = 4, 5 \times (-3) + l_{22} = 4, l_{22} = 19$$

$$l_{21}u_{13} + l_{22}u_{23} = 5 \times (4) + 19u_{23} = -3, \ 19u_{23} = -23, u_{23} = -\frac{23}{19}$$

$$l_{31}u_{12} + l_{32} = 1, 3 \times (-3) + l_{32} = 1, \ l_{32} = 10$$

$$l_{31}u_{13} + l_{32}u_{23} + l_{33} = 2$$

$$3 \times 4 + 10 \times \left(-\frac{23}{19}\right) + l_{33} = 2$$

$$l_{33} = 2 - 12 + \frac{230}{19}$$

$$= \frac{40}{19}$$

Then

$$\begin{bmatrix} 2 & 0 & 0 \\ 5 & 19 & 0 \\ 3 & 10 & 40/19 \end{bmatrix} \begin{bmatrix} y_1 \\ y_2 \\ y_3 \end{bmatrix} = \begin{bmatrix} 24 \\ 2 \\ 16 \end{bmatrix}$$

$$2y_1 = 24, y_1 = 12$$

$$5y_1 + 19y_2 = 2$$

$$19y_2 = -58, y_2 = -\frac{58}{19}$$

$$3y_1 + 10y_2 + \frac{40}{19}y_3 = 16$$

$$36 - \frac{580}{19} + \frac{40}{19}y_3 = 16$$

$$\frac{40}{19}y_3 = \frac{580}{19} - 20$$

$$= \frac{200}{19}$$

$$y_3 = 5$$

Then

$$\begin{bmatrix} 1 & -3 & 4 \\ 0 & 1 & -23/19 \\ 0 & 0 & 1 \end{bmatrix} \begin{bmatrix} x \\ y \\ z \end{bmatrix} = \begin{bmatrix} 12 \\ -58/19 \\ 5 \end{bmatrix}$$

$$x - 3y + 4z = 12$$

$$y - \frac{23}{19}z = -\frac{58}{19}$$

$$z = 5$$

$$y = \frac{115 - 58}{19}$$

$$= \frac{57}{19}$$

$$= 3$$

$$x - 9 + 20 = 12$$

$$x = 1$$

$$x - 3y + 4z = 12$$

$$y - \frac{23}{19}z = -\frac{58}{19}$$

$$z = 5$$

$$y = \frac{115 - 58}{19}$$

$$= \frac{57}{19}$$

$$= 3$$

$$x - 9 + 20 = 12$$

$$x = 1$$

$$x = 1, \ y = 3, \ z = 5.$$

Example 17: Solve the following system of equation by LU factorization method.

$$2x + y + z = 3$$

$$x + 3y + z = -2$$

$$x + y - 4z = -6$$

Solution: Let us take two matrices L and U such that LU = A or,

$$\begin{bmatrix} l_{11} & 0 & 0 \\ l_{21} & l_{22} & 0 \\ l_{31} & l_{32} & l_{33} \end{bmatrix} \begin{bmatrix} 1 & u_{12} & u_{13} \\ 0 & 1 & u_{23} \\ 0 & 0 & 1 \end{bmatrix} = \begin{bmatrix} 2 & 1 & 1 \\ 1 & 3 & 1 \\ 1 & 1 & -4 \end{bmatrix}$$

$$l_{11} = 2,\ l_{21} = 1,\ l_{31} = 1,\ l_{11}u_{12} = 1,\ u_{12} = \frac{1}{2},\ l_{11}u_{13} = 1,\ u_{13} = \frac{1}{2}$$

$$l_{21}u_{12} + l_{22} = 3,\ 1 \times \frac{1}{2} + l_{22} = 3,\ l_{22} = \frac{5}{2}$$

$$l_{21}u_{13} + l_{22}u_{23} = 1,\ 1 \times \frac{1}{2} + \frac{5}{2}u_{23} = 1,\ \frac{5}{2}u_{23} = \frac{1}{2},\ u_{23} = \frac{1}{5}$$

Then LY = B

$$\begin{bmatrix} 2 & 0 & 0 \\ 1 & 5/2 & 0 \\ 3 & 17/2 & -69/5 \end{bmatrix} \begin{bmatrix} y_1 \\ y_2 \\ y_3 \end{bmatrix} = \begin{bmatrix} 8 \\ 15 \\ 8 \end{bmatrix}$$

$$2y_1 = 8,\ y_1 = 4$$

$$y_1 + \frac{5}{2}y_2 = 15$$

$$y_2 = \frac{11 \times 2}{5}$$

$$3y_1 + \frac{17}{2}y_2 - \frac{69}{5}y_3 = 8$$

$$12 + \frac{17 \times 11 \times 2}{10} - \frac{69}{5}y_3 = 8$$

$$\frac{69}{5}y_3 = -4 - \frac{187}{5} = -\frac{207}{5}$$

$$y_3 = 3$$

$$\begin{bmatrix} 1 & -3/2 & 2 \\ 0 & 1 & 4/5 \\ 0 & 0 & 1 \end{bmatrix} \begin{bmatrix} x_1 \\ x_2 \\ x_3 \end{bmatrix} = \begin{bmatrix} 4 \\ 22/5 \\ 3 \end{bmatrix}$$

$$x_1 - \frac{3}{2}x_2 + 2x_3 = 4$$

$$x_2 + \frac{4}{5}x_3 = \frac{22}{5}$$

$$x_3 = 3$$

$$x_2 = \frac{10}{5}$$

$$= 2$$

$$x_1 - 3 + 6 = 4$$

$$x_1 = 1$$

$$x_1 = 1, x_2 = 2, x_3 = 3$$

Example 18: LU decomposition method to solve the system of equations

$$\begin{bmatrix} 1 & 2 & 2 \\ 2 & 1 & 1 \\ 2 & 2 & 1 \end{bmatrix} \begin{bmatrix} x \\ y \\ z \end{bmatrix} = \begin{bmatrix} 1 \\ 1 \\ 1 \end{bmatrix}$$

***Solution*:** Let A = LU where

$$\begin{bmatrix} 1 & 2 & 2 \\ 2 & 1 & 1 \\ 2 & 2 & 1 \end{bmatrix} = \begin{bmatrix} l_{11} & 0 & 0 \\ l_{21} & l_{22} & 0 \\ l_{31} & l_{32} & l_{33} \end{bmatrix} \begin{bmatrix} 1 & u_{12} & u_{13} \\ 0 & 1 & u_{23} \\ 0 & 0 & 1 \end{bmatrix}$$

$$l_{11} = 1, l_{11}u_{12} = 2, u_{12} = 2, l_{11}u_{13} = 2, u_{13} = 2$$

$$l_{21} = 2, l_{21}u_{12} + l_{22} = 1, 4 + l_{22} = 1, l_{22} = -3$$

$$l_{21}u_{13} + l_{22}u_{23} = 1, 2 \times 2 - 3u_{23} = 1, u_{23} = 1$$

$$l_{31} = 2, l_{31}u_{12} + l_{32} = 2, 4 + l_{32} = 2, l_{32} = -2$$

$$l_{31}u_{13} + l_{32}u_{23} + l_{33} = 1$$

$$2 \times 2 + (-2) + l_{33} = 1$$

$$l_{33} = -1$$

$$\begin{bmatrix} 1 & 0 & 0 \\ 2 & -3 & 0 \\ 2 & -2 & -1 \end{bmatrix} \begin{bmatrix} y_1 \\ y_2 \\ y_3 \end{bmatrix} = \begin{bmatrix} 1 \\ 1 \\ 1 \end{bmatrix}$$

$$y_1 = 1, 2y_1 - 3y_2 = 1, y_2 = \frac{1}{3}$$

$$2y_1 - 2y_2 - y_3 = 1, y_3 = \frac{1}{3}$$

Example 19: Solve the following system of equations using LU factorization for tri diagonal system.

$$x_1 - x_2 = 0$$

$$-2x_1 + 4x_2 - 2x_3 = -1$$

$$-x_2 + 2x_3 = 1.5$$

Solution:

$$\begin{bmatrix} 1 & -1 & 0 \\ -2 & 4 & -2 \\ 0 & -1 & 2 \end{bmatrix} = \begin{bmatrix} l_{11} & 0 & 0 \\ l_{21} & l_{22} & 0 \\ l_{31} & l_{32} & l_{33} \end{bmatrix} \begin{bmatrix} 1 & u_{12} & u_{13} \\ 0 & 1 & u_{23} \\ 0 & 0 & 1 \end{bmatrix}$$

Now,

$$l_{11} = 1, l_{11}u_{12} = -1, u_{12} = -1, l_{11}u_{13} = 0, u_{13} = -0$$

$$l_{21} = -2, l_{21}u_{12} + u_{22} = 4, l_{22} = 4 - 2 = 2$$

$$l_{21}u_{13} + l_{22}u_{23} = -2, u_{23} = -1$$

$$l_{31} = 0, l_{31}u_{12} + l_{32} = -1, l_{32} = -1$$

$$l_{31}u_{13} + l_{32}u_{23} + l_{33} = 2$$

$$l_{33} = 1$$

Then,

$$\begin{bmatrix} 1 & 2 & 2 \\ 0 & 1 & 1 \\ 0 & 0 & 1 \end{bmatrix} \begin{bmatrix} x \\ y \\ z \end{bmatrix} = \begin{bmatrix} 1 \\ 1/3 \\ 1/3 \end{bmatrix}$$

$$z = \frac{1}{3}$$

$$y + z = \frac{1}{3} \therefore y = 0$$

$$x + 2y + 2z = 1/3$$

$$\therefore \quad x = \frac{-1}{3}$$

Hence

$$x = \frac{-1}{3}, y = 0, z = 1/3$$

$$\begin{bmatrix} 1 & 0 & 0 \\ -2 & 2 & 0 \\ 0 & -1 & 1 \end{bmatrix} \begin{bmatrix} y_1 \\ y_2 \\ y_3 \end{bmatrix} = \begin{bmatrix} 0 \\ -1 \\ 1.5 \end{bmatrix}$$

$$y_1 = 0, -2y_1 + 2y_2 = -1, y_2 = -0.5$$

$$-y_2 + y_3 = 1.5, y_3 = 1$$

$$\begin{bmatrix} 1 & -1 & 0 \\ 0 & 1 & -1 \\ 0 & 0 & 1 \end{bmatrix} \begin{bmatrix} x_1 \\ x_2 \\ x_3 \end{bmatrix} = \begin{bmatrix} 0 \\ -0.5 \\ 1 \end{bmatrix}$$

$$x_3 = 1, x_2 - x_3 = -0.5, x_2 = 0.5, x_1 - x_2 = 0, x_1 = 0.5$$

$$x_1 = 0.5, x_2 = 0.5, x_3 = 1$$

Example 20: Solve the system of equations given below by the method of LU factorization.

$$2x - 3y + 10z = 3$$

$$-x + 4y + 2z = 20$$

$$5x + 2y + z = -12$$

Solution:
$$\begin{bmatrix} 2 & -3 & 10 \\ -1 & 4 & 2 \\ 5 & 2 & 1 \end{bmatrix} = \begin{bmatrix} l_{11} & 0 & 0 \\ l_{21} & l_{22} & 0 \\ l_{31} & l_{32} & l_{33} \end{bmatrix} \begin{bmatrix} 1 & u_{12} & u_{13} \\ 0 & 1 & u_{23} \\ 0 & 0 & 1 \end{bmatrix}$$

Now

$$l_{11} = 2, l_{11}u_{12} = -3, u_{12} = -3/2, l_{11}u_{13} = 10, u_{13} = 5$$

$$l_{21} = -1, l_{21}u_{12} + l_{22} = 4, l_{22} = 4 - 3/2 = 5/2$$

$$l_{21}u_{13} + l_{22}u_{23} = 2, 5/2u_{23} = 7, u_{23} = 14/5$$

$$l_{31} = 5, l_{31}u_{12} + l_{32} = 2; l_{32} = \frac{19}{2}$$

$$l_{31}u_{13} + l_{32}u_{23} + l_{33} = 1$$

$$l_{33} = -253/5$$

$$\begin{bmatrix} 2 & 0 & 0 \\ -1 & 5/2 & 0 \\ 5 & 19/2 & -253/5 \end{bmatrix} \begin{bmatrix} y_1 \\ y_2 \\ y_3 \end{bmatrix} = \begin{bmatrix} 3 \\ 20 \\ -12 \end{bmatrix}$$

$$2y_1 = 3, y_1 = 3/2, -y_1 + \frac{5}{2}y_2 = 20, y_2 = \frac{43}{5}$$

$$5y_1 + \frac{19}{2}y_2 - \frac{253}{5}y_3 = -12, y_3 = 2$$

$$\begin{bmatrix} 1 & -3/2 & 5 \\ 0 & 1 & 14/5 \\ 0 & 0 & 1 \end{bmatrix} \begin{bmatrix} x \\ y \\ z \end{bmatrix} = \begin{bmatrix} 3/2 \\ 43/5 \\ 2 \end{bmatrix}$$

$$z = 2, y + \frac{14}{5}z = \frac{43}{5}, y = 3$$

$$x - \frac{3}{2}y + 5z = \frac{3}{2}, x = -4$$

$$x = -4, y = 3, z = 2$$

12.7 GAUSS-JACOBI METHOD

This method is an iterative method and can be written as

$$x_1^{(1)} = \frac{1}{a_{11}}\left[b_1 - a_{12}x_2^{(0)} - a_{13}x_3^{(0)}\right]$$

$$x_2^{(1)} = \frac{1}{a_{22}}\left[b_2 - a_{21}x_1^{(0)} - a_{23}x_3^{(0)}\right]$$

$$x_3^{(1)} = \frac{1}{a_{33}}\left[b_3 - a_{31}x_1^{(0)} - a_{32}x_2^{(0)}\right]$$

Where

$$\begin{bmatrix} a_{11} & a_{12} & a_{13} \\ a_{21} & a_{22} & a_{23} \\ a_{31} & a_{32} & a_{33} \end{bmatrix} \begin{bmatrix} x_1 \\ x_2 \\ x_3 \end{bmatrix} = \begin{bmatrix} b_1 \\ b_2 \\ b_3 \end{bmatrix}$$

Is the matrix form of the linear equations in three unknown variables. The next iterations can be written as

$$x_1^{(i+1)} = \frac{1}{a_{11}}\left[b_1 - a_{12}x_2^{(i)} - a_{13}x_3^{(i)}\right]$$

$$x_2^{(i+1)} = \frac{1}{a_{22}}\left[b_2 - a_{21}x_1^{(i)} - a_{23}x_3^{(i)}\right]$$

$$x_3^{(i+1)} = \frac{1}{a_{33}}\left[b_3 - a_{31}x_1^{(i)} - a_{32}x_2^{(i)}\right] \qquad\qquad i = 1,2,3.....$$

The sequence $\left\{x_i^{(k)}, i = 1,2,3....\right\}$ generated above will be convergent to the solutions if

$$|a_{ii}| > \sum_{j=1,\, j\neq i}^{3} |a_{ij}|\,(i = 1,2,3)$$

i.e.

$$|a_{11}| > |a_{12}| + |a_{13}|$$

$$|a_{22}| > |a_{21}| + |a_{23}|$$

$$|a_{33}| > |a_{31}| + |a_{32}|$$

This above conditions imply that the Gaus-Jacobi iteration method converges if and only if the system of equations is strictly diagonally dominant. This method is also known as the method of simultaneous displacements. More over if the above equations are not diagonally dominant, then the equations must be written as diagonally dominant one, so that the equations will give a convergent sequence.

Example 21: Consider the system of equations and solve by Gauss-Jacobi method.

$$20x_1 + 2x_2 + 6x_3 = 28$$

$$x_1 + 20x_2 + 9x_3 = -23$$

$$2x_1 - 7x_2 - 20x_3 = -57$$

Solution: Since

$$|20| > |2| + |6|$$

$$|20| > |1| + |9|$$

$$|20| > |2| + |7|$$

Hence they are diagonally dominant

$$x_1^{(i+1)} = \frac{1}{20}\left[28 - 2x_2^{(i)} - 6x_3^{(i)}\right]$$

$$x_2^{(i+1)} = \frac{1}{20}\left[-23 - x_2^{(i)} - 9x_3^{(i)}\right]$$

$$x_3^{(i+1)} = \frac{1}{20}\left[57 + 2x_2^{(i)} - 7x_2^{(i)}\right]$$

i	$x_1^{(i)}$	$x_2^{(i)}$	$x_3^{(i)}$
0	0	0	0
1	1.4	−1.15	2.85
2	0.66	−2.50	3.39
3	0.633	−2.71	3.79
4	0.534	−2.887	3.862
5	0.5301	−2.9146	3.9139
6	0.5173	−2.9378	3.9231
7	0.5169	−2.9413	3.9299
8	0.5152	−2.9443	3.9312
9	0.5151	−2.9448	3.9320
10	0.5149	−2.9452	3.9322
11	0.5149	−2.9452	3.9323

Hence the roots are $x_1 = 0.5149$, $x_2 = -2.9452$, $x_3 = 3.9323$.

12.8   GAUSS-SEIDEL METHOD

This method is an improvement of Jacobi's method.

$$x_1^{(i+1)} = \frac{1}{a_{11}}\left[b_1 - a_{12}x_2^{(i)} - a_{13}x_3^{(i)}\right]$$

$$x_2^{(i+1)} = \frac{1}{a_{22}}\left[b_2 - a_{21}x_1^{(i+1)} - a_{23}x_3^{(i)}\right]$$

$$x_3^{(i+1)} = \frac{1}{a_{33}}\left[b_3 - a_{31}x_1^{(i+1)} - a_{32}x_2^{(i+1)}\right]$$

This method is also convergent when the system is strictly diagonally dominant.

Example 22: Using Gauss Seidel method, find the solutions of the following system of linear equations correct up to 2 places of decimal.

$$3x + y + 5z = 13$$

$$5x - 2y + z = 4$$

$$x + 6y - 2z = -1$$

Solution: First we have to make the above equations in diagonally dominant, and rewrite them as,

$$5x - 2y + z = 4$$

$$x + 6y - 2z = -1$$

$$3x + y + 5z = 13$$

Then

$$x^{(i+1)} = \frac{1}{5}\left[4 + 2y^{(i)} - az^{(i)}\right]$$

$$y^{(i+1)} = \frac{1}{6}\left[-1 - x^{(i+1)} - 2z^{(i)}\right]$$

$$z^{(i+1)} = \frac{1}{5}\left[13 - 3x^{(i+1)} - y^{(i+1)}\right]$$

We have the computational table as:

K	$x^{(K)}$	$y^{(K)}$	$z^{(K)}$
0	0	0	0
1	0.8	-0.3	2.18
2	0.244	0.519	2.3494
3	0.5377	0.527	2.172
4	0.576	0.461	2.162
5	0.552	0.462	2.176
6	0.5496	0.467	2.176
7	0.551	0.467	2.176
8	0.552	0.467	2.175

Now x = 0.55, y = 0.47, z = 2.18 correct upto two decimal places.

Example 23: Solve the following set of equations by Gauss-seidel correct to 2 places of decimal.

$$3y - 2z = 3$$
$$2x - y + 4z = 27$$
$$4x + y - 3z = 3$$

***Solution*:** We first write the equations in diagonally dominant process

i.e
$$4x + y - 3z = 3$$
$$3y - 2z = 3$$
$$2x - y + 4z = 27$$

$$x^{(i+1)} = \frac{1}{4}\left[3 - y^{(i)} + 3z^{(i)}\right]$$

$$x^{(i+1)} = \frac{1}{3}\left[3 + 2z^{(i)}\right]$$

$$z^{(i+1)} = \frac{1}{4}\left[27 - 2x^{(i+1)} + y^{(i+1)}\right]$$

Then the computational table is

K	$x^{(K)}$	$y^{(K)}$	$z^{(K)}$
0	0	0	0
1	0.75	1.0	6.625
2	5.469	5.42	5.371
3	3.423	4.581	6.184
4	4.243	5.123	5.909
5	3.901	4.939	6.03
6	4.038	5.02	5.986
7	3.984	4.991	6.006

Hence $x_1 = 4.00$, $x_2 = 5.00$, $x_3 = 6.00$ correct up to two decimal places.

Example 24: Using Gauss Seidel method, find the solution of the following system of linear equations correct up to 2 places of decimals.

$$9x_1 - 2x_2 + x_3 = 50$$

$$x_1 + 5x_2 - 3x_3 = 18$$

$$-2x_1 + 2x_2 + 7x_3 = 19$$

Solution: The system is given in the form of diagonally dominant form

$$x_1^{(i+1)} = \frac{1}{9}\left[50 + 2x_2^{(i)} - x_3^{(i)}\right]$$

$$x_2^{(i+1)} = \frac{1}{5}\left[18 - x_1^{(i+1)} + 3x_3^{(i)}\right]$$

$$x_3^{(i+1)} = \frac{1}{7}\left[19 + 2x_1^{(i+1)} - 2x_2^{(i+1)}\right]$$

$x_1^{(0)} = x_2^{(0)} = x_3^{(0)} = 0$ then the computational table is:

K	$x^{(K)}$	$y^{(K)}$	$z^{(K)}$
0	0	0	0
1	5.556	2.488	3.591
2	5.709	4.613	3.027
3	6.244	4.1674	3.308
4	6.114	4.362	3.215
5	6.227	4.284	3.269
6	6.144	4.333	3.232
7	6.159	4.307	3.243
8	6.152	4.315	3.239

Hence $x_1 = 6.15$, $x_2 = 4.13$, $x_3 = 3.24$ correct up to 2 decimal places.

Example 25: Solve the following system of equations correct to four places of decimals by Gauss – Seidel iteration method.

$$x + y + 54z = 110$$

$$27x + 6y - z = 85$$

$$6x + 15y + 2z = 72$$

Solution: We have to arrange the above equations in such a manner that the above equations are diagonally dominant.

$$27x + 6y - z = 85$$

$$6x + 15y + 2z = 72$$

$$x + y + 54z = 110$$

Then
$$x^{(i+1)} = \frac{1}{27}\left[85 - 6y^{(i)} + z^{(i)}\right]$$

$$y^{(i+1)} = \frac{1}{15}\left[72 - 6x^{(i+1)} - 2z^{(i)}\right]$$

$$z^{(i+1)} = \frac{1}{54}\left[110 - x^{(i+1)} - y^{(i+1)}\right]$$

Let us first take $x^{(0)} = y^{(0)} = z^{(0)} = 0$ then the computational table is:

K	$x^{(K)}$	$y^{(K)}$	$z^{(K)}$
0	0	0	0
1	3.148	3.541	1.913
2	2.361	3.383	1.931
3	2.695	3.465	1.923
4	2.499	3.564	1.926
5	2.427	3.575	1.926
6	2.426	3.573	1.926

Hence x = 2.43, y = 3.57, z = 1.93 correct up to two decimal places.

Example 26: Using gauss Seidel method, find the solution of the following system of linear equations correct up to 2 places of decimals:

$$3x + y + 5z = 13$$

$$5x - 2y + z = 4$$

$$x + 6y - 2z = -1$$

Solution: We arrange the above equation in the form of diagonally dominant from.

$$5x - 2y + z = 4$$

$$x + 6y - 2z = -1$$

$$3x + y + 5z = 13$$

Now

$$x^{(k+1)} = \frac{1}{5}\left[4 + 2y^{(k)} - z^{(k)}\right]$$

$$y^{(k+1)} = \frac{1}{6}\left[-1 - x^{(k+1)} + 2z^{(k)}\right]$$

$$z^{(k+1)} = \frac{1}{5}\left[13 - 3x^{(k+1)} - y^{(k+1)}\right]$$

Let

$$x_1^{(0)} = x_2^{(0)} = x_3^{(0)} = 0.$$

K	$x^{(K)}$	$y^{(K)}$	$z^{(K)}$
0	0	0	0
1	0.8	-0.3	2.18
2	0.244	0.5193	1.958
3	0.616	0.383	2.1538
4	0.522	0.464	2.194
5	0.547	0.474	2.177
6	0.554	0.467	2.174
7	0.552	0.466	2.176
8	0.551	0.467	2.176

Hence after approximation we get x = 0.55, y = 0.47, z = 2.18 correct up to 2 decimal places.

Example 27: Solve the following system of equation, correct to four places of decimals, by Gauss Seidel iteration.

$$x + y + 54z = 110$$

$$27x + 6y - z = 85$$

$$6x + 15y + 2z = 72$$

Solution: We arrange the above system to make it possible to start for Gauss Seidel

$$27x + 6y - z = 85$$

$$6x + 15y + 2z = 72$$

$$x + y + 54z = 110$$

$$x^{(k+1)} = \frac{1}{27}\left[85 - 6y^{(k)} + z^{(k)}\right]$$

$$y^{(k+1)} = \frac{1}{15}\left[72 - 6x^{(k+1)} - 2z^{(k)}\right]$$

$$z^{(k+1)} = \frac{1}{54}\left[110 - x^{(k+1)} - y^{(k+1)}\right]$$

Let $\qquad x_1^{(0)} = x_2^{(0)} = x_3^{(0)} = 0.$

K	$x^{(K)}$	$y^{(K)}$	$z^{(K)}$
0	0	0	0
1	3.148	3.5408	1.9132
2	2.432	3.5721	1.9258
3	2.4257	3.5729	1.9260
4	2.4255	3.5730	1.9260
5	2.4255	3.5730	1.9260

Hence x = 2.4255, y = 3.5730, z = 1.9260

EXERCISE

1. Solve the equations using Gauss Jordan method

$$3x_1 + 2x_2 + 3x_3 = 18$$

$$2x_1 + x_2 + x_3 = 10$$

$$x_1 + 4x_2 + 9x_3 = 16 \qquad\qquad \textbf{[Ans:}\ x_1 = 7,\ x_2 = -9,\ x_3 = 5]$$

2. Solve the equations using Gauss – elimination method.

$$6x_1 + 3x_2 + 2x_3 = 6$$

$$6x_1 + 4x_2 + 3x_3 = 0$$

$$20x_1 + 15x_2 + 12x_3 = 0$$

3. Solve the equations by Gauss – Jacobi method.

$$12.214x_1 + 2.367x_2 + 3.672x_3 = 7.814$$

$$2.412x_1 + 9.879x_2 + 1.564x_3 = 4.89$$

$$1.876x_1 + 2.985x_2 - 11.625x_3 = -0.972$$

$$\textbf{[Ans:}\ x_1 = \frac{1}{2},\ x_2 = \frac{1}{3},\ x_3 = \frac{1}{4}]$$

4. Solve the system of equations by the LU method.

$$2x_1 - 3x_2 + 10x_3 = 3$$

$$-x_1 + 4x_2 + 2x_3 = 20$$

$$5x_1 + 2x_2 + x_3 = -12$$

[**Ans:** –4, 3, 2]

Solutions and Key

1 AERO SPACE ENGINEERING

S.No	Ans	S.No	Ans	S.No	Ans
1	C	6	C	11	A
2	A	7	B	12	C
3	D	8	C	13	B
4	A	9	A	14	C
5	B	10	C	15	D

Solutions

1. The characteristic equation of A is $(3-\lambda)(3-\lambda) -1 = 0 \Rightarrow \lambda^2 -6\lambda + 8 = 0$

   $\Rightarrow (\lambda -2)(\lambda -4) = 0 \Rightarrow \lambda = 2$ or 4

2. $f(x, y) = \tan^{-1}\dfrac{x}{y} \Rightarrow \dfrac{\partial f}{\partial x} = \dfrac{y}{x^2+y^2} \Rightarrow \dfrac{\partial^2 f}{\partial y\, \partial x} = \dfrac{\partial}{\partial y}\left(\dfrac{\partial f}{\partial x}\right) = \dfrac{(x^2+y^2).1-y(2y)}{(x^2+y^2)^2} = \dfrac{x^2-y^2}{(x^2+y^2)^2}$

3. $f(x) = x\sqrt{\alpha^2 - x^2}$, satisfies all the conditions of Rolle's mean value theorem.

   Differentiating we get $f'(x) = \dfrac{\alpha^2-2x^2}{\sqrt{\alpha^2-x^2}} \Rightarrow f'(x) = \dfrac{\alpha^2-2c^2}{\sqrt{\alpha^2-c^2}}$

   Now $f'(c) = 0 \Rightarrow \alpha^2 - 2c^2 = 0 \Rightarrow c = \dfrac{\alpha}{\sqrt{2}}$

4. $\lim\limits_{x\to 0}\dfrac{\tan x-x}{x^2 \tan x}$ is in $\dfrac{0}{0}$ form hence we apply L'Hospital's rule

   We can write $\lim\limits_{x\to 0}\dfrac{\tan x-x}{x^2 \tan x} = \lim\limits_{x\to 0}\dfrac{\tan x-x}{x^3} \cdot \dfrac{x}{\tan x} = \lim\limits_{x\to 0}\dfrac{\tan x-x}{x^3}$ since $\lim\limits_{x\to 0}\dfrac{x}{\tan x} = 1$

   $\Rightarrow \lim\limits_{x\to 0}\dfrac{\tan x-x}{x^3} = \lim\limits_{x\to 0}\dfrac{\sec^2 x-1}{3x^2} = \lim\limits_{x\to 0}\dfrac{2\sec^2 x\tan x}{6x} = \lim\limits_{x\to 0}\dfrac{\sec^2 x}{3}\cdot\lim\limits_{x\to 0}\dfrac{\tan x}{x} = \dfrac{1}{3}$

5. The given equation can be written as $y = px + \dfrac{p}{p-1}$ which is in Clairaut's form. The general solution can be obtained by replacing p by c.

   Hence the general solution is $y = cx + \dfrac{c}{c-1}$

6. The given equation can be written as $(4 D^2 - 4D +1)y = 0$

   The auxiliary equation is $(4 m^2 - 4m +1) = 0 \Rightarrow (2m - 1)^2 = 0 \Rightarrow m = \dfrac{1}{2}, \dfrac{1}{2}$

   (repeated roots)

The required solution is $y = (c_1 + c_2 x)e^{\frac{x}{2}}$

7. $\text{Grad } \phi = \frac{\partial \phi}{\partial x} i + \frac{\partial \phi}{\partial y} j + \frac{\partial \phi}{\partial x} k = 2xi + \cos y \, j + k$

   $\Rightarrow \text{grad } \phi \text{ at } (0, \frac{\pi}{2}, 1) = 0.i + 0.j + k = k$

8. $L[e^{-\alpha t}] = \int_0^\infty e^{-st.} e^{-\alpha t} \, dt = \int_0^\infty e^{-(s+\alpha)t} dt = [\frac{e^{-(s+\alpha)t}}{-(s+\alpha)}]$ where t between the limits 0

   and $\infty = \frac{e^{-\infty}}{-(s+\alpha)} - \frac{e^0}{-(s+\alpha)} = \frac{1}{s+\alpha}$ [since $e^{-\infty} = 0$ and $e^0 = 1$]

9. Residue at $z = 3$ is $\lim_{z \to 3}(z-3) f(z) = \lim_{z \to 3} \frac{z^3}{(z-2)(z-1)^2} = \frac{27}{4}$

10. We have $z = \alpha e^{bt} \sin bx$ (given) Differentiating partially with respect to x we get
   $\frac{\partial z}{\partial x} = \alpha b \, e^{bt} \cos bx$

   Again partially differentiating with respect to x we get $\frac{\partial^2 z}{\partial x^2} = -\alpha b^2 e^{bt} \sin bx$

   Similarly we get $\frac{\partial^2 z}{\partial x^2} = \alpha b^2 e^{bt} \sin bx$

   Adding we get $\frac{\partial^2 z}{\partial x^2} + \frac{\partial^2 z}{\partial t^2} = 0$

11. Let $z = \alpha x + by + c$ be the solution of $pq = 1$.

   $z = \alpha x + by + c \Rightarrow \frac{\partial z}{\partial x} = p = \alpha$ and $\frac{\partial z}{\partial y} = q = b$

   Now $pq = 1 \Rightarrow \alpha b = 1 \Rightarrow b = \frac{1}{\alpha}$. Substituting in $z = \alpha x + by + c$ we get $z = \alpha x + \frac{y}{\alpha} + c$
   Which is the solution

12. $L^{-1}[\frac{1}{(s+1)^2}] = e^{-t}L^{-1}[\frac{1}{s^2}] = te^{-t}$

13. we have $y = x \Rightarrow dy = dx$, $r = xi + yj \Rightarrow dr = dx \, i + dy \, j$ and $F = x^2 i + y^2 j$
   $F. dr = (x^2 i + y^2 j), (dx \, i + dy \, j) = x^2 dx + y^2 dy = x^2 dx + x^2 dx = 2 \, x^2 dx$

   $\therefore \int F. dr = \int_0^1 2x^2 dx = \frac{2}{3}$

14. $I = \int_0^{\frac{\pi}{2}} \frac{1}{1+\tan x} dx = \int_0^{\frac{\pi}{2}} \frac{\cos x}{\cos x + \sin x} dx \Rightarrow I = \int_0^{\frac{\pi}{2}} \frac{\cos(\frac{\pi}{2} - x)}{\cos(\frac{\pi}{2} - x) + \sin(\frac{\pi}{2} - x)} dx = \int_0^{\frac{\pi}{2}} \frac{\sin x}{\sin x + \cos x}$

   $dx \Rightarrow I + I = 2I = \int_0^{\frac{\pi}{2}} \frac{\sin x + \cos x}{\sin x + \cos x} dx = \int_0^{\frac{\pi}{2}} dx = \frac{\pi}{2} \Rightarrow I = \frac{\pi}{4}$

15. $1A = \begin{bmatrix} \cos \alpha & \sin \alpha \\ -\sin \alpha & \cos \alpha \end{bmatrix} \Rightarrow A^T = \begin{bmatrix} \cos \alpha & -\sin \alpha \\ \sin \alpha & \cos \alpha \end{bmatrix}$

   $\Rightarrow A. A^T = \begin{bmatrix} \cos \alpha & \sin \alpha \\ -\sin \alpha & \cos \alpha \end{bmatrix}\begin{bmatrix} \cos \alpha & -\sin \alpha \\ \sin \alpha & \cos \alpha \end{bmatrix} =$

   $\begin{bmatrix} \cos^2 \alpha + \sin^2 \alpha & 0 \\ 0 & \cos^2 \alpha + \sin^2 \alpha \end{bmatrix} = \begin{bmatrix} 1 & 0 \\ 0 & 1 \end{bmatrix} \Rightarrow A$ is orthogonal

2 BIO MEDICAL ENGINEERING

S.NO	Ans	S.NO	Ans	S.NO	Ans
1	B	6	D	11	A
2	A	7	B	12	B
3	D	8	C	13	D
4	A	9	A	14	A
5	B	10	D	15	C

Solutions

1. We have $x = a \cos^3\theta$ and $y = a \sin^3\theta$

   Differentiating with respect to θ we get $\frac{dx}{d\theta} = a\,(3\cos^2\theta\,(-\sin\theta)) = -3a\cos^2\theta \sin\theta$ and $\frac{dy}{d\theta} = 3a\sin^2\theta\cos\theta$

   $$\Rightarrow \frac{dy}{dx} = \frac{\frac{dy}{d\theta}}{\frac{dx}{d\theta}} = -\frac{3a\sin^2\theta\cos\theta}{3a\cos^2\theta\sin\theta} = -\frac{\sin\theta}{\cos\theta} = -\tan\theta$$

2. Consider $ax^2 + by^2 = c$ $\quad$ ... (1), differentiating with respect to x we get $2ax + 2by\frac{dy}{dx} = 0 \Rightarrow \frac{dy}{dx} = -\frac{a}{b}\cdot\frac{x}{y}$

   Again differentiating with respect to x we get $\frac{d^2y}{dx^2} = -\frac{a}{b}\cdot\frac{y.1-x\frac{dy}{dx}}{y^2} = -\frac{a}{b}\cdot\frac{y+\frac{ax^2}{by}}{y^2} = -\frac{a}{b}\cdot[\frac{ax^2+by^2}{by^3}] = -\frac{ac}{b^2y^3}$ (using (1))

3. $\int \frac{xe^x}{(1+x)^2}\,dx = \int(\frac{1}{1+x} + \frac{-1}{(1+x)^2})e^x dx$ is of the form $\int[f(x) + f'(x)]dx$ where $f(x) = \frac{1}{1+x}$ Hence $\int \frac{xe^x}{(1+x)^2}\,dx = \frac{1}{1+x}e^x$

4. $\int_0^1 \frac{1-x}{1+x}\,dx = \int_0^1 \frac{2-(1+x)}{1+x}\,dx = 2\int_0^1 \frac{1}{1+x}\,dx - \int_0^1 1dx = 2\,\{\log(1+x) - x\,|^1_0 = 2\log 2 - 1$

5. $u = \sin^{-1}\frac{x^2+y^2}{x+y} \Rightarrow \sin u = \frac{x^2+y^2}{x+y} = z$ (say) Z is a homogeneous function of degree 1. By Euler's theorem $x\frac{\partial z}{\partial x} + y\frac{\partial z}{\partial y} = 1.z = z$

   $Z = \sin u \Rightarrow \frac{\partial z}{\partial x} = \cos u\frac{\partial u}{\partial x}$ and $\frac{\partial z}{\partial y} = \cos u\frac{\partial u}{\partial y}$

   Substituting in $x\frac{\partial z}{\partial x} + y\frac{\partial z}{\partial y} = z$ we get $x\frac{\partial u}{\partial x} + y\frac{\partial u}{\partial y} = \tan u$

6. Given matrix $A = \begin{bmatrix} 2 & \lambda \\ 3 & 6 \end{bmatrix}$ has no inverse if det $A = \begin{vmatrix} 2 & \lambda \\ 3 & 6 \end{vmatrix} = 0 \Rightarrow 12 - 3\lambda = 0$

   $\Rightarrow 3\lambda = 12 \Rightarrow \lambda = 4$

7. Matrix $A = \begin{bmatrix} 1 & 1 & -1 \\ 2 & -3 & 4 \\ 3 & -2 & 3 \end{bmatrix}$ applying $R_2 \to R_2 - 2R_1, R_3 \to R_3 - 3R_1$ we get

$$A = \begin{bmatrix} 1 & 1 & -1 \\ 0 & -5 & 6 \\ 0 & -5 & 6 \end{bmatrix} \text{ applying } R_3 \rightarrow R_3 - R_2 \text{ we get}$$

$$A = \begin{bmatrix} 1 & 1 & -1 \\ 0 & -5 & 6 \\ 0 & 0 & 0 \end{bmatrix} \text{ applying } R_2 \rightarrow \frac{-1}{5} R_2 \text{ we get } \begin{bmatrix} 1 & 1 & -1 \\ 0 & 1 & -6/5 \\ 0 & 0 & 0 \end{bmatrix} \text{ which is in}$$

echelon form

The number of non – zero rows is 2 Rank of A is 2

8. Put $x + y = u$ differentiating we get $1 + \dfrac{dy}{dx} = \dfrac{du}{dx} \Rightarrow \dfrac{dy}{dx} = \dfrac{du}{dx} - 1$ Substituting in the given equation we get $e^{-u} du = dx \Rightarrow \int e^{-u} du = \int 1. dx \Rightarrow - e^{-u} = x + c \Rightarrow x + e^{-u} + c = 0 \Rightarrow$ the solution is $x + e^{-(x+y)} + c = 0$

9. $\sec^2 x \tan y\, dx + \sec^2 y \tan x\, dy = 0 \Rightarrow \dfrac{\sec^2 x}{\tan x} dx + \dfrac{\sec^2 y}{\tan y} = 0$

Integrating we get $\int \dfrac{\sec^2 x}{\tan x} dx + \int \dfrac{\sec^2 y}{\tan y} dy = \log c \Rightarrow \log \tan x + \log \tan y = \log c$

$\Rightarrow \tan x \tan y = c$

10. $f(z) = \dfrac{3z+1}{(z+1)(2z-1)}$ then the residue of $f(z)$ at the pole $z = -1$ is then the residue of $f(z) = \dfrac{3z+1}{(z+1)(2z-1)}$ at the pole $z = -1$ is $\dfrac{2}{3}$

$$\lim_{z \to -1} (z + 1) \dfrac{3z+1}{(z+1)(2z-1)} = \lim_{z \to -1} \dfrac{3z+1}{(2z-1)} = \dfrac{2}{3}$$

11. The equation of the line joining the points (0.0) and (1,2) is $y = x + 1 \Rightarrow dy = dx$

$$\int_{(0,1)}^{(1,2)} (x^2 - y)\, dx + (y^2 + x)\, dy = \int_0^1 (x^2 - x - 1) + [(x+1)^2 + x]\, dx$$

$$= 2 \int_0^1 (x^2 + x)\, dx = \dfrac{5}{3}$$

12. Applying multiplication theorem on Probability we have $P\ (B/A) = \dfrac{P(A \cap B)}{P(A)}$

$\Rightarrow P(A \cap B) = P(A)\, P\ (B/A) = (0.9)(0.8) = 0.72$

13. We have mean $= np = 4$, variance $= npq = 3$, $q = \dfrac{npq}{np} = \dfrac{3}{4} \Rightarrow p = 1 - q = 1 - \dfrac{3}{4} = \dfrac{1}{4}$ $np = 4$

$\Rightarrow n. \dfrac{1}{4} = 4 \Rightarrow m = 16$

14. $P(X=1) = \dfrac{3}{10} \Rightarrow \dfrac{3}{10} = \lambda e^{-\lambda}$, $P\ (X = 2) = \dfrac{1}{5} \Rightarrow \dfrac{1}{5} = \dfrac{\lambda^2 e^{-\lambda}}{2} \Rightarrow \dfrac{\lambda e^{-\lambda}}{\lambda^2 e^{-\lambda}} = \dfrac{\frac{3}{10}}{\frac{1}{5}} \Rightarrow \dfrac{1}{\lambda} = \dfrac{3}{4} \Rightarrow \lambda = \dfrac{4}{3}$

15. Substituting in the formula Median $= \dfrac{1}{3}$ (2 Mean + Mode)

We get Median $= \dfrac{1}{3}$ ($2 \times 35.4 + 32.1$) $= \dfrac{1}{3} \times 102.9 = 34.3$

3. BIO TECHNOLOGY

S.NO	Ans	S.NO	Ans	S.NO	Ans
1	C	6	D	11	A
2	A	7	D	12	D
3	C	8	A	13	B
4	D	9	A	14	A
5	B	10	B	15	C

Solutions

1. $A = \begin{bmatrix} 5 & 2 \\ 0 & k \end{bmatrix} \Rightarrow A^2 = \begin{bmatrix} 5 & 2 \\ 0 & k \end{bmatrix}\begin{bmatrix} 5 & 2 \\ 0 & k \end{bmatrix} = \begin{bmatrix} 25 & 10+2k \\ 0 & k^2 \end{bmatrix}$

 $f(x) = x^2 - 7x + 10 \Rightarrow f(A) = A^2 - 7A + 10I = \begin{bmatrix} 25 & 10+2k \\ 0 & k^2 \end{bmatrix} - 7\begin{bmatrix} 5 & 2 \\ 0 & k \end{bmatrix} +$

 $10\begin{bmatrix} 1 & 0 \\ 0 & 1 \end{bmatrix} = \begin{bmatrix} 25 - 35 + 10 & 10+2k-14+0 \\ 0 & k^2 - 7k + 10 \end{bmatrix} = \begin{bmatrix} 0 & 0 \\ 0 & 0 \end{bmatrix}$

 $\Rightarrow 2k - 4 = 0$, $k^2 - 7k + 10 = 0$ solving we get $k = 2$

2. $\lim\limits_{x\to\infty} \dfrac{3|x|+x}{7|x|-5x} = \lim\limits_{x\to\infty} \dfrac{3x+x}{7x-5x} = \lim\limits_{x\to\infty} \dfrac{4x}{2x} = \lim\limits_{x\to\infty} \dfrac{4}{2} = 2$ [since $x\to\infty$,we have $x > 0$

 $\Rightarrow |x|=x]$

3. $y = \sqrt{\sin x + \sqrt{\sin x + \sqrt{\sin x + \ldots \text{to } \infty}}} = \sqrt{\sin x + y} \Rightarrow y^2 = \sin x + y \Rightarrow y^2 - y = \sin x$

 Differentiating we get $2y\dfrac{dy}{dx} - \dfrac{dy}{dx} = \cos x \Rightarrow (2y-1)\dfrac{dy}{dx} = \cos x \Rightarrow \dfrac{dy}{dx} = \dfrac{\cos x}{2y-1}$

4. We have $u = x^2 y + y^2 z + z^2 x$ Differentiating + partially with respect x, y, and z

 respectively we get $\dfrac{\partial u}{\partial x} = 2xy + z^2$, $\dfrac{\partial u}{\partial y} = 2yz + x^2$, $\dfrac{\partial u}{\partial z} = 2zx + y^2$

 Adding we get $\dfrac{\partial z}{\partial x} + \dfrac{\partial z}{\partial y} + \dfrac{\partial z}{\partial x} = 2xy + z^2 + 2yz + x^2 + 2zx + y^2$

 $\Rightarrow \dfrac{\partial z}{\partial x} + \dfrac{\partial z}{\partial y} + \dfrac{\partial z}{\partial x} = x^2 + y^2 + z^2 + 2xy + 2yz + 2zx = (x + y + z)^2$

5. The given equation is $\dfrac{dy}{dx} + \dfrac{y}{x} = x^3$ (Linear equation)

 We have $P = \dfrac{1}{Y}$, $Q = x^3$, Integrating factor $= e^{\int \frac{1}{x}dx} = e^{\log x} = x$

 The required solution is $x y = \int x \cdot x^3 \, dx + c$ or $xy = \dfrac{x^5}{5} + c$

6. Particular solution is $\dfrac{1}{(D^2 - 3D + 2)} e^{5x} = \dfrac{1}{(5^2 - 3 \cdot 5 + 2)} e^{5x} = \dfrac{e^{5x}}{12}$

7. The solution of the partial differential equation $p = e^q$ is where $p = \dfrac{\partial z}{\partial x}$, $q = \dfrac{\partial z}{\partial y}$ is

Let the solution be $z = a\,dx + b\,dy + c$ We have $p = e^q \Rightarrow a = e^b \Rightarrow b = \log a$

Hence, the complete solution is $z = a\,x + (\log a)\,y + c$

By the definition

8. We have $L[\sin 4t] = \dfrac{4}{s^2 + 4^2}$, by first shifting theorem we get $L[e^{-2t} \sin 4t] = $

$\dfrac{4}{(s+2)^2 + 4^2} = \dfrac{4}{s^2 + 4s + 20}$

We have $\dfrac{\partial}{\partial x}(x^y) = yx^{y-1}$ and $\dfrac{\partial}{\partial x}(y^x) = y^x \log y$

$\Rightarrow \dfrac{\partial}{\partial x}[\,f(x,y)\,] = \dfrac{\partial}{\partial x}(x^y + y^x) = y\,x^{y-1} + y^x \log y$

9. $Z = f(x^2 + y^2) \Rightarrow \dfrac{\partial z}{\partial x} = p = 2x\, f'(x^2 + y^2) \Rightarrow \dfrac{p}{2x} = f'(x^2 + y^2)$ .....(i)

and $\dfrac{\partial z}{\partial y} = q = 2y\, f'(x^2 + y^2) \Rightarrow \dfrac{q}{2y} = f'(x^2 + y^2)$ .....(ii)

from (i) and (ii) we get $\dfrac{p}{2x} = \dfrac{q}{2y} \Rightarrow p\,y - q x = 0$ i.e. $y\,p - xq = 0$

10. Here, we have mean $= np = 4$, Variance $npq = \dfrac{4}{3} \Rightarrow q = \dfrac{npq}{np} = \dfrac{\frac{4}{3}}{4} = \dfrac{1}{3}$

$\Rightarrow p = 1 - q = 1 - \dfrac{1}{3} = \dfrac{2}{3}$ substituting in $np = 4$ we get $n.\dfrac{2}{3} = 4 \Rightarrow n = 6$

11. The characteristic equation of the matrix is $\begin{vmatrix} 6 - \lambda & \sqrt{3} \\ \sqrt{3} & 4 - \lambda \end{vmatrix} = 0$

$\Rightarrow (6 - \lambda)(4 - \lambda) - 3 = 0 \Rightarrow \lambda^2 - 10\lambda + 21 = 0 \Rightarrow (\lambda - 3)(\lambda - 7) = 0$

$\Rightarrow \lambda = 3, 7$

12. Substituting in the relation Mean $-$ Mode $= 3$(Mean $-$ Median)

We get $50 - 58 = 3(50 - \text{Median}) \Rightarrow \text{Median} = 52.67$

13. We have Mean ($\bar{x}$) $= 30$, Standard deviation (σ) $= 5$

Coefficient of variation $= \dfrac{\sigma}{\bar{x}} \times 100 = \dfrac{6}{30} \times 100 = 20$

4   CHEMICAL ENGINEERING

S.NO	Ans	S.NO	Ans	S.NO	Ans
1	D	6	A	11	D
2	B	7	D	12	C
3	B	8	B	13	A
4	D	9	C	14	C
5	B	10	A	15	D

Solutions

1. $AB = \begin{bmatrix} 1 & 3 & -3 \\ 3 & 0 & 5 \end{bmatrix} \begin{bmatrix} 3 & 0 \\ -3 & 1 \\ 0 & 5 \end{bmatrix} = \begin{bmatrix} 1.3 + 3.(-3) + (-3).0 & 1.0 + 3.1 + (-3).5 \\ 3.3 + 0.(-3) + 5.0 & 3.0 + 0.1 + 5.5 \end{bmatrix} =$
$\begin{bmatrix} -6 & -9 \\ 9 & 25 \end{bmatrix}$

2. $\lim\limits_{x \to 0} \dfrac{e^{2x}-(1+x)^2}{x \log x}$ is in $\dfrac{0}{0}$ form $\Rightarrow \lim\limits_{x \to 0} \dfrac{2e^{2x}-2(1+x)}{\log x + \frac{x}{1+x}}$ $[\frac{0}{0}$ form$] \Rightarrow \lim\limits_{x \to 0} \dfrac{4e^{2x}-2}{\frac{1}{1+x}+\frac{x}{(1+x)^2}} = \dfrac{2}{2} = 1$

3. f - satisfies all the conditions of Rolle's mean value theorem $\Rightarrow$ there exists a real number c such that $f'(c) = 0$ [since $f'(x) = 3x^2 - 4$ we have $f'(c) = 0$

$\Rightarrow 3c^2 - 4 = 0 \Rightarrow c^2 = \dfrac{4}{3} \Rightarrow c = \pm \dfrac{2}{\sqrt{3}}$

4. Let $I = \int_0^{\frac{\pi}{2}} \dfrac{\sin x - \cos x}{1 + \sin x \cos x} dx$ .. (1) then we get $I = \int_0^{\frac{\pi}{2}} \dfrac{\sin\,(\frac{\pi}{2}-x) - \cos\,(\frac{\pi}{2}-X)}{1 + \sin\,(\frac{\pi}{2}-x)\cos\,(\frac{\pi}{2}-x)} dx =$

$\int_0^{\frac{\pi}{2}} \dfrac{\cos x - \sin x}{1 + \sin x \cos x} dx$ ..... (2)

Adding (1) and (2) we get $I = \int_0^{\frac{\pi}{2}} 0\,dx = 0$

5. $f(x, y, z) = 3x^2 i + 5xy^2 j + xyz^3 k \Rightarrow \text{div } f = \dfrac{\partial}{\partial x}(3x^2) + \dfrac{\partial}{\partial y}(5xy^2) + \dfrac{\partial}{\partial z}(xyz^3) = 6x + 10x y + 3xyz^2 \Rightarrow \text{div } f$ at $(1,2,3) = 80$

6. $\dfrac{dy}{dx} = \sqrt{4 - y^2} \Rightarrow \dfrac{dy}{\sqrt{4-y^2}} = dx \Rightarrow \sin^{-1} \dfrac{y}{2} = x + c \Rightarrow \dfrac{y}{2} = \sin(x + c) \Rightarrow y = 2\sin(x+c)$ is the solution

7. Consider $(D^2 + 4)\, y = \sin 3x$

The auxiliary equation is $m^2 + 4 = 0 \Rightarrow m = \pm 2$

The complementary function is $y_c = c_1 \cos 2x + c_2 \sin 2x$ particular integral is $y_p = \dfrac{1}{(D^2+ 4)} \sin 3x \Rightarrow y_p = \dfrac{1}{(-3^2+ 4)} \sin 3x = -\dfrac{1}{5} \sin 3x$ the general solution of the differential equation is $y = y_c + y_p = c_1 \cos 2x + c_2 \sin 2x - \dfrac{1}{5} \sin 3x$

8. The Laplace transform of $\sin t \cos t$ is $L(\sin t \cos t) = L\left(\dfrac{2 \sin t \cos t}{2}\right) = \dfrac{1}{2} L[\sin 2t]$
$= \dfrac{1}{2}\left[\dfrac{2}{s^2+4}\right] = \dfrac{1}{S^2+4}$

9. C is the unit circle $|z| = 1$, we have $z_0 = 0$, $f(z) = e^z$ from Cauchy's integral formula

we have $\dfrac{1}{2\pi i} \int \dfrac{e^z}{z-0} dz = f(0) = e^0 \Rightarrow \int \dfrac{e^z}{z} dz = 2\pi i$

10. A and B are independent $\Rightarrow P(A \cap B) = P(A)\, P(B)$ We have

$P(A \cup B) = P(A) + P(B) - P(A \cap B) =$

$\Rightarrow P(A \cup B) = P(A) + P(B) - P(A) \cdot P(B) \Rightarrow 0.6 = 0.35 + P(B) - 0.35\, P(B)$

$\Rightarrow 0.6 - 0.35 = P(B)\,[1 - 0.65] \Rightarrow P(B) = \dfrac{0.25}{0.65} = \dfrac{1/13}{13/20} = \dfrac{5}{13}$

11. We have mean of bin Omial distribution $= 9 \Rightarrow np = 9$ Standard deviation

$= \dfrac{3}{2} \Rightarrow \sqrt{npq} = \dfrac{3}{2} \Rightarrow \dfrac{9}{4} = npq\,]$

$\Rightarrow \dfrac{npq}{np} = \dfrac{\frac{9}{4}}{9} \Rightarrow q = \dfrac{1}{4} \Rightarrow p = \dfrac{3}{4}$

There fore $np = 9 \Rightarrow n \times \dfrac{3}{4} = 9 \Rightarrow n = 12$

12. By definition   it is two dimensional Laplace's equation]

13. We have $\nabla^2 f = \dfrac{\partial^2 f}{\partial x^2} + \dfrac{\partial^2 f}{\partial y^2} + \dfrac{\partial^2 f}{\partial z^2}$

$\Rightarrow \dfrac{\partial^2}{\partial x^2}(x^2 - y^2 + 4z) + \dfrac{\partial^2}{\partial y^2}(x^2 - y^2 + 4z) + \Rightarrow \dfrac{\partial^2}{\partial z^2}(x^2 - y^2 + 4z) \Rightarrow 2 - 2 + 0 = 0$

14. Characteristic equation of A is $= \begin{vmatrix} -1 - \lambda & 1/3 \\ -3 & -1 - \lambda \end{vmatrix} = (-1 - \lambda)(-1 - \lambda) + 1 = 0$

$\Rightarrow \lambda^2 + 2\lambda + 2 = 0 \Rightarrow \lambda = -1 \pm i$

15. In Newton – Raphs on method we use the following formula to get the next value of the root $x_{n+1} = x_n - \dfrac{f(x_n)}{f'(x_n)}$ We have $f(x) = x^2 - 13 \Rightarrow f'(x) = 2x$

Applying the above formula , we get next $x = 3.5 - \dfrac{(3.5)^2 - 13}{2 \times (3.5)} = 3.607$

5 CIVIL ENGINEERING

S.NO	Ans	S.NO	Ans	S.NO	Ans
1	A	6	B	11	A
2	D	7	A	12	C
3	D	8	B	13	A
4	C	9	C	14	C
5	B	10	B	15	D

Solutions

1. $\lim\limits_{x \to \infty} \dfrac{x = \sin x}{x + \sin x} = \lim\limits_{x \to \infty} \dfrac{1 - \frac{\sin x}{x}}{1 + \frac{\cos x}{x}} = \dfrac{1}{1} = 1$

2. The characteristic equation of A is $\begin{vmatrix} 2 - \lambda & 3 \\ x & y - \lambda \end{vmatrix} = 0 \Rightarrow (2 - \lambda)(y - \lambda) - 3x = 0$

But $\lambda = 4, 8$

Substituting $\lambda = 4$, in $(2-\lambda)(y-\lambda) - 3x = 0$ we get $-3x - 2y + 8 = 0$ .....(1)

Substituting $\lambda = 8$, in $(2-\lambda)(y-\lambda) - 3x = 0$ we get $-3x - 6y + 48 = 0$ .....(2)

Solving (1) and (2) we get $x = -4$, $y = 10$

3. Differentiating f with respect to x we get $\dfrac{\partial f}{\partial x} = \dfrac{2x}{\sqrt{2x^2 + y^2}}$

Now differentiating $\dfrac{\partial f}{\partial x}$ partially with respect to y we get

$$\dfrac{\partial^2 f}{\partial y\, \partial x} = 2x\,(2y)\left(\dfrac{-1}{2}\right)(2x^2 + y^2)^{-3/2} = -2xy(2x^2 + y^2)^{-3/2}$$

4. Grad f $= 2xyz^3 i + x^2 z^3 j + 3x^2 yz^2 k \Rightarrow$ grad f at $(2, 1, -1) = -4i - 4j + 12k$
Therefore the greatest value of directional derivative is |grad f
$| = \sqrt{16 + 16 + 144} = 4\sqrt{11}$

5. The given equation can be written as $\sin px \cos y - \cos px \sin y = p$
or $\sin(px - y) = p \Rightarrow px - y = \sin^{-1}p \Rightarrow y = px - \sin^{-1}p$ which is in Clairaut's form

Hence the solution of the given equation is $\Rightarrow y = cx - \sin^{-1}c$

6. Consider $(D^2 + 9)Y = \sin 3x$. The auxiliary equation is $m^2 + 9 = 0 \Rightarrow m = \pm 3i \Rightarrow y_c$
$= c_1 \cos 3x + c_2 \sin 3x$ $y_p = \dfrac{1}{D^2 + 9} \sin 3x = \dfrac{1}{-3^2 + 9} \sin 3x = \dfrac{1}{0} \sin 3x \Rightarrow y_p = x\left[\dfrac{1}{2D}\right] \sin 3x = x\dfrac{1}{2}$
$\left[-\dfrac{\cos 3x}{3}\right] = -\dfrac{x\cos 3x}{6}$

Hence the solution of the given equation is B. $Y = c_1 \cos 3x + c_2 \sin 3x - \dfrac{x \cos 3x}{6}$

7. $f(x)$ is differentiable everywhere in $[-2, 3]$ $f'(x) = 12x^3 - 12x^2 - 12x = 0 \Rightarrow 12x$
$(x+1)(x-3) = 0 \Rightarrow$ the critical points are $-1, 0, 2$

8. $f(x, y) = \lambda x^2 - y^2 + xy$ Differentiating partially   we get $\dfrac{\partial u}{\partial x} = 2\lambda x + y$
$\Rightarrow \dfrac{\partial^2 u}{\partial x^2} = 2\lambda$ similarly we get $\dfrac{\partial^2 u}{\partial y^2} = -2$ Adding we get $\lambda = 1$

9. The invariant points are given by $z = \dfrac{1+z}{1-z} \Rightarrow z - z^2 = 1 + z \Rightarrow z^2 = 1 \Rightarrow z = \pm 1$

10. $L(\cos^2 t) = L\left(\dfrac{1+\cos 3T}{2}\right) = L\left(\dfrac{1}{2}\right) + L\left(\dfrac{\cos 2t}{2}\right) = L\left(\dfrac{1}{2}\right) + L\left(\dfrac{\cos 2t}{2}\right) = \dfrac{1}{2s} + \dfrac{s}{2(s^2 + 4)}$

11. We have mean $= np = 2.4$, variance $npq = 1.44$
$\dfrac{npq}{np} = q = \dfrac{1.44}{2.4} = 0.6$, $p = 1 - q = 0.4$, since $np = 2.4$ we get $n(0.4) = 2.4 \Rightarrow n = 6$

12.   Total number of balls in the bag is $2 + 3 + 2 = 7$ Let S be the sample
space, then

n(S) = = Number of ways of drawing 2 balls out of 7 = $7_{c_2} = 21$

Let E = Event of drawing 2 balls, none of which is blue.

$\therefore$ n(E) = Number of ways of drawing 2 balls out of (2 + 3) balls = $5_{c_2} = 10$

Hence P (E) $=\dfrac{n(E)}{n(S)} = \dfrac{10}{21}$

13. $P(X = x) = \dfrac{e^{-\lambda}\lambda^x}{x!} \Rightarrow . \; P(X = 1) = \dfrac{e^{-\lambda}\lambda^1}{1!} = \dfrac{3}{10} \Rightarrow e^{-\lambda}\lambda = \dfrac{3}{10}$ ....(1)

And $P(X = 2) = \dfrac{e^{-\lambda}\lambda^2}{2!} = \dfrac{1}{5} \Rightarrow \dfrac{\lambda \times \frac{3}{10}}{2} = \dfrac{1}{5} \Rightarrow \lambda = \dfrac{4}{3}$

14.   The characteristic equation of A is $|A - \lambda i| = 0 \Rightarrow \lambda^3 - 2\lambda^2 - 5\lambda + 6 = 0$

   $\Rightarrow (\lambda - 1)(\lambda + 2)(\lambda - 3) = 0 \Rightarrow \lambda = -2, 1, 3$

   The other two eigen values are – 2 and 3

15.   To obtain the singularities of the function $z + 4 / \; z^2 + 2z + 5$ we take $z^2 + 2z + 5 = 0$

   $z^2 + 2z + 5 = 0 \Rightarrow (z + 1)^2 - (2i)^2 = 0 \Rightarrow (z + 1 + 2i)(z + 1 - 2i) = 0$

   $\Rightarrow$ The singularities are $-1 - 2i$ and $-1 + 2i$

6 COMPUTER SCIENCE ENGINEERING

S.NO	Ans	S.NO	Ans	S.NO	Ans
1	C	6	C	11	B
2	D	7	B	12	C
3	B	8	A	13	D
4	D	9	C	14	C
5	C	10	D	15	B

Solutions

1.   A has three elements. The number of sub sets of A is $2^3 = 8$

2.   Since the cardinality of a set is the number of elements in it the cardinality of the power set of A is $2^3 = 8$

3.   An application of the quantifiers

   We have n = 3, $n^m = 2187 \Rightarrow 3^m = 2187 \Rightarrow m = \dfrac{\log_e 2187}{\log_e 3} = 7 \Rightarrow |A| = 7$

4. $f^{-1}(-2) = \{x \in R : f(x) = -4\} = = \{x \in R : f(x) = -4\}$

   $= \{x \in R: x^2 = -4\} = = \{x \in R: x = \pm 2i\} = \phi$

   Since x = $\pm 2i$ are imaginary, does not belong to R

5.   The number of relations from A to B IS $2^{mn} = 2^{3 \times 3} = 2^9 = 512$ By Definition

6.   we have G= $\{1, -1, i. -i\}$

7.   Since G = $\{1 = i^4, \; -1 = i^2, i, -i = i^3\}$, i is generator,

   And G = $\{1 = (-i)^4, \; -1 = (-i)^2, -1, i = (-i)^3\}, -i\}, \;\; -i$ is a generator

   The generators of G are i and $-i$

8. By applying the multinomial theorem the general term is

$\dfrac{4!}{1!1!2!}(2x)^{n_1}(-y)^{n_2}(-z)^{n_3} = \dfrac{1.2.3.4}{1.1.2} \cdot 2 \cdot (-1)((-1)^2 \, x \, yz^2$

The coefficient of $x\,yz^2 = 3.4.\,2\,(-1) = -24$

9. Here we have $a_0 = 1$, $a_1 = 1$, $a_2 = 0$, $a_3 = 1, \ldots,$ A generating function for the sequence is given by $F(x) = a_0 + a_1x + a_2x^2 + a_3x^3 + \ldots = 1 + x + 0.x^2 + a_3x^3 + \ldots = 1 + x + 0.x^2 + a_3x^3 + \ldots = (1 + x + 1.x^2 + a_3x^3 + \ldots) - x^2 = (1-x)^{-1} - x^2$

10. The characteristic equation of A is $\begin{vmatrix} i-\lambda & 1 \\ 2 & -i-\lambda \end{vmatrix} = 0 \Rightarrow \lambda^2 - 1 = 0 \Rightarrow \lambda = -1, 1$

11. $\displaystyle\lim_{x\to 0}\left[\dfrac{1}{x} - \dfrac{1}{e^x-1}\right]$ is in $\infty - \infty$ form, hence we apply L'Hospital rule

$\displaystyle\lim_{x\to 0}\left[\dfrac{1}{x} - \dfrac{1}{e^x-1}\right] = \lim_{x\to 0}\left[\dfrac{e^x-1-x}{x(e^x-1)}\right]$ is on $\dfrac{0}{0}$ form

$\Rightarrow \displaystyle\lim_{x\to 0}\left[\dfrac{e^x-1-x}{x(e^x-1)}\right] = \lim_{x\to 0}\left[\dfrac{e^x-1}{(e^x-1+xe^x)}\right] = \lim_{x\to 0}\dfrac{e^x}{e^x[x+2]} = \dfrac{1}{2}$

12. We have $f(x) = 2x^2 - 7x + 10 \Rightarrow f'(x) = 4x - 7$, $f(5) = 25$, $f(2) = 4$

Applying Lagrange's theorem we get $\dfrac{f(5)-f(2)}{5-2} = f'(c) \Rightarrow \dfrac{21}{3} = 4c - 7$

$\Rightarrow 7 = 4c - 7 \Rightarrow c = \dfrac{14}{4} = \dfrac{7}{2} \in [2,5]$

13. Probability that husband is selected $= \dfrac{1}{7}$

Probability that husband is not selected $= 1 - \dfrac{1}{7} = \dfrac{6}{7}$

Probability that wife is selected $= \dfrac{1}{5}$

Probability that wife is not selected $= 1 - \dfrac{1}{5} = \dfrac{4}{5}$

Probability that only husband is selected $= \dfrac{1}{7} \cdot \dfrac{4}{5} = \dfrac{4}{35}$

Probability that only wife is selected $= \dfrac{6.1}{7.5} = \dfrac{6}{35}$

Probability that only one o them is selected is $\dfrac{4}{35} + \dfrac{6}{35} = \dfrac{2}{7}$

14. We have mean $= np = 12$, standard deviation $= 2 \Rightarrow$ variance $= npq = 4$

$\dfrac{npq}{np} = \dfrac{4}{12} = \dfrac{1}{3} \Rightarrow p = 1 - \dfrac{1}{3} = \dfrac{2}{3}$ Since $np = 12 \Rightarrow n \times \dfrac{2}{3} = 12 \Rightarrow n = \dfrac{36}{2} = 18$

15. Applying reversal law we get $(a^{-1}b)^{-1} = b^{-1}(a^{-1})^{-1} = b^{-1}a$

7 ELECTRICAL ENGINEERING

S.NO	Ans	S.NO	Ans	S.NO	Ans
1	D	6	B	11	B
2	A	7	A	12	C
3	A	8	D	13	A
4	B	9	C	14	C
5	A	10	D	15	D

Solutions

1. We have $kxyz = \det\begin{bmatrix} x & y & z \\ -x & y & z \\ x & -y & z \end{bmatrix} = xyz \det\begin{bmatrix} 1 & 1 & 1 \\ -1 & 1 & 1 \\ 1 & -1 & 1 \end{bmatrix} = 4xyz$

   $\Rightarrow kxyz = 4xyz \Rightarrow k = 4$

2. The Characteristic equation $A = \det\begin{bmatrix} -1-\lambda & \frac{1}{3} \\ -3 & -1-\lambda \end{bmatrix} = 0 \Rightarrow (-1-\lambda)(-1-\lambda)+1 = 0$

   $\Rightarrow \lambda^2 + 2\lambda + 2 = 0$

   By the quadratic formula the roots are $\lambda = -1 + i$ and $\lambda = -1 - i$.

   $\Rightarrow$ the eigen values are $\lambda = -1 + i$ and $\lambda = -1 - i$.

3. $I = \int_{-\frac{\pi}{2}}^{\frac{\pi}{2}} \sin^2 x \, dx = 2 \int_0^{\frac{\pi}{2}} \sin^2 x \, dx$ \qquad .....(1)

   [Since $\sin^2 x$ is an even function]

   $\Rightarrow I = 2 \int_0^{\frac{\pi}{2}} \sin^2\left(\frac{\pi}{2} - x\right) dx = 2 \int_0^{\frac{\pi}{2}} \cos^2 x \, dx$ \qquad ....(2)

   Adding (1) and (2) we get $2I = 2 \int_0^{\frac{\pi}{2}} 1 \, dx \Rightarrow I = \frac{\pi}{2}$

4. If $z = \cos(x^2 y) + \sin y$ then $\frac{\partial z}{\partial y} = -x^2 \sin(x^2 y) + \cos y$

5. The given equation can be written as $\frac{dy}{y} = \sec x \, dx$

   Integrating we get $\log y = \log(\sec x + \tan x) + \log c$

   $\Rightarrow y = c(\sec x + \tan x)$ which is the required solution

6. $P.I = \dfrac{1}{D^2 + 6D + 8} e^{-2x}$ \quad substituting $D = -2$ we get

   $P.I = \dfrac{1}{0} \cdot e^{-2x}$ \quad (case of failure), therefore

7. $P.I = \dfrac{x}{2D+6} e^{-2x} \Rightarrow P.I = \dfrac{x}{2(-2)+6} e^{-2x} = \dfrac{x}{2} e^{-2x}$

   $\phi(x, y, z) = xy + yz + zx \Rightarrow$

   $\text{Grad } \phi = i\dfrac{\partial \phi}{\partial x} + j\dfrac{\partial \phi}{\partial y} + k\dfrac{\partial \phi}{\partial z} = (y+z)i + (z+x)j + (x+y)k$

   $\Rightarrow \text{Grad } \phi$ at $(-1, 1, 1)$ is 2

   $\Rightarrow$ The normal derivative of ϕ at $(-1, 1, 1) = \sqrt{2^2} = 2$

8. The fixed points are of ω are $z = \dfrac{z-3}{z+1} \Rightarrow z^2 + z = z - 3]$

   $\Rightarrow z^2 = -3 \Rightarrow$ the fixed points are $\pm i\sqrt{3}$

9. We have $z = (x-a)^2 + (y-b)^2 + 1$

   Differentiating partially with respect x we get $\dfrac{\partial z}{\partial x} = 2(x-a)$

$\Rightarrow p = 2 (x-a) \Rightarrow (x-a) = \frac{p}{2}$ similarly we get $(y -b) = \frac{q}{2}$

Substituting in the given equation we get $4z = p^2 + q^2 + 4$

Which is the required partial differential equation

10. We have $Z\{5^n z^{-n}\} = \sum_{n=0}^{\infty} \frac{5^n}{z^n} = 1 + \frac{5}{z} + \frac{5^2}{z^2} + \dots = \frac{1}{1-\frac{5}{z}} = \frac{z}{z-5}$

11. $L[\cos t] = \frac{s}{s^2+4} \Rightarrow L[t \cos t] = (-1)^1 \frac{d}{ds}[\frac{s}{s^2+4}] = -\frac{s^2+4-s.2s}{(s^2+4)^2} = \frac{S^2-4}{(S^2+4)^2}$

12. let $f(z) = \cot z = \frac{\cos z}{\sin z}$ $Z = 0$ is a simple pole Let $h(z) = \cos z$, $k(z) = \sin z$

Therefore we get $Re[f(z):0] = \frac{h(0)}{k'(0)} = \frac{\cos(o)}{\cos(o)} = 1$

13. Substituting the values of mean and median in the relation

Mean – Mode = 3(Mean – Median)

We get 78 – Mode = 3 (78 −72) $\Rightarrow$ Mode = 60

14. We have $p = \frac{1}{10}$ and $n = 12$ $\Rightarrow \lambda = np = 12 \times \frac{1}{10} = 1.2$

Probability that none is defective = $P(x = 0) = \frac{e^{-\lambda}\lambda^0}{0!} = e^{-\lambda} = e^{-1.2}$

15. We have $n = 4$. Mean of the distribution = $\bar{x} = \frac{2+4+6+8}{4} = \frac{20}{4} = 5$

Variance = $\frac{\sum(x-\bar{x})^2}{n} = \frac{9+1+1+9}{4} = \frac{20}{4} = 5 \Rightarrow S.D = \sqrt{5}$

8 ELECTRONICS AND COMMUNICATIONS ENGINEERING

S.NO	Ans	S.NO	Ans	S.NO	Ans
1	C	6	D	11	B
2	D	7	A	12	D
3	A	8	B	13	C
4	B	9	A	14	A
5	C	10	C	-------	-------

Solutions

1. A is a triangular matrix , hence the diagonal elements 1,2 , 7 are the eigen values

2. Let $f(x) = \cos^5 x$, then $f(2\pi -x) = \cos^5(2\pi-x) = -\cos^5 x$

$\Rightarrow I = -2\int_0^{\pi} \cos^5 x\, dx = -2I \Rightarrow 3I = 0 \Rightarrow I = 0$

3. Separating the variables we get $\dfrac{dy}{y} = \tan x \Rightarrow \int \dfrac{dy}{y} = \int \tan x\, dx + c$

   $\Rightarrow \log y = \log \sec x + c$, Since $y=1$ at $x = 0$ (given)

   We get $\log 1 = \log \sec o \Rightarrow 0 = 0 + c \Rightarrow c = 0$

   The solution is $\log y = \log \sec x \Rightarrow y = \sec x$

4. Given equation is $p = \log(p\,x - y) \Rightarrow px - y = e^p \Rightarrow y = px - e^p$
   Which is in Clairaut's equation
   The solution is $y = cx - e^c$

5. The particular integral of $(D^2 - D - 2)\, y = \sin 2x$ is $\dfrac{1}{D^2 - D - 2} \sin 2x = \dfrac{1}{-2^2 - D - 2} \sin 2x$

   $= -\dfrac{1}{D+6} \sin 2x = -\dfrac{1}{D+6} \sin 2x = -(D-6)\dfrac{1}{(D-6)(D+6)} \sin 2x$

   $\Rightarrow -(D-6)\dfrac{1}{D^2 - 36} \sin 2x = -(D \sin 2x - 6 \sin 2x)\dfrac{1}{2^2 - 36} = Y = \dfrac{1}{40}(2 \cos 2x$

   $-6 \sin 2x) = \dfrac{1}{20}(\cos 2x - 3 \sin 2x)$

6. $f(x, y, z) = 3x^2\, i + 5\, x\, y^2\, j + xyz^3\, k \Rightarrow \operatorname{div} f = \dfrac{\partial}{\partial x}(3x^2) + \dfrac{\partial}{\partial y}(5\,x\,y^2) + \dfrac{\partial}{\partial z}(xyz^3)$

   $\Rightarrow \operatorname{div} f = 6x + 10\,xy + 3xy\,z^2 \Rightarrow \operatorname{div}$ at $(1,2,3) = 6 + 20 + 54 = 80$

7. We have $f_s(\alpha) = F_s\,[f(x)\,] = \sqrt{\dfrac{2}{\pi}} \int_0^\infty f(x) \sin \alpha x\, dx = \sqrt{\dfrac{2}{\pi}} \int_0^\infty \dfrac{1}{x} \sin \alpha x\, dx$ If we put αx

   we get $dx = \dfrac{dt}{\alpha}$

   $\Rightarrow F_s\,[f(x)] = \sqrt{\dfrac{2}{\pi}} \int_0^\infty \dfrac{1}{t} \sin dt = \sqrt{\dfrac{2}{\pi}} \cdot \dfrac{\pi}{2} = \sqrt{\dfrac{\pi}{2}},$

8. Let $z = a\,x + by + c$. be the solution We get $p = \dfrac{\partial z}{\partial x}$, $q = \dfrac{\partial z}{\partial y}$

   From the given equation $p^2 + q^2 = 1$, have $a^2 + b^2 = 1 \Rightarrow b = \sqrt{1 - a^2}$

   The complete solution is $z = a\,x + (\sqrt{1 - a^2})\,y + c$

9. $u = x^2 y + y^2 z + z^2 x \Rightarrow \dfrac{\partial u}{\partial x} = 2xy + z^2$, $\dfrac{\partial u}{\partial y} = 2yz + x^2$, $\dfrac{\partial u}{\partial z} = 2zx + y^2$

   $\Rightarrow \dfrac{\partial u}{\partial x} + \dfrac{\partial u}{\partial y} + \dfrac{\partial u}{\partial z} = 2xy + z^2 + 2yz + x^2 + 2zx + y^2$

   $= x^2 + y^2 + z^2 + 2xy + 2yz + 2\,zx = (x + y + z)^2$

10. The invariant points of ω are given by $z = \dfrac{1+z}{1-z}$

    $\Rightarrow z - z^2 = 1 + z \Rightarrow z^2 + 1 = 0 \Rightarrow z = \pm i$

11. We have $f(z) = \dfrac{ze^z}{(z-1)^3}$, $z = 1$ is a pole of order 3.

Let $g(z) = ze^z \Rightarrow g'(z) = e^z(z+1)$ and $g''(z) = e^z(z+2)$

The residue of $f(z)$ at $z = 1$ is $\dfrac{g''(1)}{2!} = \dfrac{3e}{2}$

12. Let A be the event in which red ball is drawn in the first draw

Number balls in the bag = 3 red + 4 white = 7

Number of ways of drawing a ball from 7 balls $= 7_{c_1} = 7$

Number ways of in which the first drawn ball is red $= 3_{c_1} = 3$

We get $P(A) = \dfrac{3}{7}$

Since the first drawn is not replaced, the number of balls in the bag is 6,

Let B denote the event in which second drawn ball is red

The number red balls in the bag after first draw is 2

A and B are dependent events, therefore the probability of

drawing the second red ball $= P(B/A) = \dfrac{2}{6}$

Hence we get $P(A \cap B) = P(A)\,P(B/A) = \dfrac{3}{7}\cdot\dfrac{2}{6} = \dfrac{1}{7}$

13. Applying the formula Mode = 3 Median -2 Mean

We get Mode $= 3 \times 22 - 2\times 20 = 26$

14. $\int_1^b \dfrac{dy}{y} = \log b - \log 1 = \log b$ and $\int_1^a \dfrac{dx}{x} = \log a$

We get $\int_1^a \int_1^b \dfrac{dxdy}{xy} = \log a \log b$

9 FOOD TECHNOLOGY

S.NO	Ans	S.NO	Ans	S.NO	Ans
1	A	6	C	11	D
2	C	7	B	12	B
3	D	8	D	13	C
4	B	9	A	14	B
5	B	10	B		

Solutions

1. We have $A = \dfrac{1}{3}\begin{bmatrix} 1 & 2 & 2 \\ 2 & 1 & a \\ -2 & b & -1 \end{bmatrix} \Rightarrow A^T = \dfrac{1}{3}\begin{bmatrix} 1 & 2 & -2 \\ 2 & 1 & b \\ 2 & a & -1 \end{bmatrix}$

It is given that $A\,A^T = A\,A^T = I_3$

$$\Rightarrow \frac{1}{3}\begin{bmatrix} 1 & 2 & 2 \\ 2 & 1 & a \\ -2 & b & -1 \end{bmatrix} \cdot \frac{1}{3}\begin{bmatrix} 1 & 2 & -2 \\ 2 & 1 & b \\ 2 & a & -1 \end{bmatrix} = \begin{bmatrix} 1 & 0 & 0 \\ 0 & 1 & 0 \\ 0 & 0 & 1 \end{bmatrix}$$

$$\Rightarrow \frac{1}{9}\begin{bmatrix} 9 & 4+2a & 2b-4 \\ 4+2a & 5+a^2 & b-a-4 \\ 2b-4 & 4+b-a & 4+b^2+1 \end{bmatrix} = \begin{bmatrix} 1 & 0 & 0 \\ 0 & 1 & 0 \\ 0 & 0 & 1 \end{bmatrix}$$

$\Rightarrow$ 4+2 a = 0 , 2b− 4 = 0 $\Rightarrow$ a =−2 , b =2

2. The given matrix is $\begin{bmatrix} 4 & 1-3i \\ 1+3i & 7 \end{bmatrix}$

The characteristic equation of A IS $\begin{vmatrix} 4-\lambda & 1-3i \\ 1+3i & 7-\lambda \end{vmatrix} = 0$

$\Rightarrow \lambda^2 - 11\lambda + 18 = 0 \Rightarrow \lambda = 2, 9$

3. $u = x\,y^2 + x^2 y$ x = at^2 and y = 2 at differentiating we get $\frac{dx}{dt} = 2\,at$, $\frac{dy}{dt} = 2\,a$

$\Rightarrow$ also $\frac{\partial u}{\partial x} = y^2 + 2xy$, $\frac{\partial u}{\partial y} = 2xy + x^2$

Hence $\frac{du}{dt} = \frac{\partial u}{\partial x}\frac{dx}{dt} + \frac{\partial u}{\partial y}\frac{dy}{dt} = (y^2 + 2xy)\,(2\,at) + (2xy + x^2)\,(2\,a)$

Substituting the values of x and y we get $\frac{du}{dt} = 2a^3 t^3(8+5t)$

4. $\displaystyle\lim_{x\to 0}\frac{\log(1+x^3)}{\sin^3 x}$ is in $\frac{0}{0}$ form $\Rightarrow \displaystyle\lim_{x\to 0}\frac{\log(1+x^3)}{\sin^3 x} = \lim_{x\to 0}\frac{\log(1+x^3)}{x^3} \cdot \lim_{x\to 0}\frac{x^3}{\sin^3 x}$

$= \displaystyle\lim_{x\to 0}\frac{3x^2}{\log(1+x^3)(3x^2)} = 1$

5. Put $y - x = u \Rightarrow \frac{st}{dx} - 1 = \frac{du}{dx} \Rightarrow \frac{dy}{dx} = 1 + \frac{du}{dx}$ Substituting in the given equation

we get $\frac{du}{dx} = x\tan u$ or $\cot u\;du = x\,dx$ Integrating we get $\log|\sin u| = \frac{x^2}{2} + c$ or

$\log|\sin (y - x)| = \frac{x^2}{2} + c$

6. $(m^2 - 3m + 2)\,y = 0$, given y = 0 when x = 0 and $\frac{dy}{dx} = 0$ when x = 0

The Auxiliary equation = $(m^2 - 3m + 2)\,y = 0 \Rightarrow (m-1)(m-2) = 0$

The solution is y = $c_1 e^x + c_2 e^{2x}$ .....(1)

$\Rightarrow \frac{dy}{dx} = c_1 e^x + 2c_2 e^{2x}$ .....(2)

It is given that y = 0, when x = 0 and $\frac{dy}{dx} = 0$ when = 0

Substituting the given values in (1) and (2) we get the equations $c_1 + c_2 = 0$

Abd $c_1 + 2\,c_2 = 0$

Solving these equation we get c_1 = o and c_2 = 0

The general solution of the given equation is y = 0

7.   We have $L(\sin 4t) = \dfrac{4}{s^2+4^2}$

By shifting theorem we get $L[e^{-2t} \sin 4t]$

$= \dfrac{4}{(s+2)^2+4^2} = \dfrac{4}{S^2+4S+20}$

8.   We have $pe^y = q\, e^x$

It can be written as follows:

$pe^{-x} = q\, e^{-y} = a$ (say) $\Rightarrow p = ae^x$, $q = ae^y$ putting these values in dz = p dx + q dy we get dz = ae^x dx + ae^y dy integrating we get the solution as z = $a(e^x + e^y) +$ b

where b is the constant of integration

9.   By the definition the equation is one dimensional equation

10. Since $\dfrac{dy}{dx}$ is denoted by p the given equation can be written as Y= p x + e^p

which is in Clairaut's form, the solution of which can be obtained by replacing p by c.

Hence the solution of the given equation is y = cx + e^s

11. Harmonic mean $= \dfrac{(G.M)^2}{A.M} \Rightarrow H.M = \dfrac{(15)^2}{25} = 9$

12. We P (A) $= \dfrac{1}{2}$, P (B) $= \dfrac{1}{4}$, P(C) $= \dfrac{1}{5}$

$\Rightarrow P(\bar{A}) = \dfrac{1}{2}$, $P(\bar{B}) = \dfrac{3}{4}$, $P(\bar{C}) = \dfrac{4}{5}$

The events are independent, therefore the probability that the problem is solved

$= 1 - P(\bar{A})\, P(\bar{B})\, P(\bar{C}) = 1 - \dfrac{1}{2} \times \dfrac{3.}{4.} \times \dfrac{4}{5} = \dfrac{28}{40} = \dfrac{7}{10}$

13. It is given that 9 P(X = 4) = P(X=2)

$\Rightarrow 9 \times 6_{C_4} \times p^4 q^2 = 6_{C_4} \times p^2 q^4$

$\Rightarrow 9p^2 = q^2 \Rightarrow 9\,p^2 = (1-p)^2 \Rightarrow 8p^2 +2p - 1 = 0$

$\Rightarrow p = \dfrac{-2\pm6}{16} \Rightarrow p = \dfrac{4}{16} = \dfrac{1}{4}, p = -\dfrac{1}{2}$

$\Rightarrow p = \dfrac{1}{4}$ (since p is non-negative)

14. The coefficient of correlation lies between – 1 and 1

10 INSTRUMENTATION ENGINEERING

S.NO	Ans	S.NO	Ans	S.NO	Ans
1	B	6	C	11	B
2	A	7	D	12	D
3	B	8	A	13	B
4	D	9	B	14	A
5	A	10	C	15	D

Solutions

1. We have $A = \begin{bmatrix} 1 & x & 2x \\ 1 & 3x & 5x \\ 1 & 3 & 4 \end{bmatrix} \Rightarrow$ 1.(12 x− 15x) −x (4 – 5x) + 2 x (3−3x) = − 3x −

    4x + 5 x^2 +6 x −6x^2 = −x^2 −x = 0 ⇒ −x (x+1) = 0 Since x ≠o , we get x = −1

2. $I = \int_0^1 x\,(1-x)^n dx = \int_0^1 (1-x)(1-(1-x))^n dx$

    $\Rightarrow \int_0^1 (1-x)(x^n)\ dx = \int_0^1 x^n\ dx - \int_0^1 x^{n+1} dx = \dfrac{1}{n+1} - \dfrac{1}{n+2} = \dfrac{1}{(n+1)(n+2)}$

3. We have $f(x) = x^3 - 6x^2 + 9x + 15 \Rightarrow f^1(x) = 3x^2 - 12\ x + 9$

    $f^1(x) = 0 \Rightarrow 3\ (x - 4x + 3) = 0 \Rightarrow (x-1)\ (x-3) = 0$

    ⇒ x =1 and x = 3 are the stationary points

4. $\displaystyle\int_0^3\int_0^1 \left(x^2+3y^2\right)dydx = \int_0^3\int_0^1\left(x^2+3y^2\right)dydx = \int_0^3\left[\int_0^1\left(x^2+3y^2\right)dy\right]dx = \int_0^3\left(x^2 y+y^3\right)dx$

    $\displaystyle = \int_0^3\left(x^2+1\right)dx = \left[\dfrac{x^3}{3}+x\right]_0^3 = (9+3) = 12$

5. Given equation is x dy− y dx $= a\,(x^2+y^2)$ is The given equation can be written as

    $\dfrac{x\,dy-y\,dx}{x^2+y^2}$ = a dy Integrating we get $\tan^{-1}\dfrac{Y}{X} = a\,y + c$

6. we have y = p(x − b) + $\dfrac{a}{p} \Rightarrow$ y = p x + $\dfrac{a}{p}$ − bp

    Which is in Clairaut's form. The solution of the given equation can be obtained by replacing p by c Hence the solution is y = cx + $\dfrac{a}{c}$ − bc

7. Grad $\phi\,(x, y\ z) = \nabla\phi = xyz \Rightarrow \nabla\phi = i\dfrac{\partial\phi}{\partial x} + j\dfrac{\partial\phi}{\partial y} + k\dfrac{\partial\phi}{\partial z} = i\,y\,z + j\,z\,x + k\,x\,y \Rightarrow$

    grad ϕ at (1,1, 1) is $i + j + k$

The unit normal vector in the direction of i + j +k is $\eta = \dfrac{i+j+k}{\sqrt{3}}$

The directional derivative is grad $\phi . \eta = i +j + k. \dfrac{i+j+k}{\sqrt{3}} = \dfrac{3}{\sqrt{3}} = \sqrt{3}$

8. Let the complete solution of the given equation be z = a x + by + c

Differentiating z partially we get p = $\dfrac{\partial z}{\partial x} = a$ and q = $\dfrac{\partial z}{\partial x} = b$

Substituting in $p^2 - q^2 = 1$ we get $a^2 - b^2 = 1$ or $b^2 = a^2 - 1$ or b = $\sqrt{a^2 - 1}$

The solution of the given equation is z = $a x + \sqrt{a^2 - 1}$ y + c

9. The Residue of f(z) = $\dfrac{z}{z^2+1}$ at the pole z = i is $\displaystyle\lim_{Z\to I} (z - i)\dfrac{z}{(z-i)(z+i)}$

$= \displaystyle\lim_{Z\to I} \dfrac{z}{(z+i)} = \dfrac{1}{2}$

10. We have f(z) = $\dfrac{e^{2z}}{(z-i)^3(z+i)^3}$ Since z = -1 is a pole of order 3 we have Residue f(z) at

z = 1 is $\dfrac{1}{(3-1)!} \displaystyle\lim_{z\to(-1)} \dfrac{d^2[[z+1)^3 f(z)]}{dz^2} = \dfrac{1}{2} \displaystyle\lim_{z\to-1} \dfrac{d^2 (e^{2z})}{dz^2} = \dfrac{1}{2}[\ \displaystyle\lim_{z\to-1} [4\ e^{2z}] = \dfrac{2}{e^2}$

By Cauchy's residue theorem $\displaystyle\int \dfrac{e^{2z}}{(z+1)^3} dz = 2\ \pi i\ [\dfrac{2}{e^2}] = \dfrac{4\pi i}{e^2}$

11. Mean of the distribution = np = 4

Variance = npq = $\dfrac{4}{3} \Rightarrow q = \dfrac{npq}{np} = \dfrac{4/3}{4} = \dfrac{1}{3} \Rightarrow p = \dfrac{2}{3}$

Mean = np = 4. $\Rightarrow$ n. $\dfrac{2}{3} = 4 \Rightarrow$ n = 6

12. P (A) = 0.3, P(B) = 0.2 and P (C) = 0.1 $\Rightarrow P(\bar{A}) = 0.7.$ P $(\bar{B}) = 0.\ 8$ and $P(\bar{C}) = 0.9$

A, B, C are independent events $\Rightarrow$

P $(\bar{A} \cap \bar{B} \cap \bar{C}) = P(\bar{A})\ P(\bar{B})\ P(\bar{C}) = 0.7 \times 0.8 \times 0.9 = 0;\ 504$

Therefore the probability at least one of the events happen is

$$1 - P\ (\bar{A} \cap \bar{B} \cap \bar{C}) = 1 - 0.\ 504 = 0.\ 496$$

13. The mean of the distribution is = 1 × 0.4 + 2 × 0.2 + 3 × 0.4 = 2

14. X is a Poisson variate P(X =1) = $\dfrac{3}{10}$ and P(X = 2) = $\dfrac{1}{5}$, then we have $\dfrac{e^{-\lambda}\lambda^1}{1!} = \dfrac{3}{10}$ and

$\dfrac{e^{-\lambda}\lambda^2}{2!} = \dfrac{1}{5} \Rightarrow e^{-\lambda}\lambda^2 = \dfrac{2}{5} \Rightarrow \dfrac{e^{-\lambda}\lambda^2}{e^{-\lambda}\lambda^1} = \dfrac{2/5}{3/10} \Rightarrow \lambda = \dfrac{4}{3}$

15. Let r denote the correlation coefficient σ_x denote the standard deviation of the x – series, and σ_y denote the standard deviation of y – series then we have $b_{xy} = r \frac{\sigma_x}{\sigma_y}$ and $b_{yx} = r \frac{\sigma_x}{\sigma_y}$ multiplying we get $b_{xy} \times b_{yx} = r \frac{\sigma_x}{\sigma_y} \times r \frac{\sigma_x}{\sigma_y} = r^2 \Rightarrow$

$r = \pm \sqrt{b_{xy} \times b_{yx}}$

CODE: ME MECHANICAL ENGINEERING

S.NO	Ans	S.NO	Ans	S.NO	Ans
1	A	6	C	11	A
2	D	7	B	12	B
3	C	8	C	13	C
4	B	9	A	14	D
5	B	10	C		

Solutions

1. A is singular $\Rightarrow$ det $A = 0$

$\Rightarrow \cos^2\theta - \sin^2\theta = 0 \Rightarrow \cos 2\theta = 0 \Rightarrow 2\theta = \frac{\pi}{2} \Rightarrow \theta = \frac{\pi}{4}$

2. Consider the augmented matrix $\begin{bmatrix} 3 & -1 & 4 & 3 \\ 1 & 2 & -3 & -2 \\ 6 & 5 & \lambda & -3 \end{bmatrix} \rightarrow 3R_2 - R_1,$

$R_3 - 2R_1 \sim \begin{bmatrix} 3 & -1 & 4 & 3 \\ 0 & 7 & -13 & -9 \\ 0 & 7 & \lambda - 8 & -9 \end{bmatrix} \rightarrow R_3 - R_2 \sim \begin{bmatrix} 3 & -1 & 4 & 3 \\ 0 & 7 & -3 & -9 \\ 0 & 0 & \lambda + 5 & 0 \end{bmatrix},$

therefore

the system of equations has atleastt one solution if $\lambda + 5 = 0$ i.e. if $\lambda = -5$

3. $\lim\limits_{x \to -1.5} \cos^{-1}(\frac{x+3}{3}) = \cos^{-1}[\lim\limits_{x \to -1.5}(\frac{x+3}{3})] = \cos^{-1}\frac{1}{2} = \frac{\pi}{3}$

4. $y = 5^{x^3+3x} \Rightarrow \frac{dy}{dx} = 5^{x^3+3x} \log 5 \frac{d}{dx}[x^3 + 3x]$

$\Rightarrow 5^{x^3+3x} (\log 5) [3x^2 + 3]$

5. The given equation is $z = (1 - 2xy + y^2)^{-1/2}$ .....(1)

Differentiating (1) partially with respect to x and y we get

$\frac{\partial z}{\partial x} = -\frac{1}{2}(-2y)(1 - 2xy + y^2)^{-3/2} = yz^3$

$\frac{\partial z}{\partial y} = -\frac{1}{2}(-2x + 2y)(1 - 2xy + y^2)^{-3/2} = (x - y)z^3$ Therefore we get x

$\frac{\partial z}{\partial x} - y\frac{\partial z}{\partial y} = xyz^3 - y(x-y)z^3 = xyz^3 - xyz^3 + y^2z^3 = y^2z^3,$

6. $\frac{dy}{dx} + (\sec x)\, y = \tan x\ ,\ (0 \le x < \frac{\pi}{2})$ The equation is in linear form. we have $P = \sec x$, $Q = \tan x$ $\int \sec x\, dx = \log(\sec x + \tan x)$ The integrating factor is $e^{\log(\sec x + \tan x)} = \sec x + \tan x$ The solution of the given equation is y (sec x + tan x) = $\int (\sec x + \tan x)\tan x\, dx \Rightarrow$ y (sec x + tan x) = $\int \sec x \tan x\, dx + \int \sec^2 x\, dx - \int 1.dx$ Hence the solution is y (sec x + tan x) = sec x + tan x $-$ x + c

7.   The auxiliary equation is $m^2 + 1 = 0 \Rightarrow m = \pm i$ The solution i.e . C .F Is y = $c_1\cos x + c_2\sin x$ .....(1)

   Differentiating with respect x we get $y' = -c_1\sin x + c_2\cos x$

   Since y(0) =2 and $y'(o) = -1$, from (1) we get $c_1 = 2$, $c_2 = -1$

   The solution is Y = 2 cos x $-$ sin x

8.   Divergence of F = (x + z) i + (3x + λ y)j + (x $-$5z) k is $\frac{\partial}{\partial x}(x + z) + \frac{\partial}{\partial y}$

   x + λ y) + $\frac{\partial}{\partial z}(x - 5z) = 1 + \lambda - 5 = 0 \Rightarrow \lambda = 4$

9. ϕ x ,y, z) = $x^3 - xyz + z^3 - 1 \Rightarrow$ grad $\phi = i\frac{\partial\phi}{\partial x} + j\frac{\partial\phi}{\partial y} + k\frac{\partial\phi}{\partial z} = i(3x^2 - yz) + j(-xz) + $ k $(-xy + 3z^2) \Rightarrow$ grad ϕ at (1,1,1) = 2i$-$j +2k)$\Rightarrow$ unit normal = $\frac{2I-J+2K}{3}$

10. Since L [cos 2t] = $\frac{s}{s^2 + 4}$, we get L [t cos 2t]= $-\frac{d}{ds}[\frac{s}{s^2 + 4}] = \frac{S^2 - 4}{(S^2 + 4)^2}$

11. z$-$ 1 is a pole of order. Therefore Residue of f(z) = $\frac{z-1}{(z-2)(z+1)^2}$ at z = $-$ 1, is

   $\frac{1}{1!}\frac{d}{dz}(\lim_{z\to -1}(z+1)^2[\frac{z-1}{(z-2)(z+1)^2}] = \lim_{z\to -1}\frac{d}{dz}[\frac{z-1}{z-2}] = \lim_{z\to -1}\frac{(z-2)-(z-1)}{(z-2)^2} = -\frac{1}{9}$

12. Substituting in the relation Mode = 3 Median $-$ 2 Mean we get 40 = 3.$\times$ (42) $- 2 \times$ (mean) $\Rightarrow 2 \times$ mean = 126 $-$ 40 $\Rightarrow$ mean = 43

13. p = probability of getting a male child = probability of success = $\frac{1}{2}$ And q = 1 $-$ p = $1 - \frac{1}{2}$, we have n = 4 , r = number of boys = 3 , N = 800 Probability of getting 3 boys and 1 girl= $n_{c_r}p^r\, q^{n-r} = 4_{p_3}[\frac{1}{2}]^3[\frac{1}{2}]^1 = 4 \times \frac{1}{16} = \frac{1}{4}$ No. of families having 4 children with 3 boys and one girl = 200 Hence we can expect N $\times \frac{1}{4}$ = 200 $\times \frac{1}{4}$ = 50 families

14. We have A.M = 25 , H.M = 9 G.M = $\sqrt{A.M \times H.M} = \sqrt{25 \times 9}$ = 15

CODE: MT METALLURGY MULTIPLE CHOICE QUESTIONS

S.NO	Ans	S.NO	Ans	S.NO	Ans
1	D	6	A	11	A
2	A	7	B	12	B
3	B	8	C	13	B
4	A	9	A	14	C
5	D	10	C		

Solutions

1. We have A = $\begin{bmatrix} 2 & 3+4i \\ 3-4i & 2 \end{bmatrix}$

   $\Rightarrow$ The characteristic equation A is $\begin{vmatrix} 2-\lambda & 3+4i \\ 3-4i & 2-\lambda \end{vmatrix} = 0$

   $\Rightarrow (2-\lambda)(2-\lambda)-(3+4i)(3-4i) = 0$

   $\Rightarrow \lambda^2 - 4\lambda + 4 - 9 - 16 = 0$

   $\Rightarrow \lambda^2 - 7\lambda + 3\lambda - 21 = 0 \Rightarrow (\lambda+3)(\lambda-7) = 0$

   $\Rightarrow \lambda = -3, 7$

2. The given equation is $z = (1 - 2xy + y^2)^{-1/2}$ .....(1)

   Differentiating (1) partially with respect to x and y we get

   $\frac{\partial z}{\partial x} = -\frac{1}{2}(-2y)(1-2xy+y^2)^{-3/2} = yz^3$

   $\frac{\partial z}{\partial y} = -\frac{1}{2}(-2x+2y)(1-2xy+y^2)^{-3/2} = (x-y)z^3$

   Therefore we get $x\frac{\partial z}{\partial x} - y\frac{\partial z}{\partial y} = xyz^3 - y(x-y)z^3 = xyz^3 - xyz^3 + y^2z^3 = y^2z^3$

3. $\int \frac{e^x(1+x)}{(2+x)^2}\, dx = \int [\frac{e^x(2+x)-1)}{(2+x)^2}]\, dx = \int e^x[\frac{1}{2+x} - \frac{1}{(2+x)^2}]\, dx$ [e^x [f(x) +f'(x)] dx form

   where f(x) = $\frac{1}{2+x}$] = $\frac{e^x}{2+x}$

4. $\frac{1}{(D-1)^3} xe^x = e^x \frac{1}{(D+1-1)^3} x = e^x \frac{1}{D^3}[x] = \frac{e^x x^4}{24}$

5. f(x) = $1 + x + x^2 + \ldots + x^{100} \Rightarrow f'(x) = 1 + 2x + 3x^2 + \ldots + 100x^{99}$

   $\Rightarrow f'(1) = 1 + 2 + 3 + \ldots + 100 = \frac{100(100+1)}{2} = 5050$

6. $y' + \frac{y}{x} = 2$ is in linear form p = $\frac{1}{x} \Rightarrow \int \frac{1}{x} = \log x$ Integrating factor is $e^{\log x} = x$

   The solution is xy = $\int 2x\, dx \Rightarrow xy = x^2 + c$

7. $y = xp + p^2$ is in Clairaut's form $\Rightarrow$ The general solution $Y = cx + c^2$ .....(1)

Differentiating (1) with respect to c we get $0 = x + 2c \Rightarrow c = -\dfrac{x}{2}$

Substituting in (1) we get $x^2 + 4y = 0$, which is the singular solution of the given diff. equation

8. We have $\phi = x^2 - y^2 + z - 2 \Rightarrow \nabla\phi = \text{grad } \phi = i\dfrac{\partial\phi}{\partial x} + j\dfrac{\partial\phi}{\partial y} + k\dfrac{\partial\phi}{\partial z}$

$\Rightarrow \text{grad } \phi = i\dfrac{\partial\phi}{\partial x} + j\dfrac{\partial\phi}{\partial y} + k\dfrac{\partial\phi}{\partial z} = 2xi - 2y j + k$

$\Rightarrow \text{grad } \phi$ at $(1,-1,2)$ is $2i + 2j + k$

The unit vector normal $= \dfrac{\nabla\phi}{|\nabla\phi|} = \dfrac{1}{3}\ \{\ 2i + 2j + k\ \}$

9. We have $z = f(x^2 - y^2) \Rightarrow p = f'(x^2 - y^2)\,.\,2x$

$\Rightarrow \dfrac{p}{2x} = f'(x^2 - y^2)$ similarly we get $-\dfrac{q}{2y} = f'(x^2 - y^2)$

Equating we obtain $\dfrac{p}{2x} = f'(x^2 - y^2) = -\dfrac{q}{2y}$

$\Rightarrow \dfrac{p}{2x} = -\dfrac{q}{2y} \Rightarrow py + qx = 0$ which is the required solution

10. We know that $L(\sin t) = \dfrac{1}{s^2+1}$

By shift theorem, we get $L(e^{-2t}\sin t) = \dfrac{1}{(s+2)^2+1} = \dfrac{1}{s^2+4s+5}$

Therefore we get $L(t\,e^{-2t}\sin t) = -\dfrac{d}{ds}\left[\dfrac{1}{s^2+4s+5}\right] = \dfrac{2s+4}{(s^2+4s+5)^2}$

11. Applying the formula Mode = 3 Median – 2 Mean we get

= $3\times 42 - 2$ mean $\Rightarrow$ mean = 43

12. The mean of the given series is $\bar{x} = \dfrac{2+4+6+8}{4} = 5$ The variance is $\dfrac{\Sigma(x-\bar{x})^2}{N} = \dfrac{20}{4}$

= 5, hence the standard deviation is $\sqrt{5}$

13. We have mean = np = 2.4, variance npq = 1.44

$\Rightarrow q = \dfrac{npq}{np} = \dfrac{1.44}{2.4} = 0.6 \Rightarrow p = 0.4$

Mean = 2.4 $\Rightarrow$ np = 2.4 $\Rightarrow$ n = $\dfrac{2.4}{0.4} = 6$

14. The regression coefficients of x on y and y on x are $\dfrac{9}{20}$ and $\dfrac{4}{5}$ then the value of coefficient of correlation (i.e r) is $\sqrt{\dfrac{9}{20} \times \dfrac{4}{5}} = 0.6$

GEO ENGINEERING AND GEOINFORMATICS

S.NO	Ans	S.NO	Ans	S.NO	Ans
1	C	6	C	11	A
2	D	7	B	12	B
3	C	8	C	13	C
4	A	9	A	14	C
5	D	10	C	15	D

Solutions

1. The characteristic polynomial of A is $\begin{bmatrix} 1-\lambda & 4 \\ 2 & 3-\lambda \end{bmatrix} \Rightarrow (1-\lambda)(3-\lambda)-8 = \lambda^2 - 4\lambda - 5$

2. Differentiating $y = ax^2 + bx + c$ with respect to x we get $y' = 2ax + b$, $y'' = 2a$ and $y''' = 0$

3. The equation is a linear equation i.e. of the form $\dfrac{dx}{dy} + Px = Q$ where $P = \dfrac{-1}{y}$.

   The integrating factor is $e^{-\int \frac{1}{y}dy} = e^{-\log y} = \dfrac{1}{y}$

4. *The auxiliary equation of* $y'' + y' - 6y = 0$, is $m^2 + m - 6 = 0 \Rightarrow (m-2)(m+3)=0$

   $\Rightarrow m = 2$, $m = -3$ hence the required solution is $y = c_1 e^{2x} + c_2 e^{-3x}$

5. The P.I of the given differential equation is $\dfrac{1}{2^3 + 2.2^2 + 2} e^{2x} = \dfrac{e^{2x}}{18}$

6. $u = \log \dfrac{x^2}{y} \Rightarrow u = \log x^2 - \log y = 2\log x - \log y \Rightarrow \dfrac{\partial u}{\partial x} = 2/x$, $\dfrac{\partial u}{\partial y} = -1/y$

   $\Rightarrow x\dfrac{\partial u}{\partial x} + y\dfrac{\partial u}{\partial y} = 2 - 1 = 1$

7. $\int_{x=0}^{1} \int_{y=0}^{x} e^{x+y} dy\, dx = \int_{x=0}^{1} e^x \int_{y=0}^{1} e^y dy = \int_{x=0}^{1} e^x (e^x - 1)dx = \int_{x=0}^{1}(e^{2x} - e^x)dx$

   $= (\dfrac{e^{2x}}{2} - e^x) \,|_0^1 = (\dfrac{e^2}{2} - e) - (\dfrac{1}{2} - 1) = \dfrac{e^2}{2} - e + \dfrac{1}{2} = \dfrac{(e-1)^2}{2}$

8. If $\lim\limits_{X \to 0}(\dfrac{1}{X} - \dfrac{1}{e^X - 1})$ is in $\infty - \infty$ form $\Rightarrow \lim\limits_{X \to 0}(\dfrac{e^X - 1 - X}{X(e^X - 1)})$ is in $\dfrac{0}{0}$ form applying L'Hospital rule we get $k = \dfrac{1}{2}$

9. Since f is a solenoidal we have $\nabla \bullet f = 0$

$$\Rightarrow \frac{\partial}{\partial x}[y(\alpha x^2+z)] + \frac{\partial}{\partial y}[x(y^2-z^2)] + \frac{\partial}{\partial z}[2xy(z-xy)] = 0 \Rightarrow 2\alpha xy+2xy+2xy = 0$$

$$\Rightarrow 2xy(+2) = 0 \Rightarrow \alpha+2 = 0 \Rightarrow \alpha = -2$$

10. $Z = (x^2+a)(y^2+b) \Rightarrow p = \frac{\partial z}{\partial x} = 2x(y^2+b)$ and $q = \frac{\partial z}{\partial y} = 2y(x^2+b)$

$$\Rightarrow (x^2+b) = \frac{q}{2y} \text{ and } (y^2+b) = \frac{p}{2x} \text{ substituting in the given equation we get } pq =$$

$4xyz$, which is the required partial differential equation

11. We have $P(A) = \frac{1}{2} \Rightarrow P(\overline{A}) = \frac{1}{2}$, $P(B) = \frac{1}{4} \Rightarrow P(\overline{B}) = \frac{3}{4}$, $P(C) = \frac{1}{5} \Rightarrow P(\overline{C}) = \frac{4}{5}$

The probability that the problem is not solved $= P(\overline{A})\ P(\overline{B}) = \frac{1}{2} \bullet \frac{3}{4} \bullet \frac{4}{5} = \frac{3}{10}$

Therefore the probability that the problem is solved $= 1 - \frac{3}{10} = \frac{7}{10}$

12. The residue of $f(z) = \frac{z^2}{(z-1)^2(z+2)}$ at the pole $z = -2$ is $\lim_{z\to(-2)}(z+2)\frac{z^2}{(z-1)^2(z+2)} =$

$$\lim_{z\to(-2)}(z+2)\frac{z^2}{(z-1)^2(z+2)} = \lim_{z\to(-2)}\frac{z^2}{(z-1)^2} = \frac{4}{9}$$

13. $\int_0^1 x^7(1-x)^5\ dx = \int_0^1 x^{8-1}(1-x)^{6-1}\ dx = \beta(8,6) = \frac{7!5!}{13!}$

14. We have np = 20 and variance = np q =15

$$\Rightarrow q = \frac{npq}{np} = \frac{15}{20} = \frac{3}{4} \Rightarrow p = 1-q = 1-\frac{3}{4} = \frac{1}{4}$$

Therefore np = 20 $\Rightarrow$ n$\frac{1}{4}$ = 20 $\Rightarrow$ n = 80

15. we have $u = \tan^{-1}\frac{y}{x} \Rightarrow \frac{\partial u}{\partial y} = \frac{1}{1+(\frac{y}{x})^2}\frac{\partial}{\partial y}\left(\frac{y}{x}\right) = \frac{x}{x^2+y^2}$